AF751678

Studies in the Neolithic and Urban Revolutions

The V. Gordon Childe Colloquium
Mexico, 1986

edited by
Linda Manzanilla

BAR International Series 349
1987

UNIV DUNELM
-4 NOV 1987

B.A.R.

5, Centremead, Osney Mead, Oxford OX2 0DQ, England.

930
BRI
(349)

GENERAL EDITORS

A.R. Hands, B.Sc., M.A., D.Phil.
D.R. Walker, M.A.

BAR -S349, 1987: 'Studies in the Neolithic and Urban Revolutions: The V. Gordon Childe Colloquium, Mexico, 1986'

Price £20.00 post free throughout the world. Payments made in dollars must be calculated at the current rate of exchange and $8.00 added to cover exchange charges. Cheques should be made payable to B.A.R. and sent to the above address.

MANZANILLA

© The Individual Authors, 1987

ISBN 0 86054 449 4

API 125/49

For details of all new B.A.R. publications in print please write to the above address. Information on new titles is sent regularly on request, with no obligation to purchase.

Volumes are distributed from the publisher. All B.A.R. prices are inclusive of postage by surface mail anywhere in the world.

BRITISH archaeological reports International ser., supplementary (249)

Printed in Great Britain

CONTENTS

THE NEOLITHIC REVOLUTION

THE URBAN REVOLUTION

EDITOR'S NOTE

The study of social processes has been one of the main concerns of modern archæology. It was undoubtedly Vere Gordon Childe who, fifty years ago, established theoretical issues on broad economic trends and socio-political changes in prehistory, in his book *Man Makes Himself.* As an homage to a great prehistorian, this book emerges from a meeting organised by the Institute of Anthropological Research of the National Autonomous University of Mexico, from July the 7th to 12th, 1986. On this occasion, we invited scholars who had worked in Europe, the Near East, Asia, and America, for the purpose of enriching our actual knowledge on the "Neolithic and the Urban Revolutions", half a century after Childe's first concern on these affairs. We hope that this effort will provoke further discussions on two of the most outstanding processes in history.

Acknowledgements

We would like to thank David Castledine, Nicole Spitalier, Catherine Spohn, and Pilar Agüero for the translation of the Spanish papers into English; Rosario Medina for the input into the computer; and Miguel Mireles and Miguel Navarro for the laser printing.

V. GORDON CHILDE: A MARXIST ARCHAEOLOGIST

BRUCE G. TRIGGER
McGill University

Gordon Childe's career in archæology can be divided into three stages, each of which is best known to a different audience. The first, as a culture-historian, is familiar to European archæologists; the second, as a cultural evolutionist, is known to American anthropologists; and the third, as a major social science theoretician, remains relatively unknown even to the cotery of archæologists who claim a special interest in such matters. It is frequently debated to what extent Childe was a Marxist. I believe that the extent of his commitment increased throughout his lifetime and that its impact on his archæological work altered with each of these stages.

Childe was born in North Sydney, Australia, in 1892. The rebellious son of a prosperous and conservative Church of England minister, he became a socialist while he was an undergraduate in Australia. While they were both students at Oxford University from 1914 to 1916, he and Raji Palme Dutt, later to be a leading figure in the British Communist Party, often drank together and discussed Hegel, Marx, and politics late into the night. In 1916 he returned to Australia to oppose conscription and soon became deeply involved in Labour Party politics. Between 1919 and 1921 he was private secretary to John Storey, the premier of New South Wales. After the fall of the Labour government, lacking employment and disillusioned with politics, he returned to an earlier interest in archæology. In his first, and only political, book, *How Labour Governs*, published in 1926, he reveals a deep distrust of parliamentary government, which was widespread on both the left and right of the political spectrum at that time, although his book also makes it clear that he rejected authoritarian doctrines. Apart from Dutt's brief testimony, there is nothing in his political writings or reports of his political activities at this time that indicates that his commitment to socialism was accompanied by any detailed understanding of Marxist philosophy.

There is also no evidence of specifically Marxist influence in his earliest archæological books: *The Dawn of European Civilization* (1925), *The Aryans* (1926), and *The Danube in Prehistory* (1929). These books were landmarks in the development of archæology in Britain and much of Europe. They replaced an already outmoded nineteenth-century evolutionary approach to archæology with one that was specifically culture-historical. Like most contemporary archæologists and social scientists, Childe assumed that human beings were naturally conservative and that culture change came about mainly as a result of diffusion and migration. Like other archæologists he also tended to assume that because of this, the course of history was unpredictable and to account for the accomplishments of individual cultures in terms of ethnic and even racial stereotypes. While he credited Indo-Europeans with a "peculiar vigour and genius", the Danubian people, who were assumed to be not Indo-European, were described as an inert mass that fell easy prey to conquerors and the "stagnant" megalithic cultures of western Europe were pronounced to be "not European".

The greatest creative achievement of Childe's early work was to combine the concept of the archæological culture, which had been refined by the German archæologist Gustaf Kossinna as a device for tracing the history of specific peoples in the archæological record, with the diffusionism of the great Swedish archæologist Oscar Montelius, especially the latter's belief that in prehistoric times technological skills had spread to Europe from their place of origin in the Near East. Childe's early views about diffusion were closest to those of his Oxford teachers, Arthur Evans and John L. Myres, who both stressed the creativity with which Europeans had made use of this knowledge. Childe's familiarity with a vast number of European languages and a powerful visual memory that permitted him to note and remember similarities among artifacts in museums all over Europe made possible a culture-historical synthesis of European archæology that provided a point of departure for further research throughout Europe.

Despite its continuing importance for the development of archæology, Childe quickly came to view the culture-historical approach as an intellectual dead-end. He doubted that the history of particular ethnic groups could be traced reliably far back into prehistoric times and that such studies could give much meaning to the archæological record as a whole. He came to regard the culture-historical approach as a mere duplicate of the now old-fashioned political history, which historians were abandoning in favour of social and economic history. Childe sought to emulate the latter by searching for broad economic trends in prehistory. The results of this research were drawn together in three books: *The Most Ancient East* (1928), *The Bronze Age* (1930), and *New Light on the Most Ancient East* (1934).

In the first of these books, Childe stressed the development of agriculture as a crucial event in human history and the key element defining the start of the Neolithic period. This idea had already been expounded by the well-known ethnologist Grafton Elliot Smith as early as 1915 and by the archæologists Harold Peake and Herbert Fleure in the third volume of their popular *Corridors of Time* (1927). Peake and Fleure had also adopted Pumpelly's oasis hypothesis, which proposed that massive desiccation in the Near East at the end of the last Ice Age had caused people to domesticate plants and animals in order to feed the higher densities of population that had gathered around surviving sources of water. Childe stressed, however, that individual bands could have moved elsewhere or died out rather than develop agriculture. Thus the formulation of his argument reflected the possibilism that was currently fashionable in non-Marxist geography in France and Britain.

In *The Bronze Age*, Childe studied the origins and spread of metallurgy and craft specialization. He was convinced, almost certainly wrongly, that metal-casting required full-time, although initially itinerant, specialists who, along with prospectors and miners, became the first human beings to function independently of tribal affiliations. The adoption of a metal tool-technology therefore produced a double loss of Neolithic self-sufficiency, since it required communities to become dependent on craftsmen who often were unrelated to them as well as the development of extensive trade routes that were not interrupted by periodic outbreaks of tribal warfare in order to ensure the regular delivery of supplies of copper and tin. In *New Light on the Most Ancient East*, Childe argued that two revolutions had occurred in prehistoric times in the Near East that were equivalent in their impact to the Industrial Revolution. These were the transition from food-collecting to food-producing and from self-sufficient food-producing villages to urban societies. Each of these revolutions resulted in a more productive technology and in a massive increase in population. These technologies were spread to Europe by migrations of surplus population, the exchange of manufactured goods for raw materials, and surplus craftsmen seeking employment. The result was to produce in Europe Neolithic and

Bronze Age societies that were very different from those that had developed in the Near East.

Childe's interest in economic development in prehistoric times drew its inspiration from the European archæology of that period. Yet he advanced beyond the interpretations of Smith, Peake, and Fleure in the consistency with which he applied an economic approach to the study of prehistory and in the scope of his formulations. Also, instead of seeing cultural change as the result of technological innovation, he already saw technological progress occurring within the context of broader economic and political patterns. This allowed him to explain how the same technological innovation could produce very different types of societies in Europe and the Near East.

Childe's economic approach had inherent in it the makings of a multilinear evolutionary perspective. Yet he was not primarily concerned with cultural evolution at this time. He stated categorically that "archæology's revelations...disclose no abstract evolution but the interaction of multiple concrete groups and the blending of contributions from far-sundered regions". He still regarded human beings as uninventive and relied heavily on diffusion and migration to explain cultural change. We are told at the end of *New Light on the Most Ancient East* that the main aim of the book was to justify the "general doctrine of cultural diffusion". Yet by moving towards a more materialistic explanation of cultural change, Childe was able to abandon forever the ethnic stereotypes that had featured prominently in his early work. He was later to equate the development of metallurgy with the beginning of Engels' separation of handicrafts from agriculture and thus view it in Marxist terms as marking a major social as well as a technological and economic transition. There is, however, no more evidence of a specifically Marxist input into Childe's early economic studies than there is of it into his culture-historical interpretations. Nor was his materialistic perspective complete at this time. While some economic change was viewed as a response to environmental challenges, much innovation was interpreted from a Montelian perspective as resulting from the spontaneous exercise of human intelligence to achieve greater control over the environment and make human life easier and more secure.

In 1935 Childe visited the Soviet Union for the first time. While there he met Russian archæologists, visited museums, and acquired new information about the history of eastern Europe. He also obtained some "typical Russian works on prehistory". He arrived in the Soviet Union at the end of five years of unprecedented upheaval in Russian archæology, which was intended to bring it into line with Marxist principles. The main task that was set for archæology was to account for changes in prehistoric times from the perspective of historical materialism. The primary context in terms of which change was to be understood was not technology but social organization. Change was attributed to the development of contradictions between different classes and ultimately between the forces and relations of production. These contradictions rather than *sui generis* technological change were regarded as the dynamic forces that shaped human history. Archæologists were therefore called upon not only to describe archæological remains but also to reconstruct the societies that had produced them. This required defining the mode of production and determining as much as possible about technology, social organization, and ideology from archæological data. It encouraged archæologists to study how people, especially ordinary people, had lived in prehistoric times. This led to the excavation of settlements, campsites, and workshops, which were all classes of sites that archæologists had tended to ignore in the past.

Childe did not accept the entire programme of Soviet archæology. While admiring how Soviet archæologists sought to explain cultural change in terms of internal processes and awakened to the potential value of a Marxist approach, he refused to adopt their unilinear scheme of socio-economic formations, or any other unilinear theory of social evolution. He also refused to stop viewing diffusion as a major process that promoted cultural development and to abandon the major emphasis that he placed on typology, which he saw as essential for constructing regional chronologies and tracing cultural influences between one region and another.

Yet in the decade following his first visit to the Soviet Union, Childe published three books dealing with cultural evolution: *Man Makes Himself* (1936), *What Happened in History* (1942), and *Progress and Archæology* (1944), as well as a case study, *Scotland Before the Scots* (1946). The first two were written for the general public as well as for professional archæologists and have continued to be widely read by non-archæologists. Cultural evolution was a topic of theoretical interest that had remained important in Marxist thinking but had previously been of little significance in Childe's writings.

In *Man Makes Himself* he interpreted the archæological record as evidence of a directional process whereby the increasing scientific knowledge accumulated by human beings gave them ever greater control over nature and led to the formation of new and more complex socio-political systems. Childe later regarded these views as being not sufficiently different from the idealist Montelian conception of cultural change. In *What Happened in History* Childe attempted in a much more explicitly Marxist fashion to formulate explanations of cultural change that were focused not on technological knowledge as a prime mover but on social, political, and economic institutions and the role they played in bringing about change. In accordance with the principles of dialectical materialism, he viewed every society as containing within itself both progressive and conservative tendencies that are linked by dynamic unity as well as by persistent antagonism. The latter provides the energy that in the long run brings about irreversible social change. Hence every society contains within itself the seeds for the destruction of its present state and for the creation of a new social order.

Childe did not embrace unilinear evolutionism in these works anymore than he did at any other time. He carefully documented the social and political differences between the city states that had emerged in Mesopotamia and the divine monarchy that united Old Kingdom Egypt. These differences were attributed to divergent social and political techniques for controlling agricultural surpluses that had developed in the course of their transformation from tribal to class societies. Nor did he subscribe to the naive faith in the inevitability of progress that characterized many vulgarised versions of Marxism as well as the unilinear evolutionism of the nineteenth century. He wrote under the shadow of expanding Nazi power and viewed the future darkly. Yet his pessimism led him to make an important contribution to the Marxist explanation of change by providing a more detailed analysis of the social conditions that impede social transformations than of those that promote them.

Childe argued that at any level of social development, but especially in the early civilizations, entrenched political hierarchies and inflexible systems of religious beliefs can slow or even halt social and economic change. He distinguished between progressive societies, where the relations of production favour an expansion of productive forces and there is a harmonious relationship among the means of production, social institutions, and the dominant system of beliefs, and conservative societies in which social and political factors block change. The ruling classes in the early civilizations sought to forestall technological changes that might threaten their control of society. They did this by concentrating wealth in their own hands, exercising bureaucratic control over craftsmen, inhibiting the pursuit of technical knowledge, and patronizing magic and superstition on a lavish scale, as well as by the exercise of force. They only succeeded, however, at the cost of making it more difficult for their own societies to compete with more progressive neighbouring ones. Childe thus ascribed important roles in shaping history to both the base and superstructure of societies. Yet he was careful to qualify that where the superstructure was dominant, its influence could only be negative. It has since been maintained by Soviet anthropologists that this view is in accord with orthodox Marxism.

Childe's stance also provides a definitive answer to those British Marxists, such as George Thomson, who accused him of ignoring class conflict in the early civilizations. Childe argued that social evolution occurred slowly, if at all, in those civilizations precisely because such struggles were blunted by highly effective political and religious techniques of social control. He did not ignore the concept of class struggle in early civilizations or reject it because he thought it inapplicable for studies using archæological data. On the contrary, he did not find it useful for explaining Oriental societies, which he believed had remained static for long periods of time. In his analyses of Classical societies, and in particular of the Roman Empire, he placed more emphasis on struggles among groups within societies to control wealth and power and on the shifting patterns of political control. It is possible that his differing treatment of ancient Near Eastern and Classical civilizations was based on Marx's own distinction between Oriental and Slave societies, although like other Marxists of the Stalinist period he made no explicit use of the former concept.

After World War II Childe continued to refine and develop a Marxist understanding of social change. As a result of growing disillusionment with the quality of archæological work being done in the Soviet Union, he turned away from Soviet archæology as a major source of inspiration and began to investigate the philosophical basis of Marxism itself. In the last two decades of his life, he worked to acquire a more profound and less dogmatic understanding of Marxism as an analytical tool and to apply it to the study of archæological data. As part of this effort he read widely in the field of philosophy in an effort to understand Marxism better. This led him to some conclusions that were challenging for archæology, but that were so far ahead of the thinking of most archæologists that they were ignored in his lifetime and for several decades afterwards.

Like all Marxists, Childe regarded a historical approach as uniting all of the social sciences. He argued that the significance of any generalization can only be established in relationship to specific historical contexts. This is because the rules that account for human behaviour, and human behaviour itself, change as new forms of society develop. Because of this human history is genuinely creative. It is capable of bringing into being novel and often unforeseen social orders and new forms of human self-awareness. Yet he continued to be convinced that such progress was not inevitable. Some societies remain static, while others regress or even destroy themselves. In his view, a Marxist analysis precluded the possibility of predetermined change. Functional constraints account for many similar features of social organization that are shared by unrelated cultures with analogous modes of production. Nevertheless, the specific content of cultures and of individual sequences of change is determined to so great a degree by pre-existing cultural patterns and by accidental contacts with other societies, as well as by the nature of the broader world-system in which they occur, that their precise nature cannot be predicted. As Childe noted, the precise form of the British constitution in the nineteenth century could never be deduced from the capitalist mode of production alone. His rejection of determinism presaged the views of later Marxist scholars in Western Europe, such as Jean-Paul Sartre. It also accords, however, with orthodox Marxism, which denies that general laws can explain all the diverse features of concrete human development. Childe defined knowledge as shared mental approximations of the real world that permitted human beings to act upon it and he insisted that archæologists must treat artifacts as concrete expressions of human thoughts and ideas. He also argued that human beings adapt not to real environments but to their ideas about them, even if an effective adaptation requires a reasonably close correspondence between reality and how it is perceived. Tools fulfil human needs, but these needs are not fixed. Because of this the relative efficiency of any tool or technological process cannot be measured by absolute criteria but only by chronological succession. Innovations and their applications to social needs also require new forms of thought that have ramifications extending through whole societies. Advances in technology thus reflect not simply an increase in scientific information but also the evolution of the total knowledge at the disposal of a society, including how human beings perceive themselves and their relationship to nature.

Childe was not a cultural relativist. He pointed out that the understanding of the world possessed by progressive societies has continued to expand both spatially and temporally. Comparisons with modern, small-scale societies suggested that prehistoric hunter-gatherers had only a vague and inaccurate knowledge of the past and rarely knew what was happening a few hundred kilometers away. Knowledge in these societies was different in kind as well as comprehensiveness from that of modern industrial societies, and thinking followed patterns that were significantly different from our own. Childe accepted that peoples with simple technologies do not distinguish between what is animate and inanimate as we do. Instead they view nature and society as coterminous and use social relationships as categories to organize their experience of the natural world. Notions of causality remained personal until the growing use of inanimate power to work machines engendered the idea of a mechanical causality embodied in the thinking of Isaac Newton. Childe had no qualms about pronouncing modern civilizations to be superior to all preceding ones in so far as they are able to provide a reliable guide for a far greater number of activities.

Childe elaborated his concept of knowledge in terms of the Marxist dichotomy between true and false consciousness. True consciousness is characterized by the operational correspondence between views of reality and external reality itself. In the form of technological knowledge, it exists to varying degrees in all societies. By contrast, an objective understanding of social relations or of the social significance of beliefs and values is rare in any past or modern society, although Marxist analyses maintain that such understanding will characterize the technologically-advanced, classless societies of the future. False consciousness occurs where there is no operational correspondence between what is believed and external reality. It embraces the myths that all societies create to mask and compensate for their technological incompetence and that class societies use to disguise exploitation as altruism. Childe did not claim that human beings were able to distinguish between these forms of consciousness in their own societies. Indeed, the efficaciousness of false consciousness depends upon it not being recognized as such. He also argued that a Bronze Age metallurgist might have regarded the sacrifice of a chicken as being as essential for metal-casting as was the proper mixture of tin and copper. Yet he maintained that in the long run human beings are able to distinguish what is technologically efficacious from what is not. The ultimate test of truth is thus an operational one; just as the ultimate function of true knowledge is to provide rules for action. By contrast, false consciousness perishes with the sort of society that created it; fervently believed religious concepts becoming mere superstition or folly as social conditions change. While Childe identified true knowledge mainly with technology, he believed that in the future a more objective understanding of human behaviour and society would allow human beings to control their social environment more scientifically and humanely. He hoped that history and archæology might together create a "science of progress" that would help to fill this need, even if they could never become exact sciences.

In recent years a growing number of self-styled Marxist archæologists have appeared in England and the United States. Many of them have been influenced by French "structural Marxism" and by the "critical theory" of the Frankfurt School. Many of these archæologists have attacked the distinction between true and false consciousness. They argue that all knowledge is biased by the social perspectives of those who possess it and that anything approaching objective knowledge of the human condition cannot be attained. They claim that the distinction between infrastructure and superstructure is a reflection of the biases of capitalist societies, which ignores the intricate interlacing of economic and non-economic relations. Thus they not only call the scientific credentials of Marxism into question but seek to reduce all human knowledge to the level of self-serving fantasies. Childe took similar nihilistic arguments into account. Yet he chose to stand his ground as a materialist and more specifically as a Marxist. He accepted that the world we adapt to is one that exists in our own mind. Yet by insisting that natural selection operates on the material interface between a society and its environment and by maintaining the distinction between true and false consciousness, he avoided a descent into idealism and hyper-relativism.

He also kept the realities of the archæological record and what could be done with it clearly in mind. He noted that religious beliefs, magic, and superstition leave their mark on it no less conspicuously than does technological knowledge. Yet because the possible variations in the detail of magical and religious beliefs are infinite, the archæologist has no hope of being able to infer the specific content of these beliefs in a verifiable fashion in the absence of written records or oral traditions. By contrast, the number of practical solutions to any technological problem is limited by material constraints that can be inferred with a high degree of accuracy and in much detail, using the laws of physics and chemistry. The study of prehistoric knowledge must therefore be restricted largely to technological matters and framed in terms of its practical results, not the subjective goals of those who possessed it.

Nevertheless, he also believed that the evolution and functioning of technology could only be understood if archæologists were able to reconstruct the social context within which it occurred. In the last years of his life he searched, not very successfully in his estimation, for ways to overcome the limitations of the typological approach to archæological data. Archæologists have spent the last thirty years developing new techniques for inferring social organization from archæological data.

In the course of his career as an archæologist, Childe progressed from a simplistic understanding of the role played by technology in influencing human behaviour to the sophisticated views that he held in the last decade of his life. His introduction to Marxism as a useful tool for interpreting archæological data resulted from his exposure to Soviet archæology beginning in 1935. Far from rejecting Marxism as a basis for analysing archæological data in the last two decades of his life, Childe initiated a highly productive dialogue between Marxist theories and archæological facts. This operation made a Marxist approach more vital than ever for his work and turned Childe himself into a significant Marxist thinker.

Peter Gathercole and others have suggested that Childe was hit hard by Nikita Khrushchev's revelation of Stalin's crimes and the Soviet invasion of Hungary and that the resulting disillusionment may have been a factor contributing to his suicide. Yet there is sound evidence that his disillusionment with political movements had begun far earlier as a result of his involvement with Labour politics in Australia. He had also long been aware that the Soviet Union was governed by a totalitarian regime that required even archæological researchers to tread warily. Hence he was immune to major disillusionment on this score. Moreover, his acceptance of the Marxist principle of the unity of praxis and theory did not lead him to conclude that the shortcomings of the Stalin regime were necessarily proof of the erroneousness of Marxism.

At the end of his life Childe was defeated not by disillusionment with Marxism as a research strategy but by what he saw as his own inability to develop new analytical procedures that would permit him to carry his Marxist analysis of prehistory forward. He was convinced the failure resulted from his own declining powers rather than from the intractable nature of the problem. He was assailed by the fear that his creative life was over and by a growing horror of loneliness, helplessness, and senility. Yet, as his work has been taken up, often unconsciously, by a younger generation of archæologists, it has become clear that the general perspectives that he forged in the 1950s constituted a significant point of departure for original research in the 1970s and 1980s.

*The ideas and even the wording of much of this paper are derived from three of my previous writings: *Gordon Childe: Revolutions in Archæology* (London, 1980); "Childe and Soviet Archæology", *Australian Archæology*, no. 18 (1984), pp. 1–16; and "The Role of Technology in V. Gordon Childe's Archæology", *Norwegian Archæological Review*, vol. 19 (1) (1986), pp. 1–14. The originality of this paper lies in its brevity and an accompanying dogmatic clarity.

THE PRESENCE OF VERE GORDON CHILDE IN MEXICAN ARCHAEOLOGY

JULIO CESAR OLIVE NEGRETE
Instituto Nacional de Antropología e Historia

I wish to congratulate the Institute of Anthropological Research of the National Autonomous University of Mexico under the direction of Dr. Mari Carmen Serra Puche on the organization of this conference in memory of Vere Gordon Childe and to express my gratitude to the committee headed by Dr. Linda Manzanilla for conferring on me the honor of participating in this inaugural session, which I can only explain because of the interest which for some time now I have shown in applying the theories and concepts of this illustrious Australian prehistorian and ex-Director of the London Institute of Archæology to Mexican Archæology, who Glyn Daniel asserts is the one who initiated the age of maturity in the discipline of archæology and made it accessible to the public.

How and when did the influence of this scientist reach Mexican archæology? How was it incorporated into our theory and practice and what is its significance? The answers can only be found by putting our anthropology into a world-wide context, in order to compare it and observe the internal and external tendencies it has had.

The starting point of Childe's work is his study on the birth of European civilization, published in 1925 (Daniel 1976:247; 1973:76), which he developed and perfected between that date and the Second World War, around 1939-40. By then he had elaborated the concepts and principles that contributed to the enrichment not just of archæology, but also of anthropology and history: the idea that archæology is part of history, within the preliminary consideration that the latter forms a true social science with all of its philosophical, sociological and epistemological implications; the relationship between archæology and anthropology; and, what I believe more important, the dialectical philosophy which links it to Marx and Engels and, through them, to Feuerbach and Hegel.

The lack of understanding of this philosophy has been the main factor limiting the application of Childe's theses to Mexican and also to Northamerican archæology and the reason why the discussion has descended to a fruitless criticism of Morgan's terminology; a debate has arisen over concepts such as naturalism, evolutionism, neo-evolutionism, positivism, neo-positivism and humanism, disregarding the revolutionary character of dialectical philosophy which allows us to surpass the apparent antagonisms between nature and culture through the Hegelian principle that everything real is rational and everything rational is real, "set on its feet" by the materialism of Feuerbach, Marx and Engels.

Closely united to this conception is Childe's idea of progress in history, to which archæology contributes by measuring the success of man's domination of the environment through observing material culture, which in turn reflects the accumulation of knowledge and the advancement of science as the collective heritage of the human species, which is not restricted to one group and offers individuals better alternatives for diversifying the potentiality of human existence (Childe 1973).

What was happening in Mexico in this era when European archæology was maturing? Dr. Ignacio Bernal writes: "My generation was the first in Mexico that had the opportunity to study archæology professionally, for even between 1900 and 1940 all of those dedicated to it were self-taught after a fashion, except for a few rare exceptions who studied abroad. An organized course of study did not exist, only more or less coherent classes such as those offered in the National Museum beginning shortly after 1900 and subsequently in the University. This restriction explains why men as eminent as Alfonso Caso himself had scarsely any teachers in the subject. The situation which impeded study also permitted singular aptitudes and variety. Almost everyone had studied some other profession, which therefore gave them a more extensive cultural background" (Bernal 1979).

As Dr. Bernal himself declares (verbal communication), it was logical that theory was almost never discussed either in class or in archæological practice during the time in which this first generation of professionals was formed at the National School of Anthropology from 1938 onwards.

Up until that time and perhaps as part of the phenomenon called "cultural Monroeism" (Ortega y Medina 1953), our archæology was dominated by tendencies of the Northamerican school, whose research was carried out under the sponsorship of private institutions such as museums, universities and foundations, and whose object of study —human culture of the remote past— does not form part of its own cultural tradition and is therefore extraneous to the values of its society. The concepts and techniques of Northamerican archæology had penetrated our country since the beginning of the century through the International School of American Archæology and Ethnography: the diffusion of cultural traits between cultural areas, new excavation techniques in order to obtain stratigraphic sequences and the establishment of correlations and chronologies based on ceramic typology. Gamio, on the other hand, had a definite interest in social anthropology in order to promote the integral and regional development of the country, with a nationalistic approach which has distinguished Mexican anthropology. Moreover, the latter had an ancient historical background which cannot be escaped, because of the richness of our documentary sources, and was also occupied with the reconstruction of large architectural monuments, given that archæology in Mexico has always been a state activity and has faced the challenge of defending the patrimony bequeathed by the indigenous cultures, which are considered an essential part of the national being (Lorenzo 1976, 1981; Bernal 1979; Gándara 1977; Olivé 1980).

The Northamerican trend coincided with the objective of restoring the imposing Maya monuments, carried out by the Carnegie Institution, for other reasons —to acquire private funds and prestige. Another objective in the region of Central Mexico was to establish the order of succession of the Prehispanic cultures on the basis of stratigraphy and ceramic typology, along the lines pointed out by Boas, Gamio, Spinden and Vaillant.

In the fifties the panorama began to change because of the first contributions of Mexican anthropologists, who, still under the Northamerican orientation, occupied themselves with understanding the development of the Prehispanic agrarian cultures, still using the concept of "horizon". The work of Piña Chan on the cultures that he started to call Preclassic stands out.

The development of new concerns is based of course on the achievements of the first half of the century: the foundation of the National School of Anthropology, the definition of Mesoamerica, the understanding of the calendar and the historical contents of the Mixtec codex, advances in the deciphering of the Maya inscriptions, the identification of Tula as the center of Toltec culture and the appearance of an encyclopædia on the archæology of these regions in the work on prehispanic architecture by the architect

Marquina, who, together with Mendizábal, Caso, Noguera, Jiménez Moreno and other scholars provided these contributions.

An important factor in this change was the arrival of European anthropologists, mainly of Spanish nationality, victims of Fascist persecution. They joined our research centers and teaching staffs, and provided new perspectives. Pedro Bosch Gimpera and Juan Comas Camps, founders of this Institute of Anthropological Research, deserve special mention. Bosch belonged to Childe's generation of European prehistorians, with whom he maintained scientific contact since 1928 and with whom he worked in Edinburgh. His classes in Mexico, related to universal prehistory and to ancient history of the Orient, familiarized us with the European schools of thought.

Another factor was the progress in prehistoric investigations in Mexico with the acquisition of the first scientific data on the preceramic horizon (the human fossil remains from Tepexpan and the lithic tools associated with mammoth bones at Santa Isabel Iztapan). In 1952 the Department of Prehistory was established in the National Institute of Anthropology and History, headed by the notable Mexican prehistorian Pablo Martínez del Río, who was also Director of the National School of Anthropology, where he held the chair in general prehistory and protohistory. Shortly before the foundation of this Department, José Luis Lorenzo had been sent to London to carry out postgraduate studies at the London Institute of Archæology, still under the direction of V. Gordon Childe.

Another professor at the National School of Anthropology, also of Spanish origin, the archæologist Pedro Armillas, was the first at this time to disseminate Gordon Childe's theories in his courses on Mesoamerica, directing the attention of his students to the importance of the neolithic and urban revolutions. Armillas also was a pioneer in the interest in structuring a chronological and periodification system in Precolumbian history which would reflect cultural changes. However, under the influence of evolutionism he tended toward ecological and techno-hydraulic explanations, ending by agreeing with Wittfogel's hypothesis on the importance of the hydraulic factor in the formation of the first civilizations.

José Luis Lorenzo returned to Mexico in 1953-54 where he took up prehistoric research again, occupying himself with the application of English excavation techniques to our situation and of Childe's theoretical proposals, mainly that concerning the neolithic revolution. He disseminated these ideas in his courses at the National School of Anthropology and in his prehistoric research.

The same Lorenzo and some other Mexican archæologists such as Matos (1982) consider that my master's degree thesis on Mesoamerica (1958) constituted the most serious effort of its time to handle and integrate the archæological information in this cultural superarea, based on Childe's approach, in order to place the development of ancient Mexico within a universal schematic framework which functions within a socio-cultural perspective. Matos thinks that my proposals demonstrated the limitations of the horizon concept and dynamically outlined the process of development in Mesoamerica through evolutionary and revolutionary stages; but he believes that the terminology and content originally proposed by Morgan, and whose validity I accepted following Childe, were exceeded.

In 1975 Dr. Román Piña Chan proposed the differentiation of the food-gathering from the food-producing epoch, with a division in periods which, as Matos observes, establishes a constant process of evolution without qualitative changes. This scheme was utilized in the organizaton of the collective text on Mexican archæology edited in 1975 by the National Institute of Anthropology and History entitled *México: Panorama Histórico y Cultural.*

On the other hand, if we review the monographs, archæological projects, scientific and popular books, as well as the labels in museum halls, it is observed that the criterion of periodizing by horizons still persists, based on ceramic typology and the traditional terms of Preclassic, Classic and Postclassic, with a great variety of subdivisions. In the equivalences of Northamerican authors, the term "Preclassic" is substituted by the "Formative", which conceptually refers again to historical particularism because of the assumption that each culture has its own stages of formation, florescence and desintegration.

With respect to prehistory, the periodification was established by Lorenzo in the above-mentioned work edited by the National Institute of Anthropology and History (1975). He points out the existence of a long cultural stage which he calls the Lithic, to which he attributes the assemblage of cultural characteristics of savagery, with a living pattern based on hunting and gathering. He divides this stage into horizons which he proposes to use with fluidity in order to avoid the mistake of considering the process of evolution as continuous, and disregarding the qualitative differences that occurred in the course of the process. These subdivisions are: the Archæolithic, Zenolithic and Protoneolithic, which ends in the epoch of agricultural consolidation and the initiation of pottery making. The sociopolitical change which led to food production and the establishment of villages constituted the basis of subsequent cultural development, for which reason Lorenzo still considers that the concept of the neolithic revolution is valid, as an event of qualitative change in human history.

Although research on this process began in the fifties, it was only in the mid-sixties that the archæological excavations of Richard MacNeish at Tehuacan, Puebla, provided data on the early sequences that reveal the transformation from a gathering to an agricultural economy, within a process of gradual evolution according to the aforesaid author, who points out that the characteristic elements of what a Marxist would call the neolithic revolution did not appear suddenly at Tehuacan with the nature of a new cultural complex or of a revolutionary way of life: grinding implements appeared around 6700 BC, the first cultivation not before 5000, villages around 3000, pottery about 2300, and population growth not until 500 BC (MacNeish 1964; 1973).

The contradictory theses of Lorenzo and MacNeish allow us to appreciate the ideological background underlying all theories and which explains why in the course of time Childe's ideas have been followed by only a minority of Mexican archæologists, as a reflection of what also happened in Northamerican archæology. To speak of Childe and revolutions is to invoke the ghosts of Morgan, Marx and Engels; to reject the concept of revolution in understanding the socio-economic phenomena of the remote past allays thoughts about the present and the future. To my way of thinking, it is not a question of conscious thought processes, but rather of a culturally prejuiced mentality, from which we cannot emancipate ourselves.

As in other scientific fields, objectivity in anthropology is very debatable. I am not free of it, but I venture to maintain that in this case the discussion should not deal with the duration of the period elapsed between the initiation of agriculture, its consolidation, subsequent technological and sociopolitical phenomena and the appearance of cities as the beginning of civilization.

As Dr. MacNeish establishes —and Dr. Braidwood also demonstrates in his research on the appearance of agriculture in the Near and Middle East—, in both cases it is a question of a stage covering thousands of years, for which reason Childe himself considered the neolithic revolution as the culmination of a long process, although he represented it as a single event. In Mexico Dr. Bernal also supported the thesis that one cannot speak of revolutions with respect to phenomena that develop over the course of millennia.

To me it seems that it is a question of a reductonist criterion of the concept of revolution, which is variable in the different spheres of reality. Juridical revolutions take place immediately, by changing constitutional texts; political revolutions are generated over months or years, although they might explode at a given moment; social revolutions take much more time; economic revolutions take still more and if we consider the natural order, qualitative change in biology and the appearance of new species by means of the mechanism indicated by Darwin, is a matter of generations (I do not want to refer to mutations here); on the other hand, it is correct to speak of geological revolutions which comprised extremely long periods of transformation in the earth's crust. In short, time is not a determining factor in the concept of revolution, which refers to leaps and to qualitative changes, be it in the physical, economic, social, political or juridical order, no matter how much time they may take.

This consideration forms part of the dialectic to which I have referred and according to which Childe's proposals should be evaluated; the great transformations that occurred in human existence by going from an economy that tied humanity to nature and limited its progress to another which freed it, thanks to agriculture. The following technological developments and their repercussions on production relationships resulted in the urban revolution in the Old World and in the New, without involving exact replicas.

I have heard that Childe was a multilinear evolutionist, and he himself reflects this position, contrary to that which Daniel attributes to him, who, like Dr. Bernal, classifies Childe as a moderate diffusionist because he was concerned with the dissemination of the neolithic revolution as well as the urban revolution. It is clear that Childe had the same interest as Morgan, Engels and Marx in establishing the great evolutionary stages of the whole of mankind, and in that sense he was an evolutionist, but what is not understood is his dialectial conception which dissolves the evolution-diffusion antithesis.

The dialectical approach answers another objection which has been used to exclude Childe from American archæology: the accusation that he discards the existence of ancient civilizations on this continent (Bernal 1979). The information is nor entirely precise because Childe repeatedly accepts the existence of the Maya civilization, and if he does not expressly discuss the Andean and other Mesoamerican civilizations, he does not thereby deny them. Neither is it necessary to admit the importance attributed to his diagnostic elements of civilization, such as writing, inasmuch as Childe insisted that the social totality should be considered, and not isolated features, as far as archæology permits. His general conclusion was that the development of agricultural economies from barbarism showed divergences which are explained by cultural adaptation to different physical environments and convergences, which distinguish social from organic evolution. Even though he thinks that the mechanism of social evolution is diffusion, what interest him is to show that cultural change can be described as general and intelligible rules (Childe 1966).

Late in the sixties a rebellion of students and intellectuals against academic positions took place in Mexico as well as in other countries. Here there was a strong rejection of official archæology because of its nationalism and monumentalist character.

Marxism as a universal corpus was tried as a substitute for it, with a great variety of manifestations. Bartra (1964) insisted on the validity of the ideas of Morgan, Engels and Childe, but he thought that it was necessary to stress more the importance of production relationships and he contributed to the transformation of the concept of the Asiatic mode of production, turning it into tributary despotism, in accordance with Banu's proposals. His opinions were mixed with structuralist Marxist influences, and taking Godelier as an authority, he suggested modes of production as a central theme in the transition from primitive communities to state formations. Thus there persists a minor trend which attempts to develop the ideas of Marx, Morgan and Childe, to which Matos himself, Nalda and Bate belong, as well as the above-mentioned authors.

Late in the seventies the influence of the so-called "new archæology" arrived in its diverse forms, which I will not try to characterize here, but defer to Gándara (1982). I only wish to point out that Childe's theories were relegated to a second place in theory and in practice, without ceasing to be acknowledged and popularized. An example is the excellent book by José Antonio Pérez which enabled Childe's most important works to reach archæologists and student of archæology in our country (Pérez 1980).

The Institute of Anthropological research at UNAM has continued a line of research oriented by Childe's ideas, which involves an interest in the processes of state formation in Mesoamerica; Dr. Linda Manzanilla is investigating its analogies with the Mesopotamian civilization.

Childe's ideas found a better reception in the Introductory Hall of the National Museum of Anthropology (Barba de Piña Chan and Olivé 1964) and in the Museum of Cultures, founded in 1965, where Childe's concepts are amply applied to explain prehistory and civilization in the Old an in the New Worlds.

We may conclude that Childe's presence appeared late in Mexican archæology, beginning in the fifties, sponsored by the incorporation of European anthropologists into our academic environment; that it was disseminated in courses at the National School of Anthropology and History, applied in the field of prehistoric research and extended to archæological theory on Mesoamerica since 1958, in a line which has been continued with modifications by a minority or archæologists with a Marxist orientation, and that Childe's theories have been used in the National Museum of Anthropology and to a great extent in the Museum of Cultures. In order to better understand the above, we must bear in mind that at that time, that is in the fifties, the most significant contributions of Mexican archæology had already been made, on which basis work is still being carried out.

REFERENCES

ARMILLAS, P.

1951. "Tecnología, formaciones socioeconómicas y religión en Mesoamérica", in Sol Tax (ed): *The Civilization of Ancient America*, Selected Papers of the XXIXth International Congress of Americanists, The University of Chicago Press.

1952. Mesoamérica, apuntes de clase, ENAH, México.

1957. "Cronología y Periodificación de la Historia en América Precolombina", suplemento de *Tlatoani*, México.

BARBA DE PIÑA CHAN, B. and J.C. OLIVE

1964. El hombre estudia su pasado. Guión de la Sección de Arqueología. Sala de Introducción a la Antropología, Museo Nacional de Antropología, mecanoescrito.

BARTRA, R.

1964. "La tipología y la periodificación en el método arqueológico", suplemento de *Tlatoani* no. 5, México.

BATE, L. F.

1982. "Hacia la cuantificación de las fuerzas productivas en arqueología", in *Teorías, Métodos y Técnicas en Arqueología*, reimpresiones del Instituto Panamericano de Geografía e Historia, México.

BERNAL, I.

1979. *Historia de la arqueología en México*, Porrúa, México.

BINFORD, L.

1972. *An Archæological Perspective*, New York.

BOSCH GIMPERA, P.

1948. *Historia de Oriente*, Guatemala.

1957. "Gordon Childe prehistoriador", *Seminario de problemas Científicos y Filosóficos*, suplementos, segunda serie, Instituto de Investigaciones Filosóficas, UNAM, México.

BRAIDWOOD, R. J.

1961. *Prehistoric Men*, Chicago Natural History Museum, Popular Series.

CHILDE, V.G.

1942. *What Happened in History*, Penguin Books, Harmondsworth.

1947. *Archæology as a Social Science*, Institute of Archæology, London University. 1952. *New Light on the Most Ancient East*, Grove Press Books, New York. 1954. *Los orígenes de la civilización*, Fondo de Cultura Económica, México.

1958. *Sociedad y conocimiento*, Galatea, Nueva Visión, Buenos Aires.

1958b. *L'Europe Préhistorique*, Petite Bibliothèque Payot, Paris.

1965. *Man Makes Himself*, The Fontana Library, Great Britain.

1966. *La evolución de la sociedad*, Ciencia Nueva, Madrid.

1973. *Progreso y arqueología*, La Pléyade, Buenos Aires.

1974. *Teoría de la historia*, La Pléyade, Buenos Aires.

1976. *Nacimiento de las civilizaciones orientales*, Península, Madrid.

DANIEL, G.

1973. *El concepto de prehistoria*, Labor, Buenos Aires.

1974. *Historia de la arqueología*, Alianza Editorial, Madrid.

1976. *A Hundred and Fifty Years of Archæology*, Harvard University Press, Massachusetts.

1981. *Towards a History of Archæology*, Thames and Hudson, London.

FLANNERY, K. V. (ed)

1976. *The Early Mesoamerican Village*, Academic Press, New York.

GANDARA, M.

1977. La arqueología oficial mexicana, Ma thesis, ENAH.

1982. "La vieja 'nueva arqueología'", in *Teoría, métodos y técnicas en Arqueología*, reimpresos del Instituto Panamericano de Geografía e Historia.

HARRIS, M.

1978. *El desarrollo de la teoría antropológica*, Siglo XXI de España.

LORENZO, J. L.

1961. *La Revolución Neolítica en Mesoamérica*, INAH, México.

1975. "Los primeros pobladores", in *México: Panorama histórico y cultural* vol. VI, INAH, México.

1976. *La arqueología mexicana y los arqueólogos norteamericanos*, Depto. de Prehistoria no. 14, INAH.

1981. "Notes on the History of Iberoamerican Archæology", in Glyn Daniel (ed): *Towards a History of Archæology*, Thames and Hudson, London.

MACNEISH, R. S.

1964. *El origen de la civilización mesoamericana visto desde Tehuacán*, INAH, México.

1973. "The Origins of New World Civilization", in *Readings from Scientific American*, W.H. Freeman and Co, San Francisco.

MANZANILLA, L.

1979. Comentarios en torno a un proceso histórico: la constitución de la sociedad urbana en Mesopotamia (cuarto milenio AC), Ma thesis, ENAH. (Published in 1986 at the UNAM.)

1983. "Redistribución y tributo en el Centro de México. Hipótesis y consideraciones generales", *XIX Mesa Redonda de la Sociedad Mexicana de Antropología*, Querétaro.

MATOS MOCTEZUMA, E.

1972. "Introducción y selección", in Manuel Gamio: *Arqueología e Indigenismo*, SEP, México.

1979. "Las corrientes arqueológicas en México", *Nueva Antropología* año III, no. 12.

1982. "El proceso de desarrollo en Mesoamérica", sobretiro del *Boletín de Antropología Americana*, Instituto Panamericano de Geografía e Historia.

1986. *Vida y muerte en el Templo Mayor*, Océano, México.

MARTINEZ DEL RIO, P.

1943. *Los orígenes americanos*, Páginas del Siglo XX, México.

OLIVE NEGRETE, J. C.

1958. *Estructura y dinámica de Mesoamérica*, Acta Anthropológica Epoca 2, vol. 1, no. 3, ENAH.

1980. *La antropología mexicana*, Colegio Mexicano de Antropólogos.

ORTEGA Y MEDINA, J.

1953. "Monroísmo arqueológico", *Cuadernos Americanos*, México.

PANAMEÑO, R. and E. NALDA

1979. "Arqueología: ¿para quién?", *Nueva Antropología* año III, no. 12, México.

PEREZ, J. A.

1981. *Presencia de V. Gordon Childe*, INAH.

PIÑA CHAN, R.

1953. *Las culturas preclásicas de la Cuenca de México*, Fondo de Cultura Económica.

1975. "Introducción", *México: Panorama histórico y cultural*, SEP, México.

THE NEOLITHIC REVOLUTION

CLIMATIC CHANGE AND THE ORIGINS OF AGRICULTURE

ROGER BYRNE
University of California, Berkeley

Introduction

Gordon Childe's version of the so-called "dessication hypothesis" was first published in the late 1920's, at a time when very little was known about early agriculture or its environmental context (Childe 1928). Twenty years later Braidwood and his associates began their work in Iraq and for the first time direct evidence relating to the agricultural origins problem became available. As is well known, the evidence uncovered did not support Childe's model and Braidwood went on to conclude that changes in culture not climate had been responsible for the origins of agriculture (Braidwood and Howe 1960). More recently, Braidwood's own ideas on the subject have been seriously questioned. Even so, most contemporary students of the problem appear to agree with him that climatic change was not an important causal factor in the transition to agriculture. This rejection of climate has been based on several arguments three of which are particularly important.

1. There was no significant change in climate at the time of the transition to agriculture. This is sometimes referred to as the continuity theory and has been proposed for both the Near East and Mesoamerica (Hole 1966; MacNeish 1964).

2. There was a change in climate at about the same time that plants and animals were first being domesticated, but similar changes in climate occurred earlier in the Pleistocene without any cultural response (Cohen 1977).

3. Early agriculture developed independently at different times and in several climatically different areas so no single change in climate could have been responsible (Higgs and Jarman 1972; Bender 1975).

The primary purpose of this paper is to show that these objections are to a greater or lesser degree invalid, and that climatic change did play an important causal role in the transition for hunting and gathering to agriculture.

I must emphasize at the outset, however, I am not proposing a deterministic model. Climatic change did not in itself force people to domesticate plants and animals. What I am proposing is that changes in climate during the late-Pleistocene and early Holocene brought about important changes in wild plant and animal populations and that in certain key areas this led to important changes in human subsistence activites. I am also suggesting that it was these new subsistence activities that eventually led to domestication and agriculture. This latter topic is, however, beyond the scope of the present paper. Here I only consider the changes in climate that set the stage for the origins of agriculture.

Common Denominations

It is now generally accepted that agriculture developed independently in several widely separated areas of the globe. It is also recognized that these early centers were in many ways different, both environmentally and culturally. All of this has led most students of the agricultural origins problem to reject the possibility of any generally applicable model or theory. The point I want to emphasize here is that even though the early centers were in many respects different they were also in some respects similar. These similarities, here called "common denominators", are particularly important in that they provide clues as to why plants and animals were first domesticated. Three in particular merit close attention: the chronology of domestication; the autecology of the earliest domesticates; and the present climate of the early centers.

Chronology

Prior to the late-nineteen forties virtually nothing was known about the chronology of domestication and all discussion of the subject was of necessity very theoretical. Since then numerous archæological projects have focussed on early agricultural sites, both in the Near East and elsewhere, and a considerable body of relevant evidence has been recovered. Figure 1 presents reported dates for early plant domestication in various areas of the world and although their reliability is clearly not uniform the overall pattern is probably reasonably correct.

The most obvious and important feature of the map is that during a relatively short period of time (12,000 BP to 8,000 BP) several different kinds of plants were domesticated in different areas of the globe. These early domestications were not literally synchronous but when seen against the long perspective of human prehistory the timing of events was remarkably close.

The closeness of the chronological correlation has become increasingly apparent in recent years for two main reasons. Archæological work in South America and in Asia has shown that plants were domesticated there much earlier than was formerly thought to have been the case. Also, recent botanical research on the evolutionary history of certain crop plants, as for example *Zea mays* spp. *mays*, has shown that earlier estimates as to when domestication first occurred have in some cases been too conservative (Beadle 1980; Iltis 1983).

The fundamental importance of the chronological correlation is that it demands an explanation. If one excludes the unlikely possibility that it represents a mere chance, one is forced to conclude that some common causal factor was involved. This point has been most forcefully made by Cohen (1977), for whom the common causal factor was population pressure. In this paper I shall argue that climatic change was the common causal factor and that population pressure was only of secondary importance.

Autecology of the Earliest Plant Domesticates

One of the most striking characteristics of early agriculture is that so many of the early domesticates were annuals. Annuals, or therophytes as they were called by Raunkaier, complete their life cycle in less than a year and in so doing avoid conditions unfavorable for growth. Annuals are particularly common in disturbed habitats where bare soil surfaces are more or less regularly exposed by either natural or cultural disturbance. Most weeds are annuals and by definition are adapted to cultural disturbances. Annuals are also common in desert areas where they germinate after a period of rain and complete their life cycle from seed to seed in a matter of weeks. Such plants are often referred to as desert ephemerals. A third kind of environment in which annuals form locally important components of the wild vegetation are areas with a seasonally dry climate. In Köppen terminology such areas would include mediterranean-type (Cs), savanna-type (Aw) and monsoonal (Am) climates. Most of the early domesticates appear to have evolved from wild ancestors that were adapted to these seasonally-dry climates. Although most of the earliest plant domesticates were seed-bearing annuals a number of plants with subsurface storage organs were also domesticated at a relatively early date. These plants are characterized by thickened roots, tubers, or bulbs and in most cases are geophytes in the Raunkerian system. Unfortunately, geophytes are rarely preserved in archæological contexts and as a result the history of their early domestication is not as well known as that of the seed crops. What little evidence there is indicates that at least some of them were domesticated at more or less the same time as the early seed crops.

Several authors have drawn attention to the ways in which root crops differ from seed crops (Sauer 1952; Harris 1977). The former, for example, are usually dependent upon asexual reproduction and are planted as slips or cuttings whereas the latter reproduce sexually and are planted as seeds. What is more important in the context of the present paper is that the domesticated annuals and the domesticated geophytes are both similar insofar as their life cycles represent adaptations to seasonal climates. The annual avoids the period of drought stress as a dormant seed; the geophyte stores energy in the underground storage organ so that it can grow quickly during the short growing season.

The Present Climate of the Early Centers

An obvious first question here is how can any discussion of present climate help us understand what happened 10,000 years ago? My answer is that even though the present climate of the early centers is not exactly the same as that of 10,000 years ago the present does provide a partial key to the past. The important point to note here is that all of the early centers are located in areas that today have strongly seasonal rainfall regimes.

The Near Eastern center is distinctive in that it has a Mediterranean-type climate which is characterized by winter rain and summer drought. All the other centers have summer rain and winter drought. They are monsoonal (Am) or savanna-type climates (Aw) in the Köppean sense. In the Andes and northern China, there is also a marked seasonal range in temperature. However, as I will suggest later, seasonality of rainfall was probably more important as far as the initial domestication of plants was concerned.

In summary three important conclusions can be drawn from this preliminary discussion:

1. During the period 12,000 to 8,000 BP several species of plants were domesticated in widely separated areas of the globe.
2. Most of these early domesticates were annuals or geophytes and were adapted to seasonal rainfall regimes.
3. The early agricultural centers were all located in areas that today are characterized by strongly seasonal rainfall regimes.

None of these conclusions is new or controversial. What I would suggest here, however, is that their relevance to the agricultural origins problems has been overlooked. It has been overlooked because of several widely held misconceptions as to the nature of late-Pleistocene/early-Holocene climatic change.

One important misconception is that seasonally wet and dry climates have always been part of the global pattern. The location of these "climatic types" is assumed to have changed during glacial/interglacial cycles but their continued existence has never been questioned. The alternative viewpoint I am presenting here is that seasonal rainfall regimes, such as now characterize Mediterranean and monsoonal-type climates, only developed during the late-Pleistocene and early-Holocene. Similar climatic conditions may have existed during earlier interglacials, as for example 130,000 BP to 120,000 BP, but they did not exist during the intervening glacial episodes.

On the Nature of Late-Pleistocene/Early-Holocene Climatic Change

Most of the research dealing with the climatic history of the early centers of agriculture has focussed on proxy records from the centers themselves. This regional approach has produced much of interest but it has not provided any convincing answers to the question as to why plants and animals were domesticated in the first place. What I am suggesting here is that insofar as the agricultural origins problem is a global problem any explanatory model must also be global in scope.

Ten years ago the global pattern of late-Pleistocene/early-Holocene climatic change was far from being well understood. Now thanks to the work of the CLIMAP group and others we have a fairly detailed record (CLIMAP 1976). We also have a better understanding of why and how these changes occurred. For example, it is now generally accepted that primary climatic control responsible for the glacial-interglacial cycles is the change in insolation that results from changes in earth/sun relationships. In other words, Milankovich's astronomical theory of climatic change has been confirmed (Hays *et al.* 1976).

Milankovich Confirmed

Milankovich postulated that the glacial/interglacial cycles of the Quaternary were the result of changes in earth/sun relationships. Three variable conditions were involved: the tilt of the earth's axis, the eccentricity of the earth's orbit around the sun, and precession effect which changes the timing of the aphelion and perihelion. Because these variable conditions change at a known rate Milankovich was able to calculate their combined effect in terms of changes in insolation received at different latitudes for the past million years.

In brief, the theory states that continental ice sheets can only begin to form in high latitudes when summers are cooler than average and winters are warmer than average; the assumption here being that cool summers are needed if snow is to accumulate on the ground and form ice, and that warm winters will mean more storms and more precipitation in the form of snow at high latitudes. Conversely, it is postulated that ice sheets will begin to melt when summers are warmer and winters colder, because the winter's snowfall will not survive the summer to form ice and the cooler winter will mean fewer storms and less precipitation. The important point in the context of the present paper is that increased seasonality of radiation leads to the melting of the continental ice sheets.

The striking correlation between ice volume and solar radiations is shown in Figure 2 which is from a paper by Ruddiman and MacIntyre (1981). Deglaciations are associated with warm summers and cold winters, whereas ice accumulation episodes are associated with cold summers and warm winters.

Note especially the record for the past 20,000 years. At about 20,000 BP the insolation values change from a cold summer/warm winter pattern to a warm summer/cold winter pattern. This marks the beginning of the last deglaciation. For the next 10,000 years the seasonal contrast in insolation increases until it reaches a maximum at around 10,000 BP after which it declines towards the present.

The 10,000 BP seasonality maximum is suspiciously close to the dates for the first appearance of domesticates but the question remains as to how zonal changes in radiation could create conditions favorable for domestication in localized areas.

As was emphasized earlier, the early centers of agriculture are today characterized by strongly seasonal rainfall regimes. The more appropriate question then in this context is how did increased seasonality of radiation lead to increased seasonality of rainfall? This question is a complicated one and is best dealt with in two parts. First, I will consider the situation in the Near East, and then secondly the situation in the other early centers.

Increased Seasonality in the Near East

According to the "dessication hypothesis" the melting of the Fennoscandian ice sheet led to a northward movement of the Jet Stream and fewer winter storms in the Near East. Fossil evidence from the North Atlantic now indicates that what actually happened was very different.

According to Ruddiman and MacIntyre (1981) an important development in the North Atlantic during the period 16,000 to 13,000 years BP was the formation of an extensive surface layer of brackish water derived from the melting ice sheets. Even though the layer was not very thick it had a very important effect on climate. Because of its lower density the meltwater formed a surface layer and inhabited mixing which in turn meant that penetration of heat during the summer was restricted. Conversely, the same heat was quickly lost in the fall and winter. Low winter temperatures, in part also due to lower insolation, permitted the formation of extensive sea ice as far south as 45 degrees north. This would have significantly reduced storm activity and precipitation over the ice sheets but at the same time would have increased precipitation to the south. On Ruddiman and MacIntyre's map of the deglaciation phase the storm track is indicated as heading directly towards the Mediterranean. In contrast, on their map of the ice build up phase it follows the east coast of North America.

Hydrological conditions in the eastern Mediterranean were also very different during the late-Pleistocene and early Holocene. According to Rossignol-Strick *et al.* (1982), increased run off from the Nile during the period 12,500 BP through 8,000 BP also led to the development of a brackish-water layer at the surface.

What are the implications of all this for late-Pleistocene/early-Holocene climatic change in the Near East? One reasonable assumption would seem to be that a trend towards warmer summers began around 20,000 BP and continued to the early Holocene during which time summers may have been even warmer than at present. One might also assume that this season would have been dry because of strong subsidence in the subtropical high pressure cell. Also, higher summer temperatures would have meant increased evaporation rates. Winter conditions are more difficult to estimate as far as temperature is concerned because even though insolation decreased during this time period more frequent passage of cyclonic storms could have kept temperatures relatively high. The most important change during the winter would have been the more frequent storms and a corresponding increase in precipitation. In general the period would have been characterized by increased seasonality of temperature, more intense summer drought, and higher rainfall in winter.

Increased Seasonality in the Tropics

Unlike the Near East all the other early centers of agriculture are located in areas that today are characterized by summer rain and winter drought. The key question here then is whether or not the development of the summer rainfall regime was also a late-Pleistocene/early-Holocene phenomenon. At first glance it might seem unlikely that the same kinds of changes postulated above for the Near East could also have occurred in the tropics and sub-tropics. In actual fact, however, there is good evidence that similar changes did indeed occur and that, furthermore, they occurred at more or less the same time.

The case for increased seasonality of summer rainfall is in fact an extension of the Milankovich Effect discussed above. The same increase in summer radiation that brought about the melting of the Laurentide and Scandinavian ice sheets was responsible for an intensified summer monsoon. Kutzbach and Otto-Bliesner (1982) ran a general circulation model that incorporated the summer radiation increase and concluded that the strenghtened monsoonal circulation would have meant precipitation increases of between 15 an 26 percent over a large area of northern Africa and Asia. Unfortunately, North America was nor included in their reconstruction and estimates of increased precipitation for Mesoamerica are therefore not available. However, it seems reasonable to assume that similar changes occurred there although perhaps on a somewhat lesser scale. Another limitation of the model is that it does not take into account sea surface temperatures, the assumption here being that sea surface temperatures would not be changed by seasonal changes in radiation receipt that did not involve a net annual increase or decrease.

On this latter point, it is interesting to note that Jones and Ruddiman (1982) have postulated that increased summer rainfall during the early Holocene would have led to the development of extensive areas of brackish surface water in those areas of the tropical oceans influenced by large river systems. As in the case of the North Atlantic, these brackish layers would have accentuated the seasonality that had brought about their existence in the first place. All of this does not invalidate the Kutzbach/Otto–Bliesner model but rather suggests that the estimates of summer rainfall are probably conservative, at least for those continental areas downwind of the brackish layers.

Seasonal Climates without Early Agriculture

One interesting feature of the geography of early agriculture is that the Near East was the only area of Mediterranean-type climate in which plants were domesticated at an early date. The other areas of winter-rain climate, such as southwest Australia, southernmost South Africa, central Chile, California, and the western Mediterranean, have produced no evidence of early agriculture. These have often been cited as evidence that climatic change could not have been an important causal factor in the transition to agriculture (Carter 1977). However, this argument is invalid insofar as it underestimates the distinctiveness of Near Eastern climate and climatic history.

As was mentioned earlier, seasonality would have been intensified in the Near East during the late-Pleistocene and early-Holocene because of thermal stratification in both the North Atlantic and the Mediterranean. Seasonality would have been less marked in the other winter-rain areas which are all located next to large, climatically-conservative oceans in which meltwater effects were not important. Even today, seasonality of rainfall and temperature is more extreme in the Near East than in the other areas of Mediterranean climate. In the Near East most of the rainfall occurs in a period of only three months. Also, the summers are characterized by high temperatures and low humidities, which together combine to cause extreme moisture stress. The severity of summer drought is partly attributable to the persistence of the subtropical high pressure cell and also to the large extent of desert in adjacent areas.

In the summer rain areas a similar pattern of early centers and seasonal "non-centers" is evident, but in this case the contrast is an hemispheric one. Almost all of the summer rain centers are located in the Northern Hemisphere: Mesoamerica, northern India, Southeast Asia, northern China, and southern China. Only two are located south of the equator: the eastern Andes, and New Guinea. There are extensive areas of summer rain climate in Australia, southern Africa, and eastern South America, but as yet they have not produced any evidence of early agriculture.

A most important point to note here is that the Milankovich effect operated differently in the two hemispheres. During the late-Pleistocene and early-Holocene, seasonality of radiation was at a maximum in the northern hemisphere, whereas in the southern hemisphere it was at a minimum. This was primarily due to the precession effect, that is the changing date of the perihelion. At 9,000 BP the earth was closest to the sun in July; at present the perihelion date is in January. A summer perihelion increases seasonality of radiation, whereas a winter perihelion reduces it.

Also, important in this context is the fact that seasonality of temperature in the southern hemisphere is reduced relative to the northern hemisphere because of the moderating influence of the oceans. As far as precipitation is concerned, the seasonal migration of the general circulation does result in areas with summer rain and winter drought, but there is no well developed monsoonal circulation. The only exception would be in the East Indies and northern Australia, and here the monsoon is strongly influenced by northern hemisphere developments anyway.

Increased Variability of Climate

Increased seasonality of climate was probably not the only aspect of late-Pleistocene/early Holocene climatic change that was involved in the origins of agriculture. Increased short-term variability could also have been a factor, especially on the intra-seasonal and inter-seasonal time scales. I use the qualifier "could" because the evidence for increased variability is not as good as the evidence for increased seasonality. However, on theoretical grounds to assume that such a change occurred is perfectly reasonable, especially for the northern hemisphere.

The basic point here is that during the full-glacial, the Laurentide and Scandinavian ice sheets would have had important stabilizing influence on the westerly circulation in the mid-latitudes of the northern hemisphere, and perhaps also on the Hadley circulation as well. The thermal contrast between the ice sheets and the unglaciated areas to the south would have caused the mid-latitude Jet Stream and its associated cyclonic disturbance to follow a more or less predictable path. Also, there would have been relatively little variation in the longitudinal position the troughs and ridges in the upper air westerlies.

As the continental ice sheets melted in the northern hemisphere their conservative influence on the general circulation would have progressively diminished and this in turn would have increased the variability of weather and climate at the surface. At present, for example, there is considerable variability in mid-latitude climate both on intra-seasonal and inter-seasonal time scales. Much of this variability reflects changes in the arrangement of the circumpolar vortex which in turn may be caused by relatively short term changes in the surface temperature of the oceans.

This discussion of increased variability has been restricted to the northern hemisphere because there was no equivalent effect in the southern hemisphere. The Antartic ice sheet did melt back to some extent during the last deglaciation, but change was minimal compared with what happened in the northern hemisphere.

Some Biological Consequences of Late-Pleistocene/Early-Holocene Climatic Change

In 1977 Wright reviewed the evidence for a possible connection between climatic change and the origins of agriculture and concluded that although climatic change may have been important in the Near East there was no evidence of any similar developments in the other early centers (Wright 1977). It should be emphasized, however, that Wright assumed that agriculture developed much later in the New World than in the Old World, and not surprisingly therefore he was unable to identify any common causal factors. He did, however, conclude that climatic and vegetational changes at the end of the Pleistocene were just as pronounced in the New World as in the Old World.

What I want to review briefly in the rest of this paper are some of the possible biological consequences of increased seasonality during the Pleistocene/Holocene transition.

Changes in Wild Plant Populations

In general, the lack of fossil evidence for morphological changes in plants during the late-Pleistocene and early-Holocene has led paleobotanists and others to conclude that migration rather than evolution was the more important strategy. Admittedly, this probably was the case for woody plants; however, for herbaceous taxa evolutionary changes may have been important. Two adaptations that could have occurred are:

1) An increase in annuals,

2) An increase in geophytes.

The idea that climatic changes at the end of the Pleistocene provoked the evolution of annuals was first proposed by Whyte (1977). Whyte's thesis is basically similar to the one presented here. It differs, however, in that it emphasizes increased aridity rather than seasonality, and also in that it deals only with Asia.

An increase in the seasonality of rainfall would increase the importance of annuals in two ways. Perennials would be at a disadvantage in the newly seasonal climate and there would be selection for the ability to adopt the annual habit. Also, any annuals that had been present prior to the change in climate would now have a competitive advantage and therefore become more important.

A related but more complex response to increased seasonality may have been an increase in seed size in annuals. Obviously climate is not the only determinant of seed size and many other factors are likely to be involved, such as predation, competition, dispersal strategy, edaptic conditions, etc. Even so there is some evidence that increased drought stress is positively associated with seed size (Baker 1972), and it seems reasonable to conclude therefore that a change in climate involving drier summers, or winters, could provoke the evolution of large seededness.

A second evolutionary development in areas that were becoming more seasonal could have been an increase in the importance of geophytes. Geophytes are often described as plants which store carbohydrates in order to survive the dry season. This is not strictly correct, however, because the more important function of the stored carbohydrate is to permit rapid growth when favorable conditions return. This gives the geophyte a competitive advantage over plants that have to germinate seed and put down root systems.

Changes in Wild Animal Populations

The idea that increased seasonality of climate was responsible for the extinction of the Pleistocene megafauna has been increasingly popular in recent years (Axelrod 1967; Slaughter 1967; Guthrie 1984). Needless to say I also find it convincing. I am also aware, however, of the argument that faunal changes in the early agricultural centers were not important enough to provide a strong case for any climatic effect. I would suggest, however, that this interpretation overlooks certain aspects of the faunal record.

Time and space limitations prevent a detailed discussion of this theme here, but I would like to draw attention to one aspect of the problem, namely the consequences of increased seasonality for grazing animals. As Guthrie (1984) has pointed out, caecalid grazers, such as certain equids and gomphotheres, were particularly hard hit by the environmental changes of the late-Pleistocene/early-Holocene. This was certainly true in several of the early agricultural centers.

Caecalids, unlike the ruminants, have digestive systems that are designed to process large volumes of low quality forage. It seems reasonable to conclude, therefore, that they evolved in environments where perennial grasses and forbs were always available. Such animals would be especially vulnerable to any change in climate that reduced the importance of perennials and increased the importance of annuals. In some areas, such as East Africa, the seasonality problem could be avoided by latitudinal migration. In other areas, such as the Near East, altitudinal migration would be the only option; and in this case, the availabity of summer pasture would be an important constraint.

It is tempting to conclude, therefore, that the demise of the caecalid grazers, such as horses and mammoths, led to a reduction in grazing pressure on the newly evolving annuals. Furthermore, if this was the case, it could have led to the development of monospecific stands of wild cereals, such as characterize certain *ungrazed* areas of the Near East today (Zohary 1969). As is well known in ecology, reduced predation typically leads to a reduction in species diversity (Krebs 1972). I must acknowledge that my awareness of the potential significance of this point has come from discussions with M.A. Blumler of the U.C. Berkeley Geography Department.

Conclusion

In conclusion let me first reiterate the point that I made at the beginning; namely, that I did not intend to deal with the origins of agriculture *per se*. What I have tried to do is to show how late-Pleistocene/early-Holocene climatic change could have led to an increased reliance on wild plant foods. More specifically, I have argued that, in certain key areas, increased seasonality of rainfall would have provoked the evolution of annuals and geophytes. Furthermore, the evolution of these plants would have made possible the development of essentially new subsistence strategies.

Finally, let me again refer to Gordon Childe's ideas on the origins of agriculture. It is true that the dessication hypothesis has not been supported by more recent evidence, and in this sense Childe was wrong. On the other hand, I would argue, that he was right insofar as he recognized a causal connection between deglaciation and to use his own words "the ... revolution that transformed human economy and gave man control over his own food sypply" (Childe 1951).

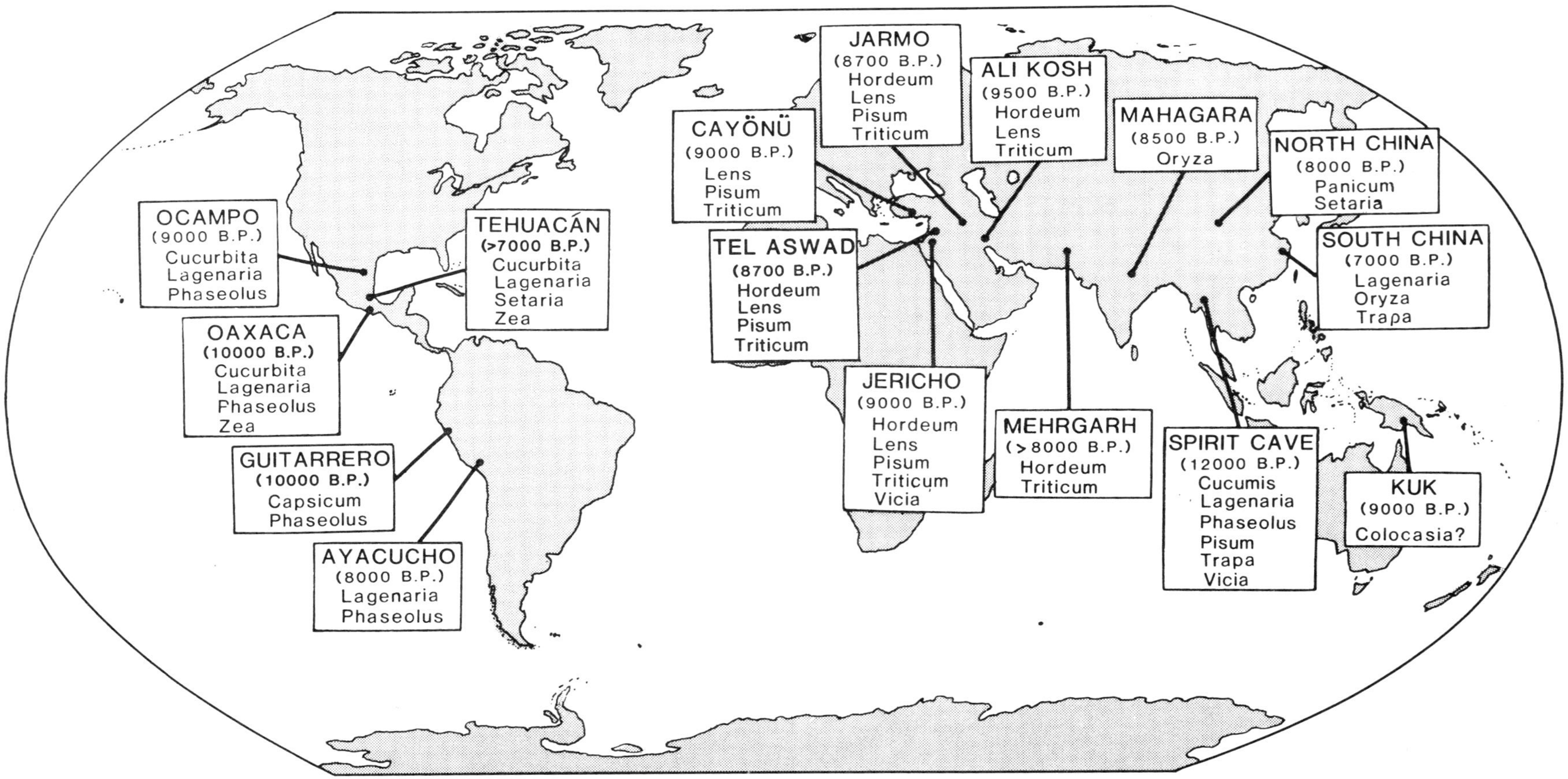

Figure 1 The Global Pattern of Early Plant Domestication
See Harlan (1976) and text

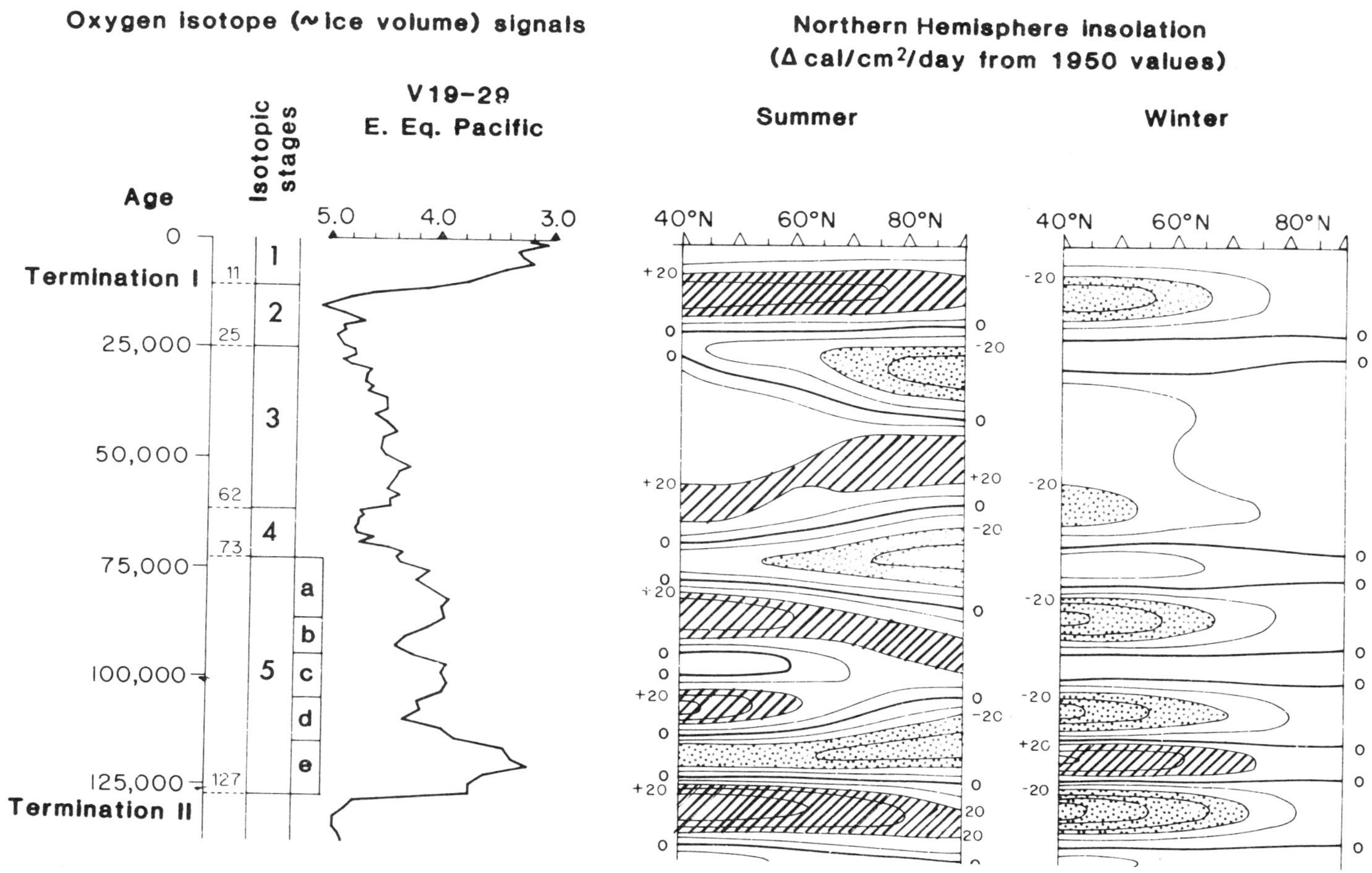

Figure 2 Late Quaternary Ice Volume and Solar Radiation
After: Ruddiman & McIntyre (1981)

REFERENCES

AXELROD, D.I.

1967. "Quaternary Extinctions of Large Mammals", *University of California Publications in Geological Science* vol. 74: 1-42.

BAKER, H.G.

1972. "Seed Weight in Relation to Environmental Conditions in California", *Ecology* vol. 53: 997-1010.

BEADLE, G.W.

1980. "The Ancestry of Corn", *Scientific American* vol. 242: 112-119.

BENDER, B.

1975. *Farming in Prehistory From Hunter-Gatherer to Food-Producer*, St. Martin's Press, New York.

BLUMLER, M.A.

1984. Climate and the Annual Habit. Unpublished M.A. thesis, Geograhy Department, University of California, Berkeley.

BRAIDWOOD, R.J. and B. HOWE

1960. *Prehistoric Investigations in Iraqi Kurdistan,* University of Chicago, Oriental Institute, Studies in Ancient Oriental Civilization No. 31.

CARTER, G.F.

1977. "A Hypothesis Suggesting a Single Origin of Agriculture", in *Origins of Agriculture,* C.A. Reed (ed.), Mouton, The Hague: 89-133.

CHILDE, V.G.

1928. *The Most Ancient East*, Kegan, Paul, Trench and Trubner, London.

1951. *Man Makes Himself*, New American Library, New York.

CLIMAP PROJECT MEMBERS

1976. "The Surface of the Ice-Age Earth", *Science* vol. 191: 1131-1137.

COHEN, M.N.

1977. *The Food Crisis in Prehistory,* Yale University Press, New Haven.

GUTHRIE, R.D.

1984. "Mosaics, Allelochemicals and Nutrients: An Ecological Theory of Late-Pleistocene Megafaunal Extinctions", in *Quaternary Extinctions: A Prehistoric Revolution*, P.S. Martin and R.G. Klein (eds.), University of Arizona Press, Tucson: 259-298.

HARLAN, J.R.

1976. "The Plants and Animals that Nourish Man", *Scientific American* vol. 235: 88-97.

HARLAN, J.R. and D. ZOHARY

1966. "Distribution of Wild Wheats and Barley", *Science* vol. 153: 1074-1080.

HARRIS, D.

1977. "Alternative Pathways Toward Agriculture", in *Origins of Agriculture,* C.A. Reed (ed.), Mouton, The Hague: 179-243.

HAYS, J.D., J. IMBRIE and N.J., SHACKLETON

1976. "Variations in the Earth's Orbit: Pacemaker of the Ice Ages", *Science* vol. 194: 1121.

HIGGS, E.S. and JARMAN, M.R.

1972. "The Origins of Animal and Plant Husbandry", in *Papers in Economic Prehistory,* E.S. Higgs (ed.), Cambridge University Press, Cambridge: 3-13. HOLE, F. 1966. "Investigating the Origins of Mesopotamian Civilization", *Science* vol. 153: 605-611.

ILTIS, H.H.

1983. "From Teosinte to Maize: The Catastrophic Sexual Transmutation", *Science* vol. 222: 886-894.

JONES, G.A. and W.F. RUDDIMAN

1982. "Assessing the Global Meltwater Spike", *Quaternary Research* vol. 17: 148-172.

KREBS, C.J.

1972. *Ecology: The Experimental Analysis of Distribution and Abundance,* Harper and Row, New York.

KUTZBACH, J.E. and B.L. OTTO-BLIESNER

1982. "The Sensitivity of the African-Asian Monsoonal Climate to Orbital Parameter Changes for 9000 BP in a Low-Resolution General Circulation Model", *Journal of Atmospheric Sciences,* vol. 39: 1177-1188.

MACNEISH, R.S.

1964. "Ancient Mesoamerican Civilization", *Science* vol. 143: 531-537.

ROSSIGNOL-STRICK, M., *et al.*

1982. "After the Deluge: Mediterranean Stagnation and Sapropel Formation", *Nature* vol. 295: 105-110.

RUDDIMAN, W.F. and A. MacINTYRE

1981. "Oceanic Mechanisms for Amplification of the 23,000-Year Ice-Volumen Cycle", *Science* vol. 212: 617-627.

SAUER, C.O.

1952. *Agricultural Origins and Dispersals,* American Geographical Society, New York.

SLAUGHTER, B.H.

1967. "Animal Ranges as a Clue to Late-Pleistocene Extinction", in *Pleistocene Extinctions: The Search for a Cause,* P.S. Martin and H.E. Wright, Jr., Yale University Press, New Haven: 155-168.

WHYTE, R.O.

1977. "The Botanical Neolithic Revolution", *Human Ecology* vol. 5: 209-222.

WRIGHT, H.E. (Jr.)

1977. "Environmental Change and the Origin of Agriculture in the Old and New Worlds", in *Origins of Agriculture,* C.A. Reed (ed.), Mouton, The Hague: 281-318.

ZOHARY, D.

1969. "The Progenitors of Wheat and Barley in Relation to Domestication and Agricultural Dispersal in the Old World", in *The Domestication and Exploitation of Plants and Animals,* P.J. Ucko and G.W. Dimbleby (eds.), Duckworth, London: 47-66.

THE NEOLITHIC REVOLUTION, A CASE ANALYSIS; A REEVALUATION OF THE CHILDEAN CONCEPT AS APPLIED TO JOMON, JAPAN

YOKO SUGIURA Y.
Universidad Nacional Autónoma de México

It was exactly fifty years ago when the book entitled *Man Makes Himself* appeared, which, together with *Social Evolution* and *What Happened in History*, form the trilogy considered by G. Daniel as one of the most important and penetrating works on historical synthesis of this century. In the above-mentioned book Childe first structures and defines his historical model, in which can be seen irrefutable influences of historical materialism or, as Childe himself calls it, "the realistic conception of history" (Childe 1951:9). Childe discusses the ability of archæology to elucidate radical changes in the economy and in the social system of production. These changes which occurred in prehistory are analogous to those that the "realistic concept of history" emphasized as factors that cause historical changes (Childe 1951:9), such as the Industrial Revolution of the eighteenth century. Childe's proposal marked a revolutionary point of departure since it has been for a long time the only established archæological model which explains historical discourse. Childe agrees with Marx, who "insisted on the preponderant importance of economic conditions, social forces of production and the application of science as casual factors in historical change" (Childe 1951:6). Although these revolutions are not "catastrophic changes", as they are usually considered, but rather processes (Childe 1951:99), we can infer on the other hand that they provoked a massive reordering in various levels of human existence.

Definition of the Neolithic Revolution proposed by Childe

The first of these revolutions took place in the Neolithic, when man acquired the ability to produce and control his own food supply by cultivating plants and by breeding domesticated animals, which in turn transformed a gathering economy into a food producing economy. This brought far-reaching changes, which had repercussions on all levels of human existence. Among these changes, Childe stands out the following aspects:

a. The size of hunting-gathering groups is a function of the amount of natural food resources and of technological efficiency. Intensification of food acquisition beyond the biotic potential can cause a gradual decrease in resources, and in some cases their extinction. This in turn is reflected in a population decrease. On the other hand, the new agricultural economy permits marked population growth (Childe 1951:69-83).

b. A food-producing economy has the ability to accumulate substantial food surplus which enables the community to survive during times of scarcity. Childe explicitely reiterates that the adoption of a mixed subsistence economy should not be confused with sedentary life, since a direct correlation between these two variables is ficticious and erroneous. In spite of this, it is implicitely deduced in his conception that there was a general trend towards sedentism among neolithic communities, given that with the "Neolithic Revolution" man could produce and manage food resources. On the other hand, this new economic capacity served as a basis for a rudimentary exchange, which later launched the second revolution.

c. A mixed subsistence base leads to the formation of autarkic communities, since the food-producing economy at this stage does not require foreign materials obtained through intercommunity exchange. Although Childe states that complete economic self-sufficiency had not yet been attained, the concept of autarky constitutes one of the medullar arguments of the Neolithic Revolution and as such it is considered as trascendental. In the first place, each community supports itself independently and is relatively isolated from other contemporary communities, due to its self-sufficient character. Hence the great cultural diversity among neolithic communities is explained. It is also deduced that intercommunity exchange plays a secondary role in a neolithic economy. Archæologists have identified objects exchanged among neolithic populations; however the products circulated were sumptuary items, or at least those not considered basic to existence. Exchange could have exercised a vital function as a mechanism for the diffusion of ideas, and of human progress in the long run. Nevertheless, since exchange was infrequent between neolithic communities, the filtration and diffusion of ideas were extremely slow, especially where unfavorable environmental conditions prevailed. For these reasons Childe emphasizes that exchange did not constitute an integral part of the neolithic economy. Only in exceptional cases does it acquire regularity and intercommunity specialization.

d. Food surpluses make it possible to maintain a segment of the population not directly dedicated to the production of its own food, and consequently social differences appear. On the other hand, cooperative activities are reflected in socio-political institutions, which in turn were consolidated and reinforced by magical-religious sanctions and ideology (Childe 1962:69).

Childe's conception of the neolithic economy and its relation to population growth and sedentism: a balance

Childe (1951:34) defines economy as the way men gain their subsistence, and he considers it the primary factor which determines the multiplication of our species and therefore its biological success. Based on this supposition, he asserts that the Neolithic Revolution or the initiation of a new food-producing economy resulted in marked population growth, which has been successful from a biological point of view, permitting the multiplication of our species (Childe 1951:35). Thus in Childe's interpretation there is a direct correlation between these two variables. On the other hand, subsistence based on a hunting and gathering economy does not have the same capacity to expand as does a food-producing economy. In other words, Childe viewed population growth as a correlative effect or a dependent variable of technological changes; in the Neolithic case this meant the incorporation of a food-producing technology and the advances in scientific knowledge.

Population growth and its relation to the food-producing economy constitutes an essential theme in discussions on social evolution (Cohen 1977; Sanders 1977, 1985; Polgar 1975; Smith and Young 1972; Spooner 1972; Boserup 1966, etc.). All agree that these two variables have a close correlation. But the central problem is the different conceptualizations of this relationship caused basically by two oppsed theoretical standpoints: the first, headed by Childe, sees population growth as the logical result of a mixed subsistence economy. Hence the demographic factor is a dependent variable. The second asserts exactly the opposite; namely, that population increase is the causal factor with regard to changes, while technology is a function of demographic pressure. The traditional conception established by Malthus and Ricardo in the nineteenth century, in which population growth is considered dependent on the biotic capacity and on inflexible technologies is implicitely accepted up to this very day. The anti-Malthusian thesis was espoused by Boserup (1965). In 1965 the book entitled ***The Conditions of***

Agricultural Growth, now considered a classic, was published. Its principal thesis is that population increase, considered an independent variable, can accelerate, and, in some cases, determine changes in agricultural technology and in social organization. That is to say, population pressure finds a solution not only in traditional regulating mechanisms such as infanticide, low fertility, migration or fissioning, but that it can lead to another alternative: the adoption of more intensive systems of soil use. Boserup's thesis is in turn supported by two principles: the elasticity of the productive capacity of the soil and the law of least effort (Spooner 1972: XVI). Other authors such as Adams (1966) and Cowgill (1975 a, b) observe that the conception of population behavior and pressure as an independent varible —a self-generating and extra-cultural factor— can lead us in the wrong direction. Other authors suggest that, to explain historical processes it is more effective to conceive of factors such as population growth, political stability and subsistence means as short-term, independent variables at most, whose alternation can cause variations on other variables, since a permanent, independent-dependent relationship between population and other variables does not exist (Hassan 1975; Dumond 1965). Faris' statements in this respect (1975) is even more decisive. He conceives the populations dynamics as subjected to particular social relations. Hence, demographic behavior changes through time, and for that reason population dynamics can not be understood without knowing first the inherent potentials of the determined social system, which conditions such dynamics.

On the other hand, ethnographic data has revealed that the take-off threshold of population growth is intrinsically related to sedentary life (Lee 1972). This last in turn has been frequently seen as one of the prerequisites for a food-producing economy. And in this sense the two variables are interrelated and form part of mutually conditioning pocesses (Kabo 1985), but the relationship is not determined. In fact, ethnographic studies make it clear that sedentism does exist in a pre-agricultural society, and that, in some cases, it is developed even without any evidence of significant new technology (Cohen 1977), if the required conditions, such as the abundance, richness and concentration of food resources, and also of food processing and storage technologies, as well as the restriction or prohibition of movement because of neighboring groups or the attraction of individuals to points to exchange, are met (Hodder 1982; Sanders 1985).

The Jomon culture; Japan as a casuistic example

Around 12,000 BP the first indications of a culture that already had quite elaborate pottery appear, thus proclaiming a new historical stage in Japan. Its basic mode of life is defined in the so-called "Jomon culture" from about 10,000 BP This culture lasts for an extraordinarily long lenght of time, and is divided into six phases: Incipient (11,000–7500 BC), Initial (7500–5300 BC), Early (5300–3600 BC), Middle (3600–2500 BC), Late (2500–1000 BC) and Terminal (1000–300 BC) (Akazawa 1982). Over these 10,000 years the Jomon culture revealed an intricate process and reached a high level of socio-economic complexity without undergoing radical changes in its subsistence pattern basically on hunting, gathering and fishing.

Process of sedentarization and demographic trends through settlement patterns

When the first evidence for ceramics appears, a cooler climate than at present still prevailed in Japan. In this Incipient phase men had still not achieved a sedentary life. They frequently still lived in rockshelters, although the occupation of new zones, such as the narrow alluvial terraces, the slopes and tops of peninsular hills and some islands, begins in the next phase. Settlements are small in size, as in the case of Ikenoki (Shizuoka), Akamidoo (Aomori), Hanawadai (Ibaragi), and Nishinozyoo (Chiba), where semi-subterranean dwelling-houses with a circular or rectangular floor plan have been located. Most of the residential sites, some associated with shell-mounds, consist of some five houses, of which about three were used at one given time (Okamoto 1974).

The trend towards sedentism becomes increasingly pronounced during the Early Jomon phase, when regional differences in the pottery tradition develop. Central and northeastern Honshu emerge as the focal region in all cultural aspects. Unlike the earlier phase, settlements are mostly located on top of plateaus or on the tips of extended hills not far from water sources and a favorable environment. The complexity and size of settlements increase considerably, as well as the lenght of occupation within determined sites. At the same time a pronounced variability in settlement size appears. As a rule some ten house-dwellings were occupied simultaneously, although more complex and larger sites covering some 50,000 square meters in area such as Aku (Nagano) existed. In this site 29 semi-subterranean houses have been located around a central space delimited with stones. Another site of similar size is the Minamibori shellmound (Yokohama), located on the edge of a hilltop, where 48 houses associated with a large number of storage pits surround a central plaza some 200 m^2 in area. The size of the houses has also increased, with an average floor space of some 22 m^2, although they range from 5 m^2 to 60 m^2 in area. Likewise construction quality improves and for the first time hearths appear inside the houses.

In the next phase the Jomon culture reaches its apogee in all senses of the word. Central and northeastern Honshu continued to play a local role. The intrinsic characteristics of the sites located on alluvial terraces, slopes and hilltops close to springs or water holes, indicate a fully sedentary life. An explosive increase in the number of sites is observed in the coastal region, as well as in the mountanious interior. The variability in site size is accentuated. Higher ranking sites are characterized by the large number of houses associated with a series of storage pits, burials and a high density of local and foreign materials, as in Miharada (Gunma), Okinohara (Niigata) and Takanekido (Chiba). House distribution continues the pattern begun in the preceding phase, arranged around a central space shaped like a horseshoe. At Miharada about 350 houses are distributed around a circular space. During the Middle Jomon phase a growing tendency to large settlements can be observed; for instance in the Tama region (Tokyo) large sites are distributed at a distance of 2 to 3 km. from each other along the Ookuri River (Kani 1982).

Along with these large settlements, medium-sized sites are found, with about ten houses arranged around a central plaza and associated with a smaller number of storage places, burials and a few sumptuary objects. Apparently sites of this category had a more limited occupation.

Another idiosyncratic feature of this phase is the notable increase in the number of large structures, ranging in size from 40 m^2 to 107 m^2, and with several hearths inside. These are distributed mainly in the northeastern region of Honshu, where heavy snowfalls occur during the winter (Togashi 1982). Some authors are inclined to think that these structures were used in food preparation (Watanabe 1977), although others suggest a socio-political function.

On the other hand, circular structures delimited by a series of large cobblestones but without evidence of domestic use appear, as well as other stone-covered circular areas with hearths. The association of the former with sumptuary ceramic and lithic objects plus the absence of articles of a utilitarian character lead us to believe that their function was principally religious or ceremonial in nature.

The extraordinary increase in the number of sites during the Middle Jomon phase reflects in turn a notable population growth. Based on data from the mid-eighth century, Koyama calculated the Middle Jomon population at about 26,000 persons, and population density at 0.9 per km^2, which indicates that the population had doubled that of the preceding phase. The difference in population density between eastern and western Japan seems clear. According to these estimates, 96% of the total population was concentrated in the west, especially in the central region of Honshu, where a density of two to three persons per km^2 is observed (Koyama 1984: 30-35). Even taking into consideration that this calculation based on data from 1974 represents a low figure, it can be regarded as valid in that it indicates the general demographic trend of the Jomon period. This implies that the focal region of Japan experienced an extraordinarily high population density during the Middle Jomon as a society with a basically gathering economy (Koyama 1978, 1984).

After Middle Jomon, the demographic trends registered decrease. Settlements concentrate in zones characterized by low elevations at the origin of *barrancas.*

Subsistence patterns, technological base and exchange network

At the end of the Pleistocene, changes in environmental conditions occur. The melting of the glaciers elevated the sea level by some 3 or 5 m. over the present level and caused the marine transgression, which reached it maximum level at about 6000 years BP Consequently, salt water penetrated into the interior of the coastal lowlands and altered litoral fisonomy, extending the coastline, bays and estuaries. On the other hand, climatic change and its concomitant effects caused the disappearance of some species of fauna and flora characteristic of the glacial period, which affected the adaptive mechanisms of the Jomon people.

The inhabitants of that time secured their food by hunting, fishing and gathering, commonly known as a gathering economy. In the anthropological literature its ability to support population growth beyond the existing biotic potential has been seriously questioned. As the central region of Honshu registered a high, though conservatively estimated population density of 2 or 3 persons per km^2 during its height and revealed a complex internal structure sustained by hunting, fishing and gathering, then we must inquire in depth into the true capacity of this type of economy and elucidate its microprocesses, which permitted the Jomon communities to achieve a considerable degree of complexity.

Hunting is an activity of great economic importance since long before the beginnings of the Jomon culture. Nevertheless, hunting and tool manufacturing techniques developed quantitatively as well as qualitatively in comparison with earlier times.

The first outstanding innovation in Jomon is the use of the bow and arrow, which permitted a notable increase in hunting capacity. The use of these implements was fundamental in the early phases, although their presence in archæological contexts diminishes notably later on. This is probably due to a change in hunting methods from an individual to a collective activity, using new techniques such as the setting of traps and the use of domesticated dogs, and does not mean that hunting declined in importance. Traps consisted of pits which varied from 1 to 2 m. in lenght and from 0.5 to 1 m. in width, and were approximately 1 m. deep. As can be seen at Kirigaoka (Tokyo), Kamiyamoto (Kanagawa), Bandooyama (Saitama) and Shyoonodaira (Nagano), they are located at a certain distance from the settlements themselves and form a unit with a considerable number of pits.

The introduction and incorporation of new hunting methods and techniques, as well as the advance of sedentism, stimulated greater settlement concentration in certain areas and greater intensification of resource exploitation (Goto 1979).

The contents of shell middens indicate that in the beginning man hunted whatever was at hand, with preference for small animals such as hares, foxes, rats, etcetera.

Subsequently, hunting for deer and wild boar with the help of dogs intensifies. Apparently man already had sufficient skills and ability to manipulate and program hunting activities; for example, bone remains from the shell midden at Kihuma (Kanagawa) according to Egami (1973) indicate that bucks occupy strikingly high proportions compared to does and fawns to assure a future food supply. About 70 species of mammals have been identified from the Jomon period by means of bone analysis, although hunting patterns differ, in accordance with the time and the region. It is worth mentioning, for example, that the axes presumably used for bear hunting are found mainly in the region with heavy snow (Nagara 1971). In time a tendency developed to select particular species of birds and mammals as prey; notably pheasant, duck and pigeon among the fowls, while the hunting of mammals centered on deer and wild boar. These last two constitute the most commonly hunted species in the forest and mountain zones (Kaneko 1973), especially in the Late Jomon phase, when deer and wild boar represent over 90% of the faunal remains. The analysis of age composition determined by the annual layer of tooth cement (Koike and Ohtaishi 1985) reveals to us that the hunting pressure on deer increased to the limits which could be sustained by the population during the Late and Terminal Jomon.

Hunting is an economic activity practiced by man since very remote times; on the other hand, evidence for fishing first appears in the Initial phase and develops to an unparalleled degree during the Jomon Middle and Late phases. The accelerated process of fishing is attributed to climatic change and to its effects on the Post-Pleistocene geographical and ecological environment. Especially in eastern Japan, the Jomon people made intensive use of these new circumstances. This is reflected in the marked tendency to settle in coastal zones, in more than 1,100 shell middens that they left from throughout the Jomon period with over 300 identified marine species, and in the high level of fishing tool technology which is almost unequaled in world prehistory. There have been several cultures in the world that supported themselves by fishing; however the Jomon culture was probably the only one that developed a complex system of aquatic resource exploitation mechanisms. And sedentary life in the coastal zones developed parallel to this new subsistence form.

Fishing activities are divided into different categories: fishing with nets, with hooks, and fishing and hunting with harpoons. The first type was practiced in central and eastern Japan, where a large quantity of weights made of clay, stone and potsherds, pumice stone floats and an object supposed to be a needle for weaving nets are found. An enormous amount of small fish bones of sardine and mackerel were also found in the shell middens, which confirms the use of nets in fishing. This method of fishing, which was still not widespread in the early phases, develops suddenly during Middle Jomon. And at the same time the species caught in nets, which consists mostly of those from the bays such as *Lateolabrax japonicus* (Akazawa 1980), *Platycephalus indicus, Mugil cephalus* and *Mylio macrocephalus*, have been identified; on the other hand deep sea fishing with nets was not yet practiced in Jomon. Freshwater species such as carp, trout, eels, and others, were also caught.

Simultaneously, fishing with hooks began to be practiced. From the beginning it was not limited to marine species, but was also used to catch freshwater fish. Hook forms vary spatially as well as temporally, although about 90% belong to the category of one-piece hooks. Usually deer antler was used to make the hooks, but wild boar bones and tusks were also utilized. The distribution of hooks is concentrated in central and northeastern Japan, especially in the Pacific coastal region, which coincides with the distribution of fishing with nets. Fishing with nets and with hooks develop at an accelerated rate in the second half of the Middle Jomon phase, when certain species were sought, preferably deep-sea fish such as *Katsuwonus pelamis, Thunnus orientalis* and

Seriola quinque radiata. Once a specific mechanism for exploitation had been established, fishing with hooks is intensified notably.

Among fishing activities, fishing with harpoons is less widespread, although it appeared at the same time as the other two types. Its distribution is limited mainly to Hokkaido and northeastern Honshu, because this method of fishing was used to catch large fish such as sharks, as well as marine mammals such as whales, seals, and dolphins which abounded in the Northern Pacific.

In short, the new subsistence pattern based on fishing is defined during Middle Jomon. In this phase, fishing reveals a high level of knowledge about the behavior of specific marine resources, a great ability to make the necessary tools, and to coordinated works. As time went by indiscriminate fishing was abandoned and the catching of certain determined species is intensified. The trend towards specialized fishing develops even more towards the end of the Jomon period, aided by a specific tool technology. It should also be pointed out that fishing relied on the use of boats since the Early Jomon.

Another specialization related to marine resource exploitation is salt production, which begins towards the final phases of the Jomon period. Thus, fishing as well as salt production lead us to infer that the production of objects for exchange was increasing during Late and Terminal Jomon (Watanabe 1974; Kaneko 1973; Goto 1979; Koyama 1984; Kusumoto 1981).

Along with fishing, the collecting of mollusks played an important role in Jomon economy. The large number of shell middens indicates an intensive exploitation of marine and lacustrine resources (Akazawa 1980). Over half of these shell middens are concentrated in the Pacific region of central Honshu, especially along Tokyo Bay. The shell middens are divided into two types according to size. The small ones are of a domestic character, varying from 2 to 3 m. in diameter, and are mainly found on hillsides and in abandoned dwelling houses. Large shell middens are much less common than the first type and are circular or shaped like a horseshoe, varying from 100 to 200 m. in diameter and are 2 to 3 m. thick. The first type appears in the Initial Jomon phase and is characterized by the great variety of remains which include not only shells but also bones and discarded artifacts. The association with house-dwellings suggests to us its relation with the domestic economy. The second type appears towards the end of Early Jomon and increases during the Middle and Late Jomon phases. Its development coincides with population and community growth, but later it suddenly disappears. Shell middens of this size have not been found either before or after this period in Japanese history.

The two types of shell mounds represent two different economic systems. For example, the analysis of seasonal dating by growth-line counting of the clam in the domestic shell-middens at Koode (Chiba) indicate that mollusk gathering is an activity carried on from the end of spring to autumn (Koike 1981, 1979). On the other hand, those of the second type reveal two aspects of preponderant importance: the possible beginnings of aquaculture and the specialization and production of dried food as reserves and as exchange objects. Possible primitive cultivation of large oysters begins in Early Jomon, with the use of sticks, boulders, fragments of grinding objects and in some cases even large potsherds (Ezaka 1973). Unlike the small shell-middens, these usually contain only one type of shell, or at most a small number of species. Their location is limited to hilltops or terraces, where the necessary conditions for drying mollusks such as sufficient sun and wind, are found. On the other hand, distance to the sea is not a primary factor, as some shell middens of this category are found several kilometers from the coast. In most cases, these giant shell mounds are not associated directly with house-dwellings, but rather are located at some distance from them. A considerable amount of ash, hearth remains and storage amphorae have been found is these middens. All of these denote that the giant shell middens functioned as specialized production areas where a reserve of food supply and exchange material in the form of dried food was produced.

Briefly, it should be emphasized that subsistence patterns and the intensified exploitation of certain aquatic and land resources developed parallel to population growth, sedentism and the appearance of more complex communities, as Cohen points out (1977: 37). The evidence from the Jomon culture also indicates that sedentism seems to have been based simply on the increasingly intensive use of vegetable foods and coastal or riverine resources. Over the course of the Jomon period, man acquired a complete system of dried shellfish food production which included knowledge and management of an optimum environment, of mollusk behavior patterns in order to increase reproduction, and to intensify gathering techniques, a whole series of specific technical skills to obtain the products and to store them (Goto 1973, 1979; Deriter 1981, 1982; Sakatsume 1970). This system was efficient enough to produce a food surplus not only for difficult times, but also to use in exchange for other needed products.

Unlike the coastal settlements, sites located in the interior and in mountanious and forest zones represent the mountain type of adaptation. The two settlement systems form part of the regional macro-system.

From the nutritional point of view, hunting, fishing and mollusk-gathering resources are of substantial importance. Nevertheless, Jomon man must have sought another food source by gathering edible nuts, fruits and plants to increase his caloric intake. Up until 1977, some 39 species had been identified from the botanical evidence (Watanabe 1977), although the number must have increased considerably in the last decade. These species reflect a marked regionalization, due to the position of the Japanese archipielago which extends from northeast to southwest. The four biotic zones which correspond with the six cultural spheres, have been identified in the Middle Jomon phase. If we assume that vegetable foods have more importance in man's caloric intake, related activites must have constituted the subsistence base. And for that reason it is logical that the cultural spheres should coincide to some degree with the biotic regions.

Of the 39 identified species, the most frequently consumed were acorns, chestnuts (*Castanea crenata* Sieb. et Zuc, *Quercus acutissima* Car, *Quercus serrata* Thunb, *Castanopsis cuspidata*), walnuts (*Juglandaceae mandschuria*) and *Aesculus turbinata* Blume. These four, together with bulbs and tubercles such as manioc, yams and arrowroot, constituted Jomon man's caloric base. The distribution of *J. mandschuria* is weighted towards eastern and northern Japan, while it decreases drastically in the south. *Quercus* come mainly from the south and decrease towards the north. *C. crenata* Sieb. et Zuc have a distribution similar to that for acorns. *A. turbinata* Blume is less frequent than the other three species and are found dispersed throughout Honshu. All of this indicates that dietary differences correspond fundamentally with the predominant vegetation type. The fact that Jomon man depended on vegetable resources for his basic food supply is by no means unique to this culture, as it is well known that current groups of hunter-gatherers support themselves mainly with gathered vegetable foods. The system for preparing *Quercus* and *A. turbinata* Blume deserves to be analyzed. Unlike *J. mandschuria* and *C. crenata* Zieb et Zuc, *A. turbinata* Blume and some species of *Quercus* require a complex procedure consisting of five stages: collection, transport, storage, elimination of the toxin, and cooking. The first three steps are commonly required in the gathering of nuts in general. Storage in baskets, in pits dug in the ground or in clay pots, often found still full of nuts, appears in the Early phase associated with house-dwellings. The high content of tannic acid in nuts of the species *acutissima* Carr, *serrata* Thumb and *mongolica* var. *grossesserrata* Rehd-et Wils, and saponyn acid and aloin found in *A. turbinata* Blume nuts require a complex process in order to eliminate the toxic elements before using them as food. Because these acids are not soluble they must first be subjected to a heating process and then mixed with an ash solution before dissolving them in water. Finally they are let to soak in clean, cold running water.

Each step of the process requires in turn a specific tool complex. In the Middle Jomon phase at least, the knowledge for adequately preparing *Quercus* existed, and later *A. turbinata* Blume began to be consumed in the Late phase. Because of the large quantities consumed and the complex system of preparation, it is thought that activities involved in the acorn and other nut-gathering economy were organized on a communal basis and carried out in determined spaces (Watanabe 1977).

On the other hand, the possible initiation of an agricultural economy during the Middle Jomon phase has frequently been debated (Ezaka 1967; Fujimori 1970). The argument is supported fundamentally by the Childean perspective in that a food-producing economy is considered a *sine qua non* which permits accelerated population growth and the development of sedentary settlements evidenced by the Jomon whose populations may exceed in some cases 200 persons, with indications of complex religious practices. The large quantity of lithic artifacts with a vaguely flat, rectangular form from the inner mountanious zone of central Honshu, which some authors assume to be agricultural implements, strengthens this hypothesis. Nevertheless, no consensus on the beginnings of agriculture before the Late Jomon phase has been reached to date. Quasi-cultivation, or in other words the cropping system of cultivation, defined by Bronson (1975:58) as the practice of deliberately growing useful plants, has also been suggested as a possible subsistence base for Jomon subsistence economy (Nakao 1966). This would have occurred in a transitional stage between the gathering economy and a food-producing economy. In order to pactice quasi-cultivation, knowledge of the management of a reduced number of tubercles and bulbs is sufficient, although a tool complex similar to that used in agriculture is also needed. Precisely, this tool-complex similarity between the agriculture system and the quasi-cultivation and extrapolations based solely on the instrument technology have been one of the causes of the confusion over the debate on the initiation of agriculture in Middle Jomon. On the other hand, it should be taken in consideration that some of these tubercles must be treated by a process similar to that described for nuts in order to eliminate their toxic contents. This implies that both the nut-collecting economy and the quasi-cutivation of tubercles form part of the same technological system.

All of this reveals that the Jomon culture was sustained by a fundamentally gathering economic system, with the combination of aquatic and land resource extraction. The Jomon economic system not only provided food for a population of considerable size, but could also produce needed surpluses for times of food scarcity and for exchange. The surpluses, then, are channelled to these two purposes, and unlikely to the maintenance of certain social sectors not engaged to the productive activites, as Childe once proposed. It should be emphasized that the Jomon subsistence system coincides with what Harris (1977) expressed, in the sense that it reached a high level of complexity, with the manipulation of vital biotic resources in order to optimize their reproduction, and the more intensive use of specific and limited resources. Also, the development of coordinated and organized activities on a communal as well as a domestic level, based on a specific exploitation-mechanism system, and the knowledge of the behavior patterns of basic resources should not be taken for granted.

Closely structured and mutually dependent exchange networks linked the coastal communities with those in the mountanious regions. The primary importance of exchange in the Jomon economy is evident in the analysis of the giant shell middens which proliferated on the Pacific coast of eastern Japan during the Middle phase and the first part of Late Jomon. In these shell middens the gathering of one or a very limited number of mollusk species in order to produce dried shellfish was done; in some cases, such as Kasori, it is calculated that almost twice the amount needed for the domestic consumption of a population was produced (Deriter 1982). This enormous quantity of dry food production was destined primarily for exchange for other needed products not obtained locally, such as obsidian, sandstone, basalt, pumice stone, asphalt, as well as mountanious products.

On the other hand, the intensification of exchange accelerated the trend towards communal specialization in exchange products, such as salt, and the fishing of specific species.

Final Considerations

In the last twenty years, the persisting idea that hunter-gatherers barely subsist by dint of unremitting effort has been refuted. Nevertheless, the supposed limitations of a gathering economy is still latent in the evolutionist scheme from Childe's time to this very day. In trying to explain macro-historical processes, the form of subsistence and social complexity has been linked in a cause-and-effect relationship. From this theoretical perspective, an economic system based on activities, such as hunting, fishing and gathering, exerts a regulating and restrictive mechanism with regard to population growth, sedentary life and socio-political complexity, since the intensification and greater efficiency in the appropiation of vital resources leads in the long run to the deterioration of the biomass, and finally to its virtual destruction. The concept of the Neolithic Revolution, proposed by Childe and endorsed by other authors, is substantiated by the process of change in the basic economic-technological level, which transformed a society of hunter-gatherers into a more complex society, since only a food-producing economy is able to control the reproductive process and increase the quality as well as the quantity of its food supply (Kabo 1985).

Until recently, groups of hunter-gatherers which had attained a degree of social complexity comparable to the Neolithic were considered exceptional cases and as such they were conceived irrelevant to explain historical processes. However in recent years more and more data has appeared which contradicts the classic and generalized idea that hunter-gatherers are "small groups in constant movement", at least in the case of terminal Pleistocene and immediately postglacial groups (Price 1981). Nevertheless, given the great theoretical importance of Childe's original concept, the presence or absence of a food-producing economy has been seen as the *sine qua non* of the process leading to the Neolithic. The proposal has been reinforced by the corroboration of data from western Asia and Europe. Unfortunately, this in turn has brought about some confusion and problems in explaining the cultural processes which lead to a complexity comparable to that of a neolithic society, but without food producing economy. Serious debates for or against the beginnings of agriculture during the Middle Jomon phase are still going on. The archæological data, supported by other scientific disciplines, makes it clear that the Jomon climax does not coincide with the introduction of the new subsistence economy. On the contrary there is a strong indication that a sufficiently complex society evolved, based on a high level of hunting, fishing and gathering economy, accompanied with food-processing and preservation techniques, storage methods, and by other factors that accelerated changes in the relations of production. It is only the second half of the Late Jomon phase, at the outset of cultural decline, when the cultivation of millet appears, although this is somewhat doubtful. Later in the Terminal phase rice cultivation was introduced from the Asiatic continent, coinciding with a marked population decrease and a certain trend towards the impoverishment of living conditions. Rice cultivation spread rapidly. The rapidity of its propagation is attributed mainly to the fact that an adequate and necessary social organization for incorporating the new economic system based on agriculture had already been established.

The lack of conclusive evidence for the existence of a food-producing economy with agriculture and breeding of domestic animals in the Middle Jomon apogee repudiated the idea that the Jomon culture could be considered a neolithic society. This is probably one of the main reasons why Jomon has been placed in a stage contemporary to the European Mesolithic, or at most in a pre-neolithic stage within the traditional evolutionist scheme (Rowley-Conwy 1984; Price 1981).

By means of a succinct description, we have tried to point out that the Jomon culture experienced a long evolutionary process. It attained a high rate of population density, full sedentism, a complex of specific tool technology and of food processing and preservation techniques and knowledge, food storage facilities, and increasingly specific skills and methods which permitted a greater intensification of the gathering of vital resources, specialization of production activities on a community level, production of surpluses, mutually dependent exchange networks, and elaborate magico-religious practices. In the course of five phases, the inherent potentialities of Jomon society provoked changes and induced to a greater complexity in the social relations of production, without completing radical changes in the level of the production forces. Obviously the post-Pleistocene conditions were the necessary factors, but these external elements have not sufficient potentialities to provoke historical changes, since they do no act as independent variables which determine such changes. What is fundamental, then, is to dilucidate the mechanisms of articulation which provoke mutually effective changes both in the specific external conditions and in the inherent potentialities of a specific social system (Faris 1975).

Another medullar aspect worth discussing is the concept of autarky, which constitutes one of the central arguments of the Childean Neolithic Revolution. Childe, as mentioned before, emphasizes the autarkic character ubiquitous in the neolithic communities. This, at the same time, hampered the development of intercommunity exchange. It has been assumed that exchange has never played a vital role in the course of the Neolithic Revolution, only restricted to the sphere of sumptuary objects.

Bronson (1975:65) emphasizes that, in some cases of pre-neolithic societies based on a marine economy, in optimum circumstances, they can be induced to adopt plant cultivation in the vecinity of the settlement in order to continue exploiting advantages of rich marine biotic potentialities. Still other hunter-gatherer groups chose, instead of cultivation, alternative mechanisms in which basic necessary products were obtained through intercommunity exchange networks. Hence, exchange involving basic goods was practiced more intensively than it has been speculated among advanced hunting, fishing and gathering groups, as in the Jomon case. Moreover, the archæological data indicates that interaction functioned by means of mutually dependent and fixed mechanisms. This exchange system accelerates, apart from other things, the process of community specialization, which although probably not full time, could produce the necessary surpluses for times of scarcity as well as for exchange; thus a certain simbiotic relationship is established between the communities.

On the other hand, Hassan (1975) argues that exchange between agricultural groups was stimulated by the limited range of subsistence activities, the development of storage techniques and by a sedentary life style, more than by the growth of surpluses itself. These three factors, which gradually led to the establishment of fixed exchange networks, can be applicable for preneolithic societies. From this perspective, it is evident that the formation of a neolithic social organization can appear in societies with an advanced, basically gathering economy, contrary to Kabo's indication that advanced groups of fishermen are located on the threshold of a dead-end street.

At a glance, the Jomon culture presents the traditionally defined characteristics of a neolithic society. However, there is a radical difference between both types of societies. The Jomon culture does not fit into Childe's scheme, since the formation of the new social order was not directly concurrent with the new subsistence system based on a food-producing economy. Neither does it conform with the scheme of population growth-greater social complexity proposed by Boserup and endorsed by other authors. There are those who qualify the concept of Neolithic Revolition proposed by Childe as extemporaneous (Forni 1985). By no means are we disputing the fundamental validity of this concept, although as some authors point out its definition has become polysemantic (Shrirelman 1985). Neither are we questioning the decisive role that a food-producing economy has exercised within the macro-historical process. What we are arguing is that the concept as defined by Childe, —that the revolutionary processes of change occur on the level of subsistence techniques— should be reevaluated due to the fact that it does not explain specific cases like the Jomon culture with a predominantly marine economy which, as Dumond (1972) suggests, tends to show a great flexibility and expansibility, combined with a mountain type economy, developed a much more complex society than that traditionally assumed in the evolutionist framework.

The Jomon case, like many others (for example, the northwestern coast and plains of the United States), lead us to rethink the Neolithic Revolution from a different perspective and would enrich the concept of the Neolithic Revolution with new casuistic data which does not agree with events in the Near East or in Europe, originally used to corroborate Childe's hypothesis.

As Shrirelman (1985) points out, the link between types of eocnomy and levels of social organization is not rigidly determined, since the latter does not have such a close correlation with economic forms as it has with its effectiveness, in other words, with the level of development in the productive forces and relations of production. This explains the reason why both types of economy can produce societies of different complexity. On the other hand, if the hypothesis that quantitative changes can provoke qualitative changes is valid, then the variables characteristic of the Jomon culture act accumulatively, so that at a given moment they can provoke qualitative transformations, even though they are not considered quantitatively radical. Thus, accumulative effects of progressive changes even in a fundamentally gathering subsistence can transform it into another level of social organization.

REFERENCES

ADAMS, R.M.

1966. *The Evolution of Urban Society*, Chicago, Aldine.

AKAZAWA, T.

1980. "Fishing Adaptation and Prehistoric Hunter-gatherers at the Nittano site, Japan", *Journal of Archæological Science* vol. 7: 325–344.

1982. "Cultural Change in Prehistoric Japan: the Receptivity Process of Rice Culture in the Japanese Archipielago", *Recent Advances in World Archæology* 1, Wendorf, F. and Close, A. (eds), London, Academic Press: 15–211.

BOSERUP, E.

1965. *The Conditions of Agricultural Growth*, Chicago, Aldine.

BRONSON, B.

1975. "The Earliest Farming: Demography as Cause and Consequence", in *Population, Ecology and Social Evolution*, S. Polgar (ed): 53–78.

COHEN, M.N.

1977. *The Food Crisis in Prehistory*, Yale University Press, New Haven and London.

COWGILL, G.

1975a. "On the Causes and Consequences of Ancient and Modern Population Changes", *American Anthropologist* 77: 505–525.

1975b. "Population Pressure as a Non-explanation", *Population Studies in Archæology and Biological Anthropology: a Symposium*, Swedlund (ed), Society for American Archæology Mem. 30: 127–131.

CHILDE, V.G.

1964. *Evolución social*, UNAM, México.

1965. *Man Makes Himself*, The Fontana Library, London.

DANIEL, G.

1965. "Preface to the Third Edition", *Man Makes Himself*, The Fontana Library, London.

DERITTER, P.

1981. "Nijon no kaizuka ni kansuru ichikosatsu", (1) *Kokogaku Journal* no. 198, Tokyo: 21–24.

1982. "Nijon no kaizuka ni kansuru ichikosatsu", (2) *Kokogaku Journal* no. 199: 18–21.

DUMOND, D.E.

1965. "Population Growth and Cultural Change", *Southwestern Journal of Anthropology* 21 (4): 302–324.

1972a. "Population Growth and Political Centralization", *Population Growth: Anthropological Implications*, B. Spooner, Cambridge, MIT Press: 286–310.

1972b. "Prehistoric Population Growth and Subsistence Change in Eskimo Alaska", *Population Growth: Anthropological Implications*, B. Spooner (ed), Cambridge, MIT Press: 311–328.

EZAKA, T.

1967. *Nijon bunka no kigen*, Kodansha, Tokyo.

1973a. "Jomon bunka no gijutsu to seikatsu", *Kodaishi hakkutsu 2: Jomon-doki to kaizuka*, Ezaka (ed), Kodansha, Tokyo: 29–49.

1973b. "Doki-bunka no kigen to tenkai", *Kodaishi hakkutsu 2: Jomon doki to kaizuka*, Ezaka (ed), Kodansha, Tokyo: 87–118.

1980a. "Kooeki: Kokuyoseki", *Kokogaku koza (9)*, Yawata, Ooba and Naito (eds), Yuzankaku, Tokyo: 257–268.

1980b. "Kooeki: Tennen Asphalto" *Kokogaku koza (9)*, Yawata, Ooba y Naito (eds), Yuzankaku, Tokyo: 279–290.

FARIS, J.

1975. "Social Evolution, Population and Production", *Population, Ecology and Social Evolution*, S. Polgar (ed), Mouton Publishers, The Hague, Paris: 235–271.

FORNI, G.

1985. "Comments, The Origins of the Food-producing Economy of V. Kabo", *Current Anthropology* vol. 26, no. 5: 609–610.

FUJITA, H.

1979. "Bungyo ron: Jomon jidai no bungyo", *Nijon no kokogaku wo Manabu* (2), Ootsuka, Tozawa, Sahara (eds), Yuhikaku, Tokyo: 197–198.

FUJITA, E.

1970. *Jomon Nooko*, Gakuseisha, Tokyo.

GOTO, Kazuhito

1979. "Shugyoro no gijyutsu no hensen", *Nihon no kokogaku wo manabu: 2*, Ootsuka, Tozawa y Sahara (eds), Yuhikaku, Tokyo: 108–136.

GOTO, Kazutami

1973. "Jomon jidai no Tokyowanengan no kaizuka bunka no kenkyu", *Bosochiho no kenkyu*, Yuzankaku, Tokyo.

HARRIS, D.

1977. "Alternative Pathways Toward Agriculture", *Origins of Agriculture*, C. Reed (ed), The Hague, Mouton: 179–243.

HARUNARI, T.

1982. "Jomon shakairon", *Jomonbunka no kenkyu 8*, Yuzankaku, Tokyo.

HASSAN, F.A.

1975. "Determination of the Size, Density and Growth Rate of a Hunting-gathering Population", *Population, Ecology and Social Evolution*, S. Polgar (ed), The Hague, Mouton: 27–52.

HODDER, I.

1982. *The Present Past*, B. T. Batsford LTD, London.

ISHIDA, E. and SHIMIZU, S.

1968. *Symposium Nijon-nookoobunka no kigen*, Kadokawa-shinsho, Tokyo.

KABO, V.

1985. "The Origin of the Food-producing Economy", *Current Anthropology* vol. 26, no. 5: 601–616.

KAGAWA, M.

1974. "Primitive Agriculture in Japan: Latest Jomon Agricultural Society and Means of Production", *Asian Perspectives* 16: 1–15.

KANEKO

1973. "Gyoro no tenkai. Shuryo no tenkai", *Kodaishi-hakkutsu (2): Jomon doki to kaizika*, Ezaka (ed), Kodansha, Tokyo: 119–135.

KANI, M.

1982. "Tama New Town no Jomon-shuraku", *Kokogaku Journal* no. 203: 15–19.

KOBAYASHI, T.

1973. "Tama New Town no Senjyusha-shutoshite Jomonjidai no Settlement system".

KOIKE, H.

1981. "Iseki no Kaizuka Kara Saguru Seikatsureki", *Kagaku Asahi* vol. 41, no. 6: 56–59.

KOIKE, H. and N. OHTASHI

1985. "Prehistoric Hunting Pressure Estimated by the Age Composition of Excavated Sika Deer (*Cervus nippon*) Using the Annual Layer of Tooth Cement", *Journal of Archæological Science* vol. 12: 443–456.

KOYAMA, S.

1978. "Jomon Subsistence and Population", *Senri Ethnological Studies 2*, Osaka: 1–65.

1984. *Jomonjidai*, Chuko shinsho, Tokyo.

KUSUMOTO, M.

1981. *Jomon seikatsu no saigen*, Chikumabooks 25, Chikuma shobo, Tokyo.

LEE, R.B.

1972. "Population Growth and the Beginnings of Sedentary Life among the Kung Bushman", *Population Growth: Anthropological Implications*, B. Spooner (ed), MIT Press, Cambridge: 329–342.

MIYAMOTO, N.

1979. "Jyu seikatsu", *Nijon no kokogaku wo manabu 2*, Ootsuka, Tozawa, Sahara (eds), Yuhikaku, Tokyo: 254–272.

MIZUNO, M.

1969. "Jomonjidai shuraku kenkyu eno kisotekisoosa", *Kodaibunka* 21: 3–4.

1980. "Mokko no sekai", *Kokogaku koza 9*, Yawata, Ooba and Naito (eds), Yuzankaku, Tokyo: 181–192.

NAGARA, Y.

1971. "Shuroo", *Kokogaku koza 9*, Yawata, Ooba, and Naito (eds), Yuzankaku: 27–37.

NAKAO, S.

1966. *Saibai shokumotsu to Noko no kigen*, Iwanami-shinsho, Tokyo.

OKAMOTO, I.

1974. "Shuro-shakai no hatten to Jomon jidai", *Nihon no rekishi I: Nihon no akebono*, Shueisha: 93–132.

OTOKITA, T.

1980. "Gyoroo", *Kokogaku Koza 9*, Yawata, Ooba and Naito (eds), Yuzankaku, Tokyo: 16–26.

POLGAR, S. (ed)

1975. *Population, Ecology and Social Evolution*, The Hague, Mouton.

1975. "Population, Evolution and Theoretical Paradigms", *Population, Ecology and Social Evolution*, Polgar (ed), The Hague, Mouton: 1–25.

PRICE, T.D.

1982. "Complexity in 'Non-complex' Societies", *Archæological Approaches to the Study of Complexity*, Van der Leew (ed), University of Amsterdam: 54–97.

ROWLEY-CONWY, P.

1984. "Postglacial Foraging and Early Farming Economies in Japan and Korea: a West European Perspective", *World Archæology* vol. 16, no. 1: 28–42.

SATO, O.

1982. "Tama-gawa churyuiki no Jomon-jidai shuraku", *Kokogaku Journal* no. 203: 10–14.

SAKATSUME, N.

1970. *Kaizuka ni manabu*, Gakuseisha, Tokyo.

SHIMIZU, J.

1980. "Senkoo", *Kokogaku koza 9*, Yawata, Ooba y Naito (ed), Yuzankaku, Tokyo: 225–234.

SHRIRELMAN, V.A.

1985. "Comments, The Origins of the Food-producing Economy by V. Kabo", *Current Anthropology* vol. 26, no. 5: 612.

SMITH, P.S.

1972. "Changes in Population Pressure in Archæological Explanation", *World Archæology* 4: 5–18.

SMITH, P.E.L. and YOUNG, T.C.

1972. "The Evolution of Early Agriculture and Culture in Greater Mesopotamia: a Trial Model", *Population Growth: Anthropological Implications*, B. Spooner (ed), Cambridge, MIT Press: 1–59.

SPOONER, B.

1972. "Introduction", *Population Growth: Anthropological Implications*, B. Spooner (ed), MIT Press, Cambridge: 15–27.

TOGASHI, Y.

1982. "Jomon shuraku no oogata jyukyo", *Kokogaku Journal* no. 203: 24–26.

TOZAWA, M.

1979. "Jomon-nookoron", *Nijon no Kokogaku ni Manabu 2*, Ootsuka, Tozawa and Sahara (eds), Yuhikaku, Tokyo: 173–191.

UMEZAWA, T.

1980. "Jomon no Shuraku sooron", *Kokogaku Journal* no. 203: 2–4.

WAJIMA, S. *et al.*

1968. *Kanto heiya ni okeru Jomon kaishin no saikosuijyun ni tsuite*, Shigen kagaku kenkyujyo 70.

WATANABE, M.

1973. "Shyoku seikatsu no hensen", *Kodaishi hakkutsu 2: Jomon doki to kaizuka*, Kodaksha, Tokyo: 136–139.

1973. "Maisoo no hensen", *Kodaishi hakkutsu 2: Jomon doki to kaizuka*, Kodansha, Tokyo: 140–151.

1975. *Jomon jidai no gyogyoo*, Yuzankaku, Tokyo.

1977. *Jomon jidai no shokubutsushoku*, Yuzankaku, Tokyo.

RICE CULTIVATION AND ITS PLACE IN SOUTHEAST ASIAN PREHISTORY: SOME FIRST INDICATIONS

CHARLES HIGHAM
University of Otago, New Zealand

A symposium dedicated to the memory of Vere Gordon Childe will bring from each participant, some self-reflection on their intellectual indebtedness to him. When I began studying archæology at the Institute of Archæology, Childe was on the point of retirement as its director. At that juncture, he was a dominant contributor to the explanation of culture change. In *Piecing Together the Past*, he set out with masterly clarity, his integrative theoretical approach to the data, and his major texts on prehistoric Europe and the Near East put his approach into practice (Childe 1925, 1934, 1956). The last three decades have seen a transformation in the discipline in terms of geographical spread of research and the identification of variables relevant in the interpretation of instances of culture change which particularly interested Childe. The present paper has taken as its theme, the Neolithic Revolution in a river valley in Southeast Asia, but it could equally have considered the Urban Revolution in the same area. It will be seen that some of the variables invoked were not considered by Childe, but that is as he would have predicted and wished. A worker who made such a contribution would be best placed to appreciate the plethora of further issues generated.

The quest for the initial steps towards rice cultivation in Southeast Asia have been dogged by failure. In 1972-3, excavations at Banyan Valley Cave in the rugged hills of Northern Thailand revealed rice husks in association with polished stone axes and pottery together with characteristic Hoabinhian flaked stone tools. At the time, it was anticipated that this discovery would place early rice cultivation in an upland, evergreen rainforest context. Radiocarbon dates, however, were to demonstrate a very late occupation, even into the first millennium AD, while Yen's consideration of the rice husks indicated a wild rather than a domestic variety (Yen 1977). Penny (1982) then sought evidence for early rice domestication in the piedmont between the Petchabun Mountains and the Khorat Plateau, but to no avail. This lack of evidence is vexing, because the plateau itself is one of several lowland areas studded with prehistoric settlement excavations which reveal the exploitation of rice, maintenance of domestic cattle and pigs as well as fishing, foraging and collecting in the aquatic habitats and their woodland margins (Higham and Kijngam 1984). These settlements cover up to 3 or so hectares, and the occupants clearly participated in a series of exchange networks which incorporated exotic valuables (fig 1). In the Red River Valley, the sites are ascribed to the Phung Nguyen and Dong Dau phases, and in Northeast Thailand, Ban Chiang and Ban Na Di are perhaps the best documented. Like sites are found in the middle and lower Mekong Valley and in the valley of the Chao Phraya River. They tend to cluster in the middle courses of tributary streams where the flood risk would have been minimal and conditions suited to rice cultivation, secured (White 1982). While we are unsure of the chronology of initial expansion, dates in the third millennium BC are considered reasonable. The inhabitants were, to judge from their physical remains, healthy and robust (Houghton and Wiriyaromp 1984). The burial data reflect socially ranked groups whose relative status was signalled by *inter alia*, imported stone bracelets, marine shell jewellery and in due course, bronze ornaments and silk clothing.

The proliferation of these settlements in what appears to have been a hitherto unoccupied landscape has been ascribed to settlement fission following local population build up to and beyond thresholds which invoked scalar stress (Johnson 1982; Higham and Kijngam 1984). Formative phases in the adoption of a subsistence base incorporating rice and maintenance of domestic stock remain unknown, but demand scrutiny.

Prehistoric Coastal Settlement Round the Gulf of Siam

The Bang Pakong is one of the major rivers which flows into the upper Gulf of Siam. Its estuary is located *c.* 80 km. east of Bangkok, and its headwaters originate in the hills which flank the strategic gap between the Chao Phraya Lowlands and the Cambodian Basin. It is a valley which can reasonably be expected to provide a long prehistoric and early historic sequence, a situation confirmed by an intensive site survey undertaken early in 1984. As part of the Bang Pakong research programme, the site of Khok Phanom Di was excavated over a seven month season during 1985, and the findings, which are clearly still of a preliminary and tentative nature, seem to provide the basis for a model which might clarify our thoughts on how rice entered the diet, and why such pervasive population growth and expansion of human settlement occurred thereafter. By projecting our results onto a broader canvas, it is also possible to review the subsequent impact of rice cultivation within mainland Southeast Asia.

The terrain underlying the present Gulf of Siam is covered by a shallow sea. This means that any rise in the mean sea level was translated into extensive inundation and conversely, the exposure of marine sediments. The marine transgressions which covered much of Sundaland are now responding to analysis, particularly in the Strait of Malacca. It has been found that until *c.* 10,000 years ago, the sea level was between 40–60 m. lower than at present, and that it rose from a level of -13 m. to +5 m. between 8,000 and 4,000 BP. Consequently, from a date of about 4,000–6,000 BP, a situation arose where raised beaches above modern sea level were formed, and it becomes possible to recover evidence for a marine adaptation (Geyh, Kudrass and Streif 1979). Such changes make it imperative to follow the environmental adjustments associated with the establishement of human settlements. Khok Phanom Di is a large (*c.* 5 ha.) site now situated about 22 km. from the present coast. While the excavations were in progress, Maloney was taking cores for the analysis of sediment and pollen remains from the vicinity of the site (Maloney 1985). He has confirmed that the soils of the Bangkok Plain comprise in essence, older marine clays, younger marine clays, acid sulphate soils and the present tidal flats. The first two have not yet been dated in detail, and doubtless there were numerous local changes reflecting isostatic adjustments and sedimentation regimes as well as sea level changes. Maloney's preliminary review of a core taken 200 m. southwest of Khok Phanom Di has revealed marine clay to a depth of 2 to 3.5 m. below the present ground surface. He has tentatively equated this with the older series of acid sulphate soils of the Bangkok Plain. This underlies clays laid down under brackish to freshwater regimes, the pollen from which indicates the presence of a brackish water lagoon. The transition to the brackish water clays probably equates with the initial occupation of the site nearby. Subsequently, there were several minor fluctuations in the sea level, indicated by the pollen and sedimentological data. During periods of higher sea, Khok Phanom Di may have been situated on an island or offshore rivermouth bar. During all but the latest occupation phase, the occupants had access to open tidal flats, the mangrove shore and freshwater swamp forest. Grass pollen, including rice-type specimens, were present in the pollen spectrum. Maloney has also identified possible riverine deposits which suggest that the course of the ancestral Bang Pakong River may have been close at hand. Hylleberg and Anuwat have added to our preliminary assessment of the palæoenvironment by identifying some of the microscopic gastropods to freshwater species. This indicates the local presence of probably ephemeral stands of fresh water. In a word, Khok Phanom Di was situated in a habitat giving access to a mangrove-fringed shore, an estuary, and seasonal freshwater ponds behind the barrier. The fluctuations in this habitat with time will only become clear with the course of further research.

It is important to stress some characteristics of such a coastal-estuarine ecosystem. The Bangkok Plain is subtropical and heavily influenced by the southwest monsoon. This brings rain between May and November. The rest of the year is characterised by dry conditions as the wind pattern shifts, introducing dry, cool air from Northeastern Asia. The present reaction to this sharp seasonality places a premium on systems of food storage. Rice is readily stored in granaries, and fish are converted into a fermented meal and stored in pottery jars. This process explains the importance of salt down to the present day. The abundant wet season feed for cattle is stored on the hoof. Although fish and shellfish are less readily available during the dry season, many species can be obtained through an increased expenditure of effort. Freshwater gastropods such as pila for example, can be collected from lake margins or ricefields during the rainy season, but one has to dig for them in the dry months because they æstivate in subterranean burrows.

The effects of the monsoon are considerably offset by proximity to the sea. The extensive intertidal zone, which reflects the flat terrain characteristic of the Gulf of Siam, yields harvest of sedentary shellfish and crustacea irrespective of season. The resources of the mangrove swamp are likewise unaffected by seasonality, while the presence of a major river nearby has important implications for the supply of fresh water as well as transport facilities and food. A subtropical coastal habitat is not, however, necessarily an optimal habitat and it is necessary to look at the general characteristics of the shoreline ecosystem in more detail before reviewing what was found during the excavation of Khok Phanom Di. Essentially, there are three different situations to which the prehistoric inhabitants of the site were exposed. These are the standstill after a period of regression, a situation during or immediately following a regression, and effects of a transgression. It is recalled that Maloney has identified evidence in a core only 200 m. from the site, that there was indeed a transgression of limited duration during the period of occupation.

Chappel and Thom (1977) have described the likely course of events under such conditions. During the standstill, they envisage a mangrove dominated fringe giving way to a raised beach backed by tidal creeks and a high tidal flat. During a regression, the high tidal flat would be transformed into an area under temporary freshwater ponds, while the exposed terrain would comprise evaporite basins interspersed by dunes, giving way to the mangrove belt at the high tide mark. Most food resources would be found along the active beach fringe. A transgression would involve the formation of a beach ridge behind which there would be a shallow lagoon favouring the growth of mangroves. The extent to which this idealised portrait is the case for Khok Phanom Di will be determined by pollen and sediment studies now under way. For the purposes of the present paper, it is relevant to note that under regressive conditions, former tidal creeks and flats would be converted into seasonal freshwater ponds likely to sustain such annual wild grasses as rice. Again, the lack of predictable fresh water in the absence of a river, would impose severe limitations on human settlement along such a coastline. It has to be noted, therefore, that the coastal habitat probably incorporated a few very rich estuarine oases, interspersed by long stretches of inhospitable dunes or mangrove forest.

Let us now turn to the excavation of Khok Phanom Di (fig 2). It is currently situated *c.* 22 km. from the shore, the surrounding sediments comprising marine clays. It covers *c.* 5 ha. and rises 12 m. above the surrounding plain. It is, then, an extremely large settlement given its assumed age, and no comparable sites in terms of size, depth and location have been identified round the shores of the gulf. The excavators resolved to depart from the traditional long and narrow excavation format, which had been dictated at earlier excavations by the maximum span of a bamboo roof. In the event, we opened a 10 by 10 m. square in the hope that it would allow the recovery of complete burials, their spacial distribution, and the recognition of structures and activity areas. After seven months of excavation, natural soil was reached at a depth of 6.8 m. and for the present purposes, it is possible to recognise three major stratigraphic zones which are named zones A, B and C. Zone A, the earliest, comprises a complicated series of lenses. In the main, they are made up of either ash, occupation material including pottery and biological remains, discrete shell middens and small lenses of red soil. There are numerous small hearths filled with ash and charcoal as well as postholes some of which retain the very wood of the prehistoric structures. Zone A was about 3 m. thick, and within it, there were 104 inhumation burials. The earliest six were found isolated from one another and with the exception of one shell bead, were devoid of grave goods. The remaining 98 burials were set out in rows and superimposed, the distribution revealing several nucleated clusters separated by areas with no burials at all (fig 3). Zone B offers a sharp contrast. The northeast quadrant was dominated by a rectangular raised platform surmounted by clay walls. The platform comprised 43 successive floors made of fill covered by smooth plastered surfaces. The posthole configurations indicate several rebuilding phases. No occupation material was found on any of the surfaces and the platform was probably part of a larger structure, because the plastered floors proceeded beyond the uppermost clay wall foundations and beyond the edges of the square. The distribution of burials remained nucleated, one of the latest such groups comprising burials cut through the latest platform floor. In front of the platform mound and lying along the same orientation as its walls, however, there was a row of burials ringed by a set of postholes which indicate an above ground mortuary structure. It is apparent that the distribution of burials was nucleated as in zone A, and indeed individual clusters reveal continuity between the two zones. This is matched by the enduring similarities in the mortuary rite.

Zone C, which occupies the uppermost metre of deposit, is a much darker soil than zone B. While zone B comprises a sandy matrix interspersed with a few shell middens and ash lenses, zone C is rich in pottery, animal bone and the implements used in the manufacture of pottery. No burials were encountered.

Let us begin with a review of the biological remains. Zones A and B are rich in faunal material. With the exception of a few macaque, pig and deer bones, there is scarcely any evidence for the exploitation of mammals, either domestic or wild. The overwhelming majority of remains are aquatic, particularly fish, shellfish, crab, and turtle. The main species of shellfish are adapted to a mangrove habitat, though some, too, are found on sandy tidal flats. On two occasions, human remains were found with the partially digested remains of stomach contents in place. An initial investigation has revealed the presence of fish bones and scales. A sample of all archæological layers was processed through a flotation chamber. The residue contained numerous small fish bones. With zone C, the presence of marine species suddenly halted, giving way to mammalian remains associated with a swamp-woodland, including pig, monkey and a large, probably wild, water buffalo.

The flotation procedure was designed in order to recover the remains of rice and other plants. In addition to these samples, rice chaff found while excavating and blocks of cultural deposit were taken for excavation in the laboratory. Rice was also found represented as chaff impressions on the clay adhering to the exterior surfaces of broken potsherds, and in the case of at least zones A and B, rice chaff was used as a tempering agent when preparing clay for pottery manufacture. An initial glance at the fæcal remains also revealed the presence of rice remains. At present, we are unsure whether the rice comes from a wild or domestic plant, or perhaps from both. Jill Thompson is currently considering this issue.

During the build up of zones A and B then, subsistence activities were concerned with fishing, shellfish collecting, and obtaining turtle, crab and other marine resources. The harvesting of rice was also undertaken, but we must defer the question of whether it was harvested as a wild plant or deliberately propagated until the remains have been considered in more detail. Such rich resources, of which only rice was seasonal and could be easily stored, are compatible with either a fully sedentary or at least a logistically mobile subsistence strategy. There are very few estuaries with gently shelving beaches round the Gulf of Siam, and it is reasonable to assume that the position of Khok Phanom Di was unusual.

When we turn to material culture, we find that the site was located so as to command at least two major sources of raw materials: clay and marine shell. While the area was deficient in high quality stone, its coastal-estuarine situation conferred considerable advantages as a nodal point for exchange. From the initial occupation to the final abandonment, the occupants made pottery vessels. We have found numerous clay anvils for shaping pots and burnishing stones with worn facets. Clay cylinders representing the earliest stage in shaping individual vessels have been found from the lowest to the latest levels. The numerous superimposed ash spreads have been interpreted as the remains of the bonfire method of firing vessels. Some inhumation burials included anvils, burnishing pebbles, prepared clay cylinders for fashioning vessels and complete pots. In contrast to the preponderance of cord-marked sherds in the zone A occupation layers, most of the grave wares were a lustrous black burnished style with incised decoration infilled with red pigment. There is no hint of developmental stages in the pottery-making tradition. Pottery was fully operational from the moment the site was occupied (Vincent 1987).

The same situation applies to fishing technology. Clay net weights and some fish hooks were present from the earliest contexts. There is also one example of a bone harpoon. Access to marine shell was also important: local manufacture of shell jewellery is attested by the discovery of shell from which tabs have been drilled in order to shape disc beads. Shell jewellery is also a component of the grave goods in both zones A and B.

Participation in a regional exchange network is evidenced too by the presence of polished stone adzes. These are present from the earliest occupation. A cache of ten was found in the earliest level, and single specimens are present, albeit rarely, in burials. Many adzes were sharpened to such an extent that they were worn to a very small size. There is a variety of stone, implying multiple sources. The number of postholes found and volume of ash and charcoal in hearths and firing areas hints at a considerable use of such adzes in the modification of the local vegetation.

Only the most general statements can be made about the mortuary remains, because the important task of sexing and ageing the skeletons has yet to commence. The six lowest skeletons comprise three adults, a child and two infants. They were interred in shallow graves with heads pointing to the east. The child was buried in a flexed position on the back. One of the adults was accompanied by a solitary shell bead, but all the others were buried without grave goods. Grave 151, that of an infant, was found with the body covered in red ochre and wrapped in white, unwoven fabric resembling bark cloth. This "tapa" covering became common in later zone A burials. There is a gap of about 40 cm. between these early burials and the balance from zone A. The latter set exhibit a marked preponderance of very young infants, there being at least 52 infant burials, five children and 30 adults. The mortuary ritual involved the excavation of a shallow grave orientated so that the head pointed to the east. Two bodies were placed on wooden biers, a practice which might have been more generally recognised had wood been more likely to survive.

The corpse was covered above and below with red ochre and together with any grave goods, was wrapped in binding sheets made of an unwoven, probably beaten bark fabric. The presence of grave goods varied with individuals. Twenty-one infants were buried in scoops with none of the ritual activity described above. Twelve were associated only with ochre and "tapa". The remaining complete burials were richer in terms of offerings. All had ochre over the bones, and in addition, two were accompanied by pots, three by pots and shell beads, two by bracelets, two by shell beads and bracelets, one by pots, beads and bracelets. These richer infant burials tended to be late within zone A. Grave goods associated with adults include red ochre, pottery vessels, shell beads, burnishing stones, a polished stone adze and clay cylinders of prepared clay. In contrast to the infants, no adults were interred with bone bracelets. Ochre was universally applied to adult bodies. Of the 32 complete examples, 7 were associated with pebbles bearing burnishing facets. It would be interesting to ascertain the sex of this set. None of the interments stands out, as do the few infant burials mentioned above, on the basis of grave wealth or ritual of burial though further analysis is necessary, particularly of the pottery fabric, before this can be demonstrated beyond reasonable doubt. The two richest graves from this assemblage are numbers 72 and 91. The former contains three pots, three burnishing stones and some shell beads placed between the knees. There is a concentration of fish bones over the left shoulder, and the body was covered in red ochre. The latter contained pots, as well as two burnishing stones and a string of shell beads over the chest. Most adults were interred with either one or two pots or with pots and shell beads.

There was a major stratigraphic break between zones A and B, as well as a 40–50 cm. hiatus in the presence of burials. The upper zone revealed a more homogeneous buildup without the intense differential lensing found between 4–6.8 m. Moreover, the northeast corner of the excavation square contained the raised platform which has already been described. The proportion of infant burials fell from 55% in zone A to 30%. There was a wider range of grave goods in zone B. In addition to those described for zone A, we also encounter clay anvils, large shell bracelets, shell discs, shell ear ornaments, a fishhook and sections of turtle carapace. The shell beads and pots reveal typological changes from the preceding forms, but the basic rite remained the same with the head orientated to the east. Again, nucleation into discrete groups was a feature of the burial layout. Two late groups post-dated the last addition to the platform, but the major concentration of zone B burials was distributed in front of and parallel with it. This group revealed marked disparity in the grave goods found with both adults and children. The poorest burials were associated with relatively shallow graves and either an absence of material goods or at best, one or two pottery vessels and some red ochre. These contrast with two outstandingly rich adult burials and three equally well endowed child burials.

Burial 15 was placed in a grave which stood out on account of its size: it was 3 m. in length, 1 m. wide and 80 cm. deep. It contained the remains of a woman aged *c.* 45 years at death. Her body was covered by a pyramid of shaped clay cylinders representing an early stage of pottery manufacture. A pot had been balanced on top of this pyramid, reaching almost to the top of the grave. Four or five further pots of considerable quality were placed over the woman's legs, while a particular feature of the burial was the richness of the shell jewellery. About 250,000 shell-disc beads were encountered, disposed as strands both above and under the chest. She also wore several necklaces made of large 'I' shaped beads linked by the smaller disc variety. Two horned shell discs covered each shoulder. She wore a shell bracelet and shell disc ear ornaments. At her feet was a clay potter's anvil and two burnishing pebbles which had been placed in a shell container. Her body was covered in red ochre. An adjacent grave of the same date contained an infant buried in an identical manner and accompanied by nearly as rich an array of grave goods. Again, the ochre-covered body lay under a pile of clay cylinders. A bracelet had been placed over the left wrist, and the burnished pots over the legs lay beside a miniature potter's anvil. The upper part of the body was covered in the same types of shell beads as were found with burial 15. The grave too, was far larger than necessary to contain the body. Burial 43 was only partially present in the excavation area. At present, we can confirm that it was accompanied by a considerable quantity of shell beads as well as two large shell discs.

A second rich infant burial was again found to be roughly contemporary with and in the same row as the three rich interments just described. This one, however, was found within a circular grave. The body was contained within two pots, one of which stood out on account of its superb decoration and almost eggshell-thin fabric. Two pots were found beside the feet, and a quantity of shell beads were present over the chest area. This third child burial was rather earlier than the other two. The child was buried with three of four pots and a considerable number of 'I' and disc shaped shell beads. The row of burials found in front of the platform was distinguished from zone A graves by the presence of postholes set round individual graves and indeed, possibly round the entire grave cluster as well. This finding suggests the presence of above ground mortuary structures. Indeed the lack of any intercutting graves surely indicates a planned cemetery in which the location of earlier graves was known. Within this area, there were several discrete shell middens and spreads of ash and charcoal which lap up to the top of individual graves as if reflecting feasting before or after the burial ritual.

Although the analysis of mortuary behaviour has hardly begun, it is still possible to make one or two tentative conclusions. The earliest six graves were not grouped, as occurred later. Nor were they endowed with a range of grave goods. Like those found in the rest of zone A, they were set within an occupation-industrial area. Some were cut through or set within shell middens or occupation layers. There is variation in the number of grave goods for zone A as a whole, but no individual or group of graves stands out as a being unsually rich. This situation contrasts with the establishment of a specific cemetery in zone B. Much attention was given to the construction and maintenance of a platform structure, and graves show a sharp dichotomy between considerable wealth associated with a group of two adults and three children, and the relative lack of wealth with the balance. The essential burial ritual, however, was similar throughout the zones in that the same orientation was employed, ochre was used to cover the corpse, pots and shell beads were commonly placed with the dead and graves were nucleated.

The last burial was cut from just below the distinct stratigraphic change involving a transition from virtually sand to a dark occupation layer rich in material culture and organic remains. No further marine shells or crabs were found, and mammalian species became common. Stone bracelets were favoured, pottery was made *in situ* as evidenced by numerous clay cylinders and anvils. There were many stone adzes, but no evidence for prehistoric metallurgy. This change is correlated with a lowering of the sea level and a considerable consequential adjustment in subsistence and exchange activities. Analysis of the pottery should inform us on the issue of continuity in terms of people occupying Khok Phanom Di in late zone B and in zone C.

At this juncture, it would be most desirable to be able to consider a complete series of radiocarbon dates from Khok Phanom Di. These indicate that the deposits between about 2 and 4 m. below datum were probably laid down between 3,000–1,500 BC. Until the remaining sample results are available, it will not be possible to predict the overall chronology with any confidence. There are however, two further clues. In 1979, Damrongkiadt Noksakul excavated a test square at the site and obtained a set of radiocarbon dates determined on the basis of human bone from burials. Two burials found about 2 m. above natural soil were dated to *c.* 4,000 BC (uncorrected). The remaining dates from higher burials were so mutually contradictory as to be of little use. If these two dates are supported by further determinations, then initial occupation may well lie within the millennium 5,000 to 4,000 BC. These early dates must be placed alongside the determinations of Van de Kevie (1971), which suggest that the deposition of brackish soils equated tentatively with part of the occupation span of Khok Phanom Di took place between *c.* 4,000–1,350 BC (corrected). A tentative prediction, therefore, is a date of *c.* 4,000 BC for the initial occupation, with zone B dating to *c.* 3,000–1,250 BC and zone C from 1,250 BC.

A Model for the Inception of Rice Exploitation and Settlement Expansion

Given the preliminary stage in our enquiries, it could be argued that any overview of the site in a regional context is premature. This is only partially true. Models based on the site's unusual size and location were formulated even before excavation commenced, and these are being refined as laboratory analyses proceed. Thus, an earlier draft of this paper noted an absence of rice in the lowest 40 cm. of deposits, but this had now been ruled out by Vincent's analysis of the pottery fabric and the recognition of rice in a coprolite found at the very base of the site.

Recent excavations in Northeast Thailand, the Chao Phraya Valley, Central Cambodia and the Middle Country above the Red River Delta are unanimous in disclosing prehistoric village settlements whose occupants practiced a mixed economy of hunting, gathering, fishing, stock raising and the cultivation of rice. As far as we can judge on present evidence, such expansionary settlement was under way by the third and possibly the late fourth millennia BC. The most logical source for such settlement ultimately lies in local hunter-gatherer groups. On *prima facie* grounds, it seems likely that trends to agriculture and expansionary settlement occurred among complex sedentary coastal communities, of which the best documented is Khok Phanom Di rather than among the occupants of small inland rock-shelters. It was located within a mangrove-estuarine habitat where marine resources were overwhelmingly dominant. The rice remains may have come from a wild or cultivated plant. The impact of human harvesting on wild rice is known to result in modifications to the seeding rate and the number of spikelets per panicle even over five generations, so clearly much detailed research on the pollen cores and actual remains of rice from the site is necessary before a clear picture of the status of the rice in the successive layers is known. Again, it should not be forgotten than in historic times, the inhabitants of the margins of the Tonle Sap have harvested wild floating rice by tapping the seed heads into their boats (Delvert 1961). For the purposes of the present discussion, we will regard the earlier occupation contexts as reflecting an essentially coastal gathering orientation, wherein a very high proportion of the food was obtained by fishing, collecting marine and estuarine resources such as crabs and shell fish, harvesting nuts and rice and either hunting or herding pigs. As such, it is felt legitimate to consider their culture within the general framework of complex sedentary hunter-gatherers. Price and Brown (1985) have defined complexity in this context as involving a dense co-residing population, large settlement size, permanent habitations, evidence for differential wealth, energy expended in the burial rite and pattern in the location of interment, and logistic rather than residential mobility. The preferred approach to explaining the development of such complex hunter-gatherer groups avoids prime movers, such as population pressure on available food resources, and rather stresses interactions between a range of cultural and environmental variables.

In terms of the environment, our present knowledge of prehistoric Khok Phanom Di recalls several characteristics noted by Yesner (1980) as being relevant for maritime hunter-gatherers as a whole. These begin with a high biomass. It has already been emphasized that the site had access to the sedentary and productive resources of the intertidal flats and mangrove swamps. These shellfish, fish and crabs are predictable and can survive high culling rates. There were also the riverine resources and the plants of the backswamps. Among the latter, we include wild rice. These resources are stable and predictable given a consistent sea level. It is also noted that such rich estuarine enclaves are rare. Indeed, only four major rivers flow into the Gulf of Siam.

Yesner also noted common cultural variables among those who occupy these optimal enclaves. One is that their rarity fosters a high degree of territoriality on the part of the favoured occupants. Another is that young and old can contribute to the diet by collecting shellfish, thereby reducing constraints on population growth felt by mobile forager groups. A third is the development of high population density within highly circumscribed areas. Since a marine diet based on shellfish, fish and crabs is low in vitamins and calories, inclusion of vegetables and plant supplements is important.

With reference to monsoonal subtropical Southeast Asia, it is relevant to add the reduced impact of seasonality which characterises the maritime tract when compared with the more stressful conditions inland, where the full impact of a long dry season is felt. If the contrast between coastal and interior resources encourages exchange, then those dominating both the coast and the river valleys, with access to boats, are well placed to occupy nodal positions in any emerging exchange network. Consummation of these theoretical predictions about Khok Phanom Di is seen in its outstanding size. If the dates proposed above are correct, then there is no known site in Southeast Asia of similar antiquity which approaches the 12 m. depth and 5 ha. area of this mound. The actual occupation followed soon after the land was exposed by a falling sea. There probably remain earlier sites which followed the coastline as it receded and before it stabilised at Khok Phanom Di, for the initial occupants were already conversant with pottery making, had access to numerous polished stone adzes and had a developed fishing technology.

The lower levels of the site suggest a sedentary population in that there are no clear lenses revealing site abandonment, and the buildup of layers is patterned: most hearths were found, often superimposed, in one area; pottery firing in another and shortly after initial settlement, specific areas were used over time for disposing of the dead. The adzes were made of exotic stone hinting at an early inception of exchange networks which in itself probably involved locally specialised craft production, particularly in the manufacture of pottery vessels and shell jewellery.

Sedentary communities as at Khok Phanom Di, set in rich habitats, permit population growth, though given the high incidence of infant mortality in zone A, this may have taken some time to manifest itself. Now high co-residential populations invoke the problem recently addressed by Johnson (1982) and described as scalar stress. This involves a series of population thresholds which when approached, invite a series of new responses. The simplest early response to high population aggregates is fission whereby a segment of the community moves elsewhere. This is straightforward in the absence of social circumscription, and was doubtless a commonly taken option. When alternative coastal enclaves are already taken up, however, and the more marginal hinterland also suffers a packing of communities, options are narrowed and may invoke conflict or new forms of social organisation, not least ranking and increased use of ritual behaviour to resolve conflict. The point about ranking, of course, is that it is a means of reducing stress through vesting leadership and status in one or few individuals. It hardly needs to be emphasised that ranking is also a convenient mechanism for the ordering of exchange relationships and storage and distribution of foodstuffs. In turn, it accentuates a demand for prestige goods which act as emblems of the highly ranked individuals.

The question of food storage brings us to rice, a plant which was harvested at Khok Phanom Di and to judge from the context of the rice remains, was locally husked.

Processing rice is not arduous, since it can be cooked as whole grains. Grinding stones at Khok Phanom Di are notable for their absence. Once the husks are removed, rice can be stored and consumed without further processing. Indeed, brown rice is more nutritious than the white polished grain. We are left in no doubt that rice was consumed from the initial occupation of the site. Maloney and Brown (1986) and Maloney (1987) have provided most important insight into the issue of rice exploitation. Their two pollen spectra reveal a sharp increase in the amount of charcoal, accompanied by a slight increase in the fraction of grass pollen, at depths of about 5 m. below the present ground level. These they interpret as burn-offs which occurred at or soon after the initial occupation. There was no major increase in grass pollen until considerably later in the pollen spectrum, at which point the charcoal fraction surged and was this time, linked with a major and enduring increase in grass pollen, including some derived from rice field weeds. It is felt probable that the latter event occurred towards the end of zone B or solely within zone C, although the correlation of the pollen cores and the cultural stratigraphy has not yet been effected.

At present, it is felt likely that the zone A data reflect a sedentary population, which as it grew in numbers, encountered scalar stress. Let us assume that initially, at least, the last problem was resolved by fission. Similar coastal enclaves would soon be taken up if they weren't already. The next best option for settlement was probably along the river valleys and paticularly the tributary streams which provided routes, water, aquatic resources and suffered minimal flood risk. In Northeast Thailand, it is just these positions which witnessed early occupation at, for example, Ban Chiang and Non Nok Tha. The earliest biological remains from Ban Chiang reveal a broadly based gathering, fishing and collecting pattern allied with the maintenance of domestic cattle, pig and dog and the harvesting of rice. This adaptation appears to have been successful biologically, as there are numerous such sites and their inhabitants were robust, well fed and healthy. They are held to represent a successful adaptation to the inland stream valleys by sedentary communities which numbered rice harvesting among their many subsistence activities.

This habitat, however, is marginal only in relation to the concentrated biomass characteristic of the coastal estuarine enclaves where the impact of the long dry season was lessened by proximity to the sea. It is not hard to imagine that the tracts just behind the optimal coastal zone filled fairly rapidly, since the sedentary coastal settlements are seen as the original donors. The point is that for sites like Khok Phanom Di, fission is an option only until the forces of social circumscription are invoked. The converse is continuing population growth and the possibility that more intense social ranking systems and attendant ritual developed to cope with the resultant scalar stress.

Evidence for this proposition could take the form of growing disparity in the wealth of grave goods among individuals buried at about the same time, differential energy expended in burying the higher-ranked individuals, evidence for population growth, more restricted distribution of valuables and perhaps, intensified production of more marginal food sources. Such intensification, for example, could take the form of expanding through swamp-forest clearance, the area within which rice flourished. The local intensification of rice production could have involved no conceptual break with the preceding system of harvesting of relatively small-scale burns to encourage grass growth, but rather greatly magnified interference with competitors to enlarge suitable habitats. Such improvement of the natural habitat for rice, however, could have increased territoriality.

The above model sees sedentism, population growth and fissioning to more marginal habitats, followed by increased social circumscription in the optimal zone and increased social ranking as factors involved in trends towards intensified exploitation of rice. How do the data from Khok Phanom Di fit the model, and how might it be tested in the future? Initial settlement is seen as reflecting the pull of predictable and concentrated marine food resources to hunter-gatherers already probably logistically mobile and accustomed to a marine habitat. They secured food through exploiting optimum resources, a strategy greatly assisted by marine transport. As settlement developed through zone B, population grew and the co-residing group underwent a process of fissioning whereby certain emigrant groups occupied more marginal inland locations, including tributary stream valleys. Forest clearance there expanded or opened up areas suitable for the proliferation of rice, but this plant was one, possibly minor component in a very flexible and wide ranging subsistencc strategy. Domestic cattle, which were not suited to the optimal marine zone, and domestic pigs were maintained in the marginal zones. The large co-residential group at Khok Phanom Di encouraged some ranking, as did the development of exchange networks to supply interior communities with desirable marine products, not least shell jewellery. This is reflected in some disparity in individual wealth among the zone A burials of adults and children. This phase of fissioning and trends towards complexity probably took place between *c.* 3,500–2,500 BC. By the sharp change seen with the start of zone B, increased circumscription and the developing exchange networks, as well as the growing population, entailed investing more energy in ritual and ascriptive ranking. Burials display a sharp dichotomy between rich and poor, and there is some evidence that those skilled in primary production of quality ceramics were among the most highly ranked echelon. Indeed, a child too young to have been skilled in pottery making was buried with much shell jewellery together with a miniature potter's anvil. Increased ritual and attention to the burial rite is indicated by the transition of the part of the site excavated from an occupation area with burials to a specific cemetery complete with a platform. Four of the five outstandingly rich burials lie in the shadow of this structure. It is considered likely that rice was by now, at least encouraged by the expansion of its preferred habitat. This settlement in the optimal zone remained reliant on the predictable marine resources, however, particularly fish, shellfish, crabs and turtles. Turtle remains now figure prominantly in some burials. This phase dates to the period 3,000–1,250 BC.

So far, the environment has been seen as a contributing but not a determining variable. The transition to zone C saw a major and pervasive change in the habitat. The sea fell sufficiently to rule out the exploitation of coastal food supplies. The cemetery was abandoned and probably, the site as a whole was deserted. Re-occupation saw a continuation of the site as a pottery making centre but now the animal remains were adapted to the forest and river. Rice-chaff tempered pottery vessels indicate that rice was locally available. This phase probably dates from *c.* 1,250 BC.

This preliminary model visualises rice harvesting in both the marginal and optimal zones as a component part of a flexible multi-facetted subsistence strategy. During the period under review, rice was significant to the prehistorian more for its potential for intensification than the realisation of more extensive or productive cultivation. Indeed, the first major intensification probably came with the formation of centralised chiefdoms in the period 500–0 BC.

Some aspects of the model set out above fit well with the mortuary, structural and biological data from Khok Phanom Di, largely because there were seminal in its formulation. Other aspects will be tested. Thus, Maloney is now assembling the first pollen spectra associated with an archæological site in Thailand, and his finding on the question of forest clearance or other environmental modifications in favour of rice will be of critical importance. Again, his work on the diatoms and sediments will illuminate the changing incidence of brackish and freshwater habitats available to the prehistoric occupants. The abundant rice remains will be analysed for clues as to their domestic or wild status and the impact of human interference. The human remains will permit the identification of the people's health and changing diet, while the remains of their fæces and stomach contents will prove invaluable in reconstructing individual diets. The location, size and subsistence economies of the sites found in the prehistoric hinterland of Khok Phanom Di, not to mention those drowned by the rising sea level, will be of crucial relevance to the issue of fission and settlement of marginal environments, if it occurred. Should these basic elements be proven valid by further research, then there is no reason why similar trends were not operating in analogous coastal tracts. The initial expansion of people numbering rice cultivation among their subsistence pursuits in the Red River Valley for example, is most logically sought along the immediately adjacent coast than as the result of some shadowy migration. Indeed, migrations were one of Gordon Childe's absorbing interests which have become rather unfashionable of late. In mainland Southeast Asia, we think we are confronted by expansionary trends following scalar stress and settlement fission, but isn't that in essence, the same process, but taken back a stage to try and understand the underlying causes?

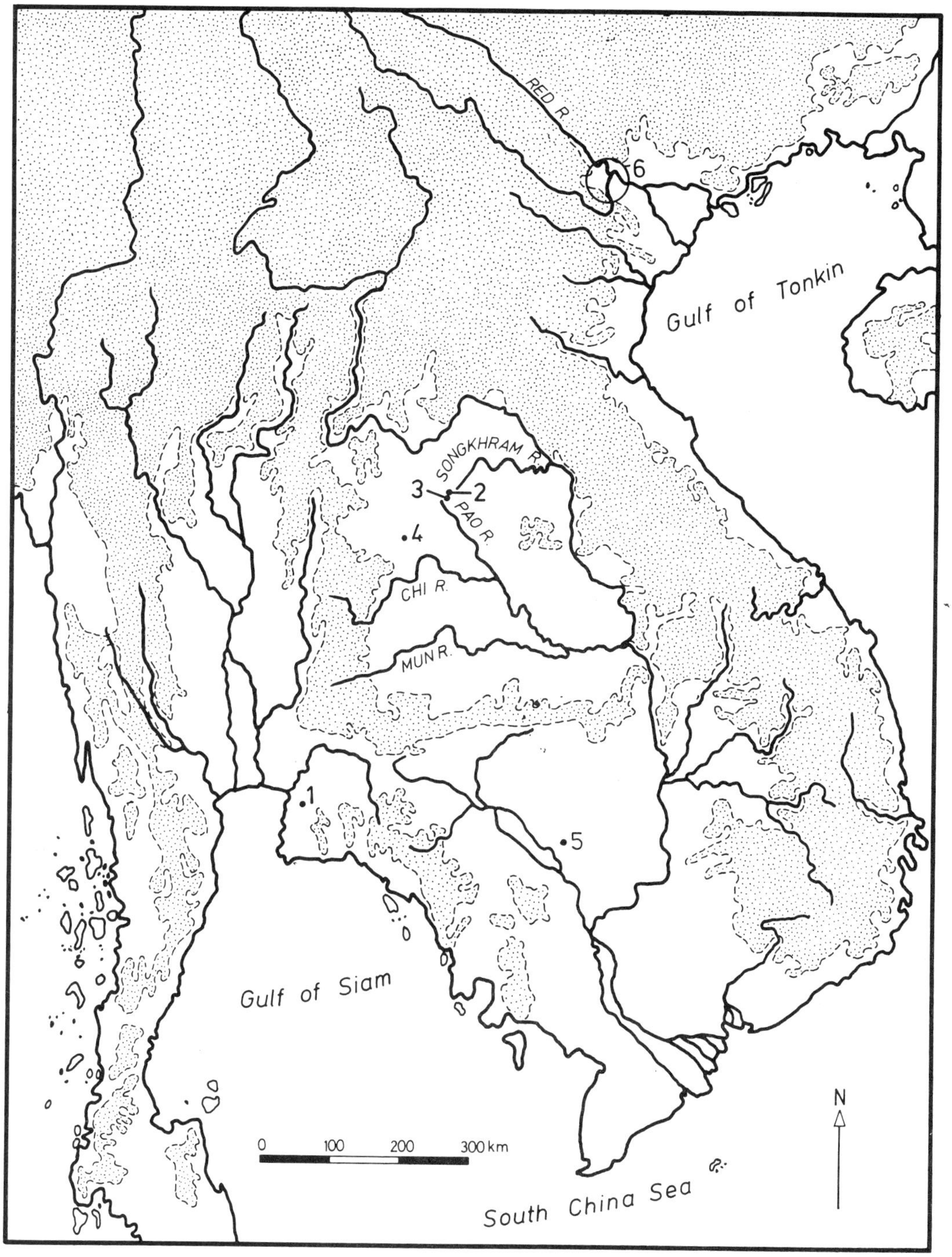

Figure 1: Mainland Southeast Asia showing sites mentioned in the text: 1. Khok Phanom Di, 2. Ban Chiang, 3. Ban Na Di, 4. Non Nok Tha, 5. Samrong Sen, 6. Phung Nguyen sites.

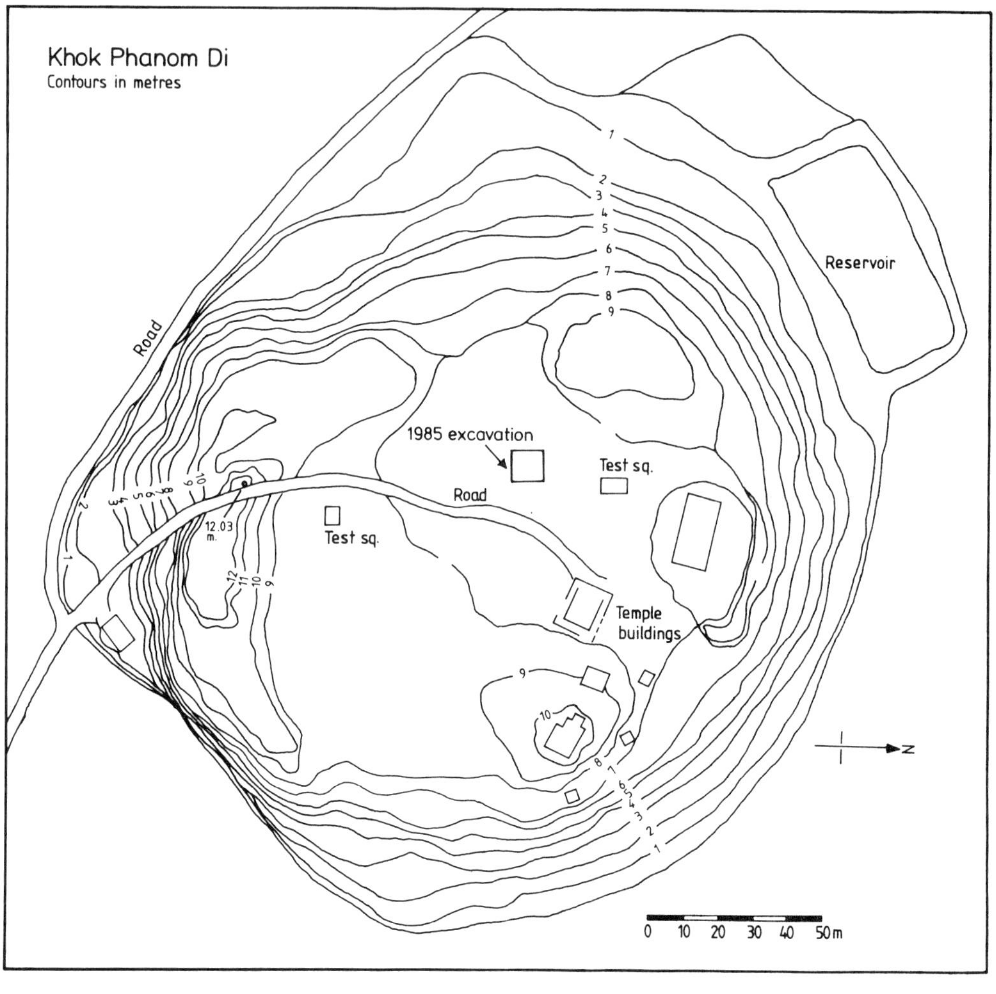

Figure 2: Khok Phanom Di, showing the location of the 1985 excavation.

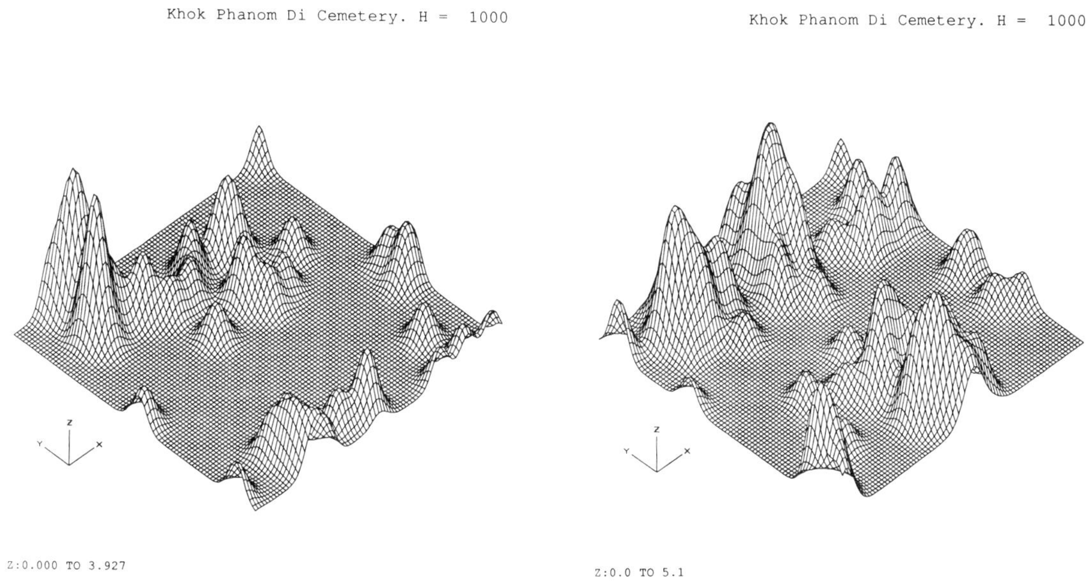

Figure 3: The relative concentrations of burials in zones A (right) and zone B at Khok Phanom Di. The height of each peak represents the degree to which burials are concentrated in that part of the excavation square. It is noted that to a considerable extent, burial clusters were duplicated from one zone to the next. (Courtesy G. Standring)

REFERENCES

BINFORD, L.R.

1968. "Post-Pleistocene Adaptations", Binford S.R. and L.R. Binford (eds.), *New Perspectives in Archæology*, Aldine, Chicago.

BROWN, J.A. and T.D. PRICE

1985. "Aspects of Hunter-Gatherer Complexity", Price, T.D. and J.A. Brown (eds.), *Prehistoric Hunter-Gatherers. The Emergence of Cultural Complexity*, Academic Press, New York: 3–20.

CHAPPEL, J. and B.G. THOM

1977. "Sea Levels and Coasts", Allen, J., J. Golson and R. Jones (eds.), *Sunda and Sahul*, Academic Press, London: 275–291.

CHILDE, V.G.

1925. *The Dawn of European Civilization*, Routledge and Kegan Paul, London.

1934. *New Light on the Most Ancient East*, Routledge and Kegan Paul, London.

1956. *Piecing Together the Past*, Routledge and Kegan Paul, London.

DELVERT, J.

1961. "Le Paysan Cambodgien", *Le Monde d'Outre Mer Present.* Première Série No. 10, École Practique des Hautes Études, Sorbonne.

GEYH, M.A., KUDRASS H.R. and H. STREIFF

1979. "Sea Level Changes in the Late Pleistocene and Holocene of the Strait of Malacca", *Nature* 278: 441–3.

HIGHAM, C.F.W. and A. KIJNGAM

1984. *Prehistoric Investigations in Northeast Thailand*, British Archæological Reports, (International Series 231), (i–iii), Oxford.

HOUGHTON, P. and W. WIRIYAROMP

1984. "The Human Remains from Ban Na Di", C.F.W. Higham and A. Kijngam (eds), *Prehistoric Investigations in Northeast Thailand*, British Archæological Reports. (International Series 231), (i–iii), Oxford.

JOHNSON, G.A.

1982. "Organisational Structure and Scalar Stress", A. C. Renfrew, M.J. Rowlands and B.A. Segraves (eds.), *Theory and Explanation in Archæology*, Academic Press, New York.

MALONEY, B.K.

1987. in Higham, C.F.W., R. Bannanurag, B.K. Maloney and B.A. Vincent: "Khok Phanom Di, the results of the 1984–5 season". *Bull. Indo-Pacific Preh. Assoc.*, in press.

MALONEY, B.K. and J.M. BROWN

1986. "Two Pollen Diagrams from Khok Phanom Di: a Preliminary Report", *Indian Ocean Newsletter*, in press.

PENNY, J.S.

1982. "Petchabun Piedmont Survey", *Expedition* 24(4): 65–72.

PRICE, T.D. and J.A. BROWN (eds.)

1985. *Prehistoric Hunter-Gatherers: The Emergence of Cultural Complexity*, Academic Press, New York.

VAN DER KEVIE, W.

1971. "Acid Sulphate Soils in Central Thailand FAO/UNDP. Regional Seminar on Soil Survey and Soil Fertility Research", New Delhi, Feb. 1971.

VINCENT, B.A.

1987. in Higham, C.F.W., R. Bannanurag, B.K. Maloney and B.A. Vincent: "Khok Phanom Di, the results of the 1984–5 season". *Bull. Indo-Pacific Preh. Assoc.*, in press.

WHITE, J.C.

1982. "Prehistoric Environment and Subsistence in Northeast Thailand", *Southeast Asian Studies Newsletter* 9: 1–3.

YEN, D.

1977. "Plant Remains from Northern Thailand", in Allen, J.J., Golson, J. and R. Jones (eds.), *Sunda and Sahul*, Academic Press, London.

YESNER, D.R.

1980. "Maritime Hunter-gatherers: Ecology and Prehistory", *Current Anthropology* 21: 727–750.

Acknowledgements

The excavation of Khok Phanom Di was co-directed by the author and Mrs. Rachanie Bannanurag. We are most grateful to the Director-General of the Fine Arts Department, Khun Thaweesak Senanarong and the Director of the Research Division, Pisit Charoenwongsa for their support at all levels and times. We further thank the Director-General, Dr. Chumpol Swasdiyakorn and the staff of the Social Sciences Division, National Research Council of Thailand, for their constant courtesy and help. We acknowledge with gratitude the assistance provided by the Ford Foundation, the Wenner-Gren Foundation, the Shell Company of Thailand through the courtesy of M.R. Sarisdiguna Kitayakara and Mr. Precha Phonpraserth, the Universities of Cambridge and Otago, the University Grants Committee of New Zealand, the Centre for Field Research and the British Academy. Specialised analyses are under way, and I acknowledge a deep debt to my associates Damrongkiadt Noksakul and his team of students for contributing so much in the field, to Brian Vincent (ceramics), Bernard Maloney and Judith Brown (pollen), Jacqui Pilditch (jewellery), Amphan Kijngam (fish bone), Philip Houghton, Praphid Choosiri, Nancy Tayles, Carolyn McGill and Katherine Roy (human bone), Douglas Yen and Jill Thompson (plant remains), Alan Grant (mammalian bone), Barbara West (bird and micro-mammalian bone), Glen Standring (burial distribution), Pirapon Pisnupong (stone artefacts), Rose Daamen (shell), Gordon Hillman (fæcal and stomach contents), Anat Bamrungwongse (bone artefacts), and the staff at the radiocarbon dating laboratories in Wellington and Canberra.

DARWINIAN EVOLUTION AND CULTURAL CHANGE: THE CASE OF AGRICULTURE

DAVID RINDOS

Introduction

In this paper it is my intention to sketch out some of the fundamental premises underlying a Darwinian approach to the understanding of cultural change. I will give major attention to a major error that has been repeatedly made in the anthropological literature in the explanation of cultural evolution. We err when we assume that cultures respond to environmental changes in an adaptive manner —that cultural change is generated by adaptively directed cultural activities. The error may be rectified only by a deeper understanding of the Darwinian concepts of undirected variation and natural selection. I will point out that the common understanding of these terms is deficient in depth and subtlety and that this leads to error and contradiction in our descriptions and reconstruction of cultural events.

The Darwinian Perspective

The fundamental premise of Darwinism is that evolutionary change is the result of the natural selection of heritable individual variation. As Ernest Mayr has repeatedly pointed out (e. g. 1942, 1969, 1982: 519–520), evolutionary change is the result of a two step-process: (1) the production of undirected variation and (2) its sorting by means of differential success of individual variant forms over time. Darwinism, unlike all competing theories for organic change, places major emphasis upon the ***undirected nature of the processes generating variation*** in heritable traits. This does not imply the absurd claim that variation is generated "randomly" in the mathematical sense of that term; that is, that the process is totally stochastic and not bounded in any sense. Clearly, no human will exhibit variation in wing structure or photosynthetic pathways. Instead, variation at any moment in time will always be bounded in very important ways by the nature of the organism itself; that is, by its evolved history. If we view culture as an inheritance system —functionally equivalent although only vaguely analogous to the genetic inheritance system— we may accept that the characteristics found at any given time will limit the variant forms that potentially might be generated in that system. This limitation will, of course, place constraints upon the further evolution of that system. Hence, culturally transmitted limitations upon human behaviors may have long term effects upon the evolution of that system. The history of culture is the future of culture in a very real sense.

We may see this phenomenon in a clear manner when we consider the early beginnings of plant domestication. As I have pointed out (1984), a simple relationship exists between humans and plants on which they feed. Over long periods of time, human feeding behavior will affect the local flora in such a manner as to place certain morphological traits of members of a plant species at a competetive advantage over others in terms of attracting human consumers as dispersal agents for the plant. The further evolution of the plant species, therefore, will be altered by the feeding and dispersal behavior of humans. Yet this evolution, in any specific case, takes place in the context of two interdependent historical processes —that of the plant genome including mutation and recombination, and that of the learned subsistence behavior of the humans. A subsistence pattern that lacks a particular behavior, for example the processing of tubers for starch, will place limits upon the development of specific domestication events quite as effectively as the non-appearance of an "appropriate" genetic event within the plant. Hence, we must recognize that speaking of the general evolutioary pressures within the development of agricultural systems may only sensitize us to the types of symbioses that might have occurred; it cannot replace the careful study of the particulars that have occurred in any specific system.

In this context, the properties of the cultural transmission system that governs variable human behaviors must be taken into account when we seek to understand the origin and evolution of cultural traits. Here, much useful work has already been done by Boyd and Richerson (see especially 1985) and Cavalli-Sforza and Feldman (1981). Cultural change may be analogized with much profit to genetic evolution (Campbell 1965). For example, innovations appear in populations, and spread if favored or are abandoned if not, by processes showing striking similarities to natural selection and random drift. Nevertheless, we must recognize that these sorts of general descriptions are only a heuristic model and that the explanation of specific cultural changes must take into account the specifics of the system being considered.

As an example, we may consider work that has been done on domestication and the origins and further development of agricultural system in the central United States known as the American Bottom (Rindos and Johannessen 1983). Following the long incidental domestication of several species of nut trees, a starchy seed complex of native plants began to grow in importance in this region during the Middle Woodland period (*c.* 150 BC–AD 300). The contribution of this complex of seed plants to the diet increased, in both absolute and relative terms, until the latest phases of the Mississipian period (AD 1000–1500). At about AD 800, maize begins to appear throughout the plant record recovered from the region, with the evidence indicating a widespread and remarkably abrupt adoption of this imported plant. The fact that the starchy seed complex remains abundant in the palæoethnobotanic record despite the introduction of maize indicates that this new crop was not grown *instead of* the existing plant complex, but rather was grown *in addition to* plants that had been involved in an evolved agricultural system with humans for as much as one-thousand years. Furthermore, the speed at which maize appears in the record is remarkable. Seeds and cobs, often in very substantial quantities, have been recovered from 50% to 90% of the features analyzed at sites of the Emergent Mississipian period (AD 800–1000). In the sites dated to the immediately previous phases, maize is virtually unknown, having been recovered in minute quantities from only 2% of the features analyzed. The data clearly indicate that maize appearing in the record marks the introduction of a new crop, and not the appearance of agriculture itself. The diffusion of this new, hard-seeded variety of maize could not have been successful unless humans were already practicing the necessary agricultural behaviors. The spread of this new crop was not in any sense a "random" event based upon any inherent qualities of the maize plant, but was rather preconditioned by the existence of an already functioning indigenous agricultural system.

Of course, the Darwinian approach is well within the anthropological perspective: to describe these human characteristics in terms of the generation of potentially selectable traits is to recognize implicitly that decision-making, experimentation or bias are, themselves, culturally bound and determined. Given the widespread anthropological acceptance of the fact that no "absolute" rationality exists which may form the basis for the hierarchical ranking of fundamental human thought processes, we may nevertheless accept that human actions, no matter their motivation or rationalization, may have effects upon the future development of human cultures.

Adopting a Darwinian perspective involves a reorientation of our thought processes. Rather than concentrating upon the *origin* of the particular variant trait that was to form the basis for future developments, we stress the *effect* the possession of this trait, in its incipient form, was to have upon individual humans. Here our attention is directed to the fact that all cultural behaviors have some influence, no matter how small, upon human survival and reproductive success. This will be true even if changes in these behaviors are totally and completely independent of any change in gene frequencies in the populations under consideration. Put in other terms, cultural behaviors may affect human demography. Hence, the concept of "pure" or "demographic" Darwinian fitness is totally applicable to the effects of cultural traits upon human populations.

Let us return to the introduction of maize into the American Bottom. We have already noted that maize was introduced into a cultural and agricultural system that had been in operation for at least one thousand years. Nevertheless, following its adoption major changes may be seen in the archæological record. These include increases in population, nucleation, social complexity and centralization. We have argued (Rindos and Johannessen 1983) that some of these changes were precipitated by the introduction of maize which, by means of its higher yield potential, permitted increases in populations in the region. Of greater importance, however, qualities of the varieties of maize that came into use may have had major effects upon the reliability and predictability of the agricultural system itself, and hence upon the social structuring of the region.

As population increased as a function of the higher average yield obtainable from maize, so did the need for farm land. Increased farm land could be obtained only by clearing the native forests. Our evidence indicates that this occurred, at least in part, by clearing off the incidentally domesticated "nut groves" that had arisen over many centuries in regions of human habitation. A major change occurs in the charcoal record when maize appears as a major component of the diet. Before the introduction of maize, most firewood was from genera such as *Populus* and *Salix*; after maize becomes abundant, *Carya* (hickory) and *Juglans* (walnut) charcoal appear in the record for the first time in large quantities. It appears that the growing need for land to grow maize to support a burgeoning population brought about the destruction of what had previously been an important dietary resource.

Much data indicates that while maize is higher yielding in the American Bottom region than the native seed crop complex, this increased yield is likely accompanied by a radical increase in variability of yield. That is, even though average yield obtainable from maize cultivation increases dramatically, the year to year variance on that yield increases even more. Any change in absolute yield is "progress" only at the moment of its first occurrence —over relatively short periods of time, increases in yield will literally be eaten up by the increased populations they generate. Yet, as population increases, the negative effects of increased variability on yield become increasingly severe: one cannot consume during times of scarcity the no longer existant surplus that had been generated during times of abundance.

The most obvious way to deal with the interacting factors of increasing population, increasing potential yield and increased variance on that yield would be attempts to buffer the system by increased association and trade between regions. Here if a crop is bad in one locality, maize could be imported from other localities during the crisis period. Note that this is a type of activity that requires no foresight, merely a response to a specific condition of immediate reduced food availability. Furthermore, over time such arrangements could grow and have consequences that were totally unforseeable at the moment that the exchange systems were initially established. Hence, the increased centralization and integration characteristic of Mississippian culture may be traced to attempts to deal with the new variability in yield characteristic of a maize-based agricultural system.

No question can exist that agriculture, and changes in agricultural system may have major effects upon local demography and these in turn may affect social systems in radical ways. However, it is significant that differential fitness, the "currency" of evolutionary change requires that more than one state may be observed in the population of concern. If all individuals have exactly the same trait set, evolutionary change is, by definition, impossible. Cultural processes may be of major importance in determining which traits may be acceptable to members of that culture. Therefore, a Darwinian view of cultural evolution does not restrict itself solely to the demographic aspects of cultural behaviors, but must also look inside the system and consider the impact of variable culturally defined behaviors upon the further evolution of the system. I have earlier dealt with this distinction between the "demographic" and "symbolic" aspects of cultural transmission systems under the rubric of cultural selection of the first and second types (CS 1 and CS 2; Rindos 1985). If culture is, indeed, an inheritance system, as we believe it to be, those factors serving to increase the variance within the system will increase the rate of evolution of the phenotypes generated within the cultural environment ***even if the specific variants produced are not intrinsically directional*** (see Rindos 1986). From the Darwinian perspective, individual "inventions" or cultural "choices" are viewed as ways in which heritable variation may be generated, and these variant forms are then available to selection.

We have generally seen the meaning of human innovations solely in terms of their *causal* role in cultural evolution. I would claim that these innovations are better viewed as events that *facilitate* evolution. Cultural processes such as innovation or discovery are processes that permit but do not directly cause cultural change. The true reason for change only may be detected in the social and demographic consequences of these changes in behavior.

Any general ability for human cultures to adapt to the environment —to produce new and variable behaviors in concert with the flexibility to accept certain types of innovation— should be seen as a consequence of the selective processes that acted during the evolution of the human cultural capacity. This entails simply the claim that the evolved capacity of our species to react to the environment in a highly pleiotropic manner is a consequence of selection for increased phenotypic flexibility within our lineage — that is, for the ability to ***quantitatively increase, not qualitatively change*** the amount of heritable variation produced in the phenotype (Rindos 1986). The specific changes that occur, however, are not the result of a cultural capacity that allows humans the ability inevitably to generate directed or adapative variation. Instead these changes arise by means of the selection of the indirected (in an evolutionary, not a cultural, sense) variation produced within cultures. Put in other terms: innovations are not adaptations without selection's action. We cannot use the evolved capacity of humans to respond to the environment as *explanation* for specific changes. We cannot explain the particular by invoking the general; instead, we should look directly to the evolved system for insights.

Consider another example from the American Bottom. It is likely that the "introduction" of maize around AD 800 into the region actually marked the appearance of new varieties of flinty, early maturing maize which was stored as a dry grain. Previously, it is likely that maize was grown for its immature ears that were stored in a processed form. Here, culinary and storage techniques and traditions interacted to affect the growth potential of the agricultural system. If a crop is stored in a processed form, the seed needed to plant the next year's crop must be estimated and left unharvested until it is mature. In a year with a particularly good crop, much food could be preserved and this could serve as a buffer upon unpredictable reductions in other resources. Of course, the same would be done if the crop is stored as a dry grain. However, the potential growth rates of these two systems would not be the same.

Consider years of abundance: the stored crop is not consumed in its totality. If the stored crop were in a processed form, some might even end up going to waste. But if the stored crop were a dry grain, an alternative exists —planting of some of the left-over grain, increasing the area under cultivation by a slight amount. In this manner, mean productivity could slowly rise as the result of a succession of favorable years. Storage of a crop in its reproductive form may increase the potential rate of growth of the system. However, this same identity of "food" and "seed" would have negative effects during bad seasons. Under conditions of moderate stress, some of the seed grain for the next season's planting could be consumed. Here, the farmer is willing to take a gamble that next year's crop would be sufficiently good that it is worth taking the risk of consuming some portion of the seed in the face of present needs. Of course, this is a risky tactic to take in that it reduces the long-term stability of this particular system. And it is important to note that such a reduction in stability is far less likely to occur when the crop is stored in a processed form: a cultural and culinary marker sets the processed, "edible" food apart from the "seed" for the plant. In the most extreme form, the edible part of the plant is totally differentiated from the reproductive part both culturally and biologically (for example, manioc and many other tropical tubers). Here the much discussed stability of tropical tuber systems may also be seen as a function of cultural classification systems concerning what is edible.

The distinction between culture seen as explanation and culture seen as the result of natural, evolutionary processes is neither trivial nor purely definitional. Instead, adopting the Darwinian perspective totally changes the manner in which we approach the historical, ethnographic and archæological records. It is not an easy point of view to adopt. Much of the adversion to treating cultural processes as natural arises from a cultural bias (Rindos 1985), and is largely centered upon the issue of the potency of human intentionality in explaining cultural change (Rindos 1984). Translated into evolutionary terms, the common-sense view of cultural change holds that *the variations are generated as specific responses to adaptive needs —that is, that the variations arising in cultural systems are directional and therefore, the variations themselves, not selection, order evolutionary processes in human culture.*

Cultural traits, of course, may be adaptive but this is a result of the fact that they gave a relative advantage to the individuals with these traits. Here we analyze cultural traits in exactly the same terms used for the evolution of adaptations within any species. But we must also be willing to go beyond this analysis. Cultural traits may spread not because they are adaptive to some need but because they indirectly increase the relative fertility (fitness in the pure Darwinian sense) of individuals within a cultural setting. If, for example, religious affiliation is a relatively heritable trait, then we can expect that over time (and all other things being equal) the proportion of individuals within any group belonging to "procreative" religions such as Islam, Roman Catholicism, Orthodox Judaism, and Mormonism will rise. It is absurd to call any such change in the proportions of individuals belonging to these sects a result of an "adaptation" by its members to any enviromental "need" or, put in other terms, that a positive correlation between relative fitness and environmental adaptation must always exist.

In my discussion of the spread of agricultural systems (Rindos 1984), I have maintained the position that fitness and "adaptiveness" may even show a negative correlation. Here, the relative fitness induced into a culture by means of its agricultural behavior is seen as being simply the result of a higher realized rate of population growth —agricultural behaviors increase the carrying capacity of the local environment for humans and, hence, the proportion of individuals per unit area who will have agricultural modes of subsistence will inevitably grow to be larger than the proportion of individuals with most other subsistence strategies. It is important to realize that this will hold true even if agricultural behaviors bring with them a decrease in robustness, a decline in life expectancy, an increase in morbidity and mortality or a higher infant mortality rate (accompanied, of course, by an even higher fertility rate). While all of these factors may easily be judged as indicators of decreased adaptation, the higher growth rate of agriculture will nevertheless favor it over other forms of human subsistence.

A very striking example of how fitness and adaptiveness may become decoupled may be seen by considering the relative fitness of competing agricultural traditions. Begin with the assumption that agricultural behaviors bring with them a decrease in overall "adaptiveness" of human populations (as measured by the types of parameters mentioned above). Consider two competing agricultural populations, one of which has a more adaptive agricultural tradition than the other measured in terms of variables such as stability in average yield. It is easily demonstrated (Rindos 1984, Chapter 6) that, on the average, decreased stability in productivity will increase the probability that a given agricultural tradition will spread. Here, instability in agricultural production serves as a driving force in spreading a particular agricultural tradition by literally driving out individuals to colonize new regions. Occasional episodes of lowered productivity, induced not by the environment but rather by the plants and techniques of the agricultural system itself, will cause the spread of that tradition to occur at a rate greater than that of other, more adaptive and stable, agricultural traditions. Hence, not only is agricultural more fit while being less adaptive, but a positive selection for instability in production will tend to increase the maladaptiveness of agricultural systems over time.

Do we have any evidence that the theoretical model presented above has any valadity, or is it merely a rather unpleasant anti-adaptationistic scenerio —a sort of Malthusian "Justo-So" story? A recent volume, edited by M. Cohen and G. Armelagos (1984) presents abundant evidence indicating that the archæological record is largely in accord with predictions that may be drawn from the coevolutionary model for the origins and spread of agriculture. Under the coevolutionary model, it would be predicted that major population stress would originate with highly developed agricultural systems (the phase of "agricultural domestication" described in Rindos 1984 and the discussion of population pressure in chapter 5). It would be further intensified as agricultural systems developed and spread by means of the positive selection for optimally unstable systems during diffusion episodes.

These predictions stand in stark contrast to those of the two other contemporary schools modeling the origins of agriculture. The first school represents the mainstream of anthropological and archæological theory. Despite great differences in approach by specific authors, it is unified by an adaptationistic, cultural ecological perspective. An equilibrium based analysis of cultural function and change is one of the major assumptions underlying work done by members of this school. I include here such authors as Flannery (1965, 1968, 1973), Binford (1968), Bray (1976, 1977), Harris (1969, 1972, 1977), Reed (1977) and Wright (1977). As Roosevelt (1984: 569) points out, an equilibrium centered view of cultural change has clear implications in terms of stress on humans that might be found in the archæological record: "the equilibrium theory predicts that physiological stress should occur only rarely and that cultural adaptation should increasingly buffer people from stress". Under this model, the transition to sedentism would be expected to be accompanied by a "decrease in mortality". Cultural ecology with its emphasis upon adaptation and homeostasis sees cultural change as motivated by human adaptation; *increased adaptiveness* should therefore result from realized cultural changes.

The other contemporary school for agricultural origins applies a population pressure model. This is best exemplified in the work of Cohen (1977), although others such as Abernathy (1976), Grigg (1976), Smith (1972) and Spooner (1972) have advanced similar views. The theory (and I will not criticize it here on theoretical grounds) holds that cultural change is the result of human adaptations to their own increasing numbers. Population pressures generated by the slow increase of human populations forces new adaptations to be adopted if humans are simply to "stay in the same place"; cultural change is the result of the need to *maintain adaptiveness.* Here the predictions for the archæological record would be "that physiological stress should be recurrent and persistent, with particularly severe stress possibly occuring during incipient agriculture" (Roosevelt 1984: 569). The demographic determinists, unlike the cultural ecologist, also would hold that sedentism should be accompanied by increases in mortality.

In Roosevelt's review of the archæological indicators of skeletal stress, she was able to make the following generalizations:

> Although there is a relative lack of evidence for the Paleolithic stage, enough skeletons have been studied that it seems clear that seasonal and periodic stress regularly affected most prehistoric hunting-gathering populations... What also seems clear is that severe and chronic stress...is not characteristic of these populations. There is no evidence of frequent, severe malnutrition, so the diet must have been adequate...
>
> During the periods when effective agriculture first came into use there seems to be a temporary upturn in health and survial rates in a few regions: Europe, North America, and the Eastern Mediterranean...it seems that cultivated plants probably increased the storable calorie supply, removing for a time any seasonal or periodic problems in food supply. In most regions, however, the development of agriculture seems not to have had this effect, and there seems to have been a slight increase in physiological stress.
>
> Stress, however, does not seem to have become common and widespread until after the development of high degrees of sedentism, population density and reliance on intensive agriculture. At this stage in all regions the incidence of physiological stress increases greatly, and average mortality rates increase appreciably... Stature in many populations appears to have been considerably lower than would be expected if genetically-determined height maxima had been reached, which suggest that the growth arrests documented by pathologies were causing stunting...
>
> It seems that a large proportion of most sedentary populations under intensive agriculture underwent chronic and life-threatening malnutrition and disease, especially during infancy and childhood. The causes of the nutritional stress are likely to have been the poverty of the staple crops in most nutrients except calories, periodic famines caused by the instability of the agricultural system, and chronic lack of food due to both population growth and economic expropriation by elites. (Roosevelt 1984: 572–573).

As Roosevelt notes in summary, "the origin of agriculture, then, cannot accurately be attributed to the existence of unusually high levels of (population) pressure at the time". Furthermore the increased stress occuring with the appearance of developed agricultural systems is in total contradiction to the expectations of the cultural ecologists. Note, however, that the general pattern and the specific timing of appearance of indicators of greatly hightened stress is totally congruent with the predictions to be deduced from the coevolutionary model.

Both cultural ecologists and demographic determinists share a common assumption: cultural change is mediated by adaptive needs. In fact, from a Darwinian perspective, the two positions are demarcated by only the slightest difference in emphasis. The cultural ecologists sees cultural change as the means to successfully maintain a high level of adaptation to changes in environmental conditions; the demographic determinists sees cultural change as successfully maintaining an increased level of limited adaption for a short period of time, after which growing population pressure necessitates further readjustments. Yet both center their arguments on adaptation as a dictated response by culture to changes in realized environmental conditions. Here, the cultural ecologist faced with an environment that was slowly and continually degrading would be in the same position as the demographic determinist. And in fact, what is the difference between an assumption of a continually degrading environment and the assumption that population numbers inexorably grow? In mathematical terms the two would be equivalent —per capita yield will decrease even in the face of successful adaptations that might raise it for a short period of time. In either case the future is apparent— no matter what is done, new stresses will continue to arise. Further cultural change is assured by the need to accomplish further adjustments by means of new cultural adaptations that will, at least temporarily, solve adaptive problems.

The stress laid upon adaptation as the prime mover in sociocultural change is the major (at times it appears the only) avenue to anthropological and archæological investigation. The work of almost all recent scholars treating agricultural origins begins, continues and ends with the assumption that adaptation is the only way to explain this major shift in human subsistence patterning. Consider Binford's early statement:

> Culture is all those means whose forms are not under direct genetic control...which serve to adjust individuals and groups within their ecological communities. If we seek understanding of the origins of agriculture...we must analyze these cultural means as adaptive adjustments... (S)elective pressures favoring new cultural forms result from non-equilibrium conditions in the local ecosystem. Our task, then, becomes the isolation of the variables *initiating directional change in the interal structuring of ecological systems.* (Binford 1968: 323, emphasis added).

Or taking a more recent treatment by Cohen:

> I suggest therefore that the development of agriculture was an *adjustment* which human populations were *forced to make in response* to their own increasing numbers... (Agriculture) has only one advantage...that of providing more calories...(I)t will thus be practiced only when *necessitated* by population pressure. (Cohen 1977: 144–15, emphasis mine).

Such examples could be multiplied at great lenght (and are considered in just such a manner in Rindos 1985). For our purposes here, however, it is more useful to consider the source of this prejudice (and prejudice it is because no convincing argument justifying the practice has yet to be advanced) in a bit more depth.

Natural Selection and Cultural Change.

In taking a Darwinian perspective towards human culture, rather than seeing culture in typological terms —that is as an entity into itself having its own properties and finalities— we see what we call culture as the summed and averaged beliefs, attitudes and behaviors of the members of a specific society at a specific time. Here, a tremendous store of variation is seen to be present. Variant forms of almost all cultural traits are expected to be expressed at any time. Nevertheless, over long periods of time, the mean and mode values of these variant forms will change. The process underlying all such change in heritable traits, even if they are culturally transmitted, is selection (Dunnel 1980). This selection may be working on phenotypically transmitted traits that are coded as parts of a culturally defined inheritance system, but it is selection none the less.

The critical issue in this context is whether natural selection is the process responsible for the changes that have occurred over time in human subsistence patterns. The centrality of natural selection in Darwinism arises from its ability to bring about evolution. From the Darwinian perspective, undirected variation is important for its role in fueling the engine of evolutionary change by generating new forms which may then be subject to selection. Indeed, we may claim confidently that without a true concept of undirected variation, natural selection is not only unnecessary but is actually impossible. Natural selection, within Darwinian theory is the only directional force in evolution. If variation is less than undirected, then natural selection cannot be seen as a *creative* force in evolution and must perforce maintain its simple, pre-Darwinian role of removing those variant forms that accidently deviate from the true type of the species (see discussion in Rindos 1984). Only if we see variation as being produced randomly with respect to selective pressures may we claim that the directionality that may be observed in evolution over time is the result of natural selection (but recall the remarks above on inherent limitations upon variation).

The issue of *undirected* variation is critical to the development of a scientific understanding of human cultural evolution. Viewing variation as undirected changes the whole way in which we do the business of attempting to explain cultural evolution. Here, the spread of behavior throughout a society, or the spread of a particular type of behavior (rather than another) throughout the species, is the result of the fitness induced by that behavior (and again, note I am using "fitness" in a broad sense that includes both demographic and symbolic aspects and goes beyond simple genetic contribution to future generations). Rather than seeing change as consequent to the adoption of a particular form of behavior, emphasis is placed upon the *historical consequences* of a particular variant form of behavior for the humans exhibiting that behavior and for the cultures that are thereby altered.

REFERENCES

ABERNATHY, V.

1979. *Population Pressure and Cultural Adjustment*, New York, Huam Science Press.

BINFORD, L. R.

1968. "Post-Pleistocene Adaptations", in *New Perspectives in Archæology*, edited by S. R. Binford and L. R. Binford, Chicago, Aldine: 313–341.

BOYD, R. and P. J. RICHERSON

1985. *Culture and the Evolutionary Process*, Chicago, University of Chicago Press.

BRAY, W.

1976. "From Predation to Production: The Nature of Agricultural Evolution in Mexico and Peru", in *Problems in Economic and Social Archæology*, edited by G. de G. Seveking, T. H. Longworth and K. E. Wilson, London, Duckworth: 73–95.

1977. "From Foraging to Farming in Early Mexico", in *Hunters, Gatherers and the First Farmers Beyond Europe*, edited by S. V. Megaw, Leicester, Leicester University Press: 225–249.

CAMPBELL, D. T.

1965. "Variation and Selective Retention in Sociocultural Evolution", in *Social Change in Underdeveloped Areas: A Reinterpretation of Evolutionary Theory*, edited by R. W. Mack, G. Blanksten and H. R. Barringer, Cambridge, Schenkman: 19–48.

CAVALLI-SFORZA, L. L. and M. W. FELDMAN

1981. *Cultural Transmission and Evolution: A Quantitative Approach*, Princeton NJ, Princeton University Press.

COHEN, M. N.

1977. *The Food Crisis in Prehistory: Overpopulation and the Origins of Agriculture*, New Haven, Yale University Press.

COHEN, M. N. and G. J. ARMELAGOS

1984. *Paleopathology at the Origins of Agriculture*, Orlando, Academic Press.

DUNNELL, R. C.

1980. "Evolutionary Theory and Archæology", in *Advances in Archæological Method and Theory* (Vol 3), edited by M.B. Schiffer, New York, Academic Press: 38–99.

FLANNERY, K. V.

1965. "The Ecology of Early Food Production in Mesopotamia", *Science* 147: 1247–1256.

1968. "Archæological Systems Theory and Early Mesoamerica", in *Anthropological Archæology in the Americas*, edited by B. Meggers, Washington DC, Anthropological Society of Washington: 67–86.

1973. "The Origins of Agriculture", *Annual Review of Anthropology* 2: 271–310.

GRIGG, D. B.

1976. "Population Pressure and Agricultural Change", *Progressive Geography* 8: 135–176.

HARRIS, D.

1969. "Agricultural Systems, Ecosystems and the Origins of Agriculture", in *The Domestication and Exploitation of Plants and Animals*, edited by P. J. Ucko and G.W. Dimbleby, London, Duckworth: 3–14.

1972. "The Origins of Agriculture in the Tropics", *American Scientist* 60: 180–193.

1977. "The Origins of Agriculture: Alternative Pathways Towards Agriculture", in *Origins of Agriculture*, edited by C. A. Reed, The Hague, Mouton: 173–249.

MAYR, E.

1942. *Systematics and the Origin of Species*, New York, Columbia Univesity Press.

1969. *Principles of Systematic Zoology*, New York, McGraw Hill.

1982. *The Growth of Biological Thought*, Cambridge MA, Harvard University Press.

REED, C. A.

1977. "Origins of Agriculture: Discussion and Some Conclusions", in *Origins of Agriculture*, edited by C. A. Reed, The Hague, Mouton: 879–953.

RINDOS, D.

1984. *The Origins of Agriculture: an Evolutionary Perspective*, New York, Academic Press.

1985. "Darwinian Selection, Symbolic Variation and the Evolution of Culture", *Current Anthropology* 26: 65–88.

1986. "The Genetics of Cultural Anthropology: Towards a Genetic Model for the Origin of the Capacity for Culture", *Journal of Anthropological Archæology* 5.

RINDOS, D. and S. JOHANNESSEN

1983. "Diversity and Uniformity in Mississippian Plant Usage", Paper for the Symposium 'Mississippian Cultural Development and Variation in the Central and Upper Mississippi River Drainage', 48th Annual Meeting, Society for American Archæology, Pittsburg PA, 28 April 1983.

ROOSEVELT, A. C.

1984. "Population, Health, and the Evolution of Subsistence: Conclusions from the Conference", in *Paleopathology at the Origins of Agriculture*, edited by M. N. Cohen and G. A. Armelagos, Orlando, Academic Press: 559–584.

SMITH, P. E. L.

1972. "Changes in Population Pressure in Archæological Explanation", *World Archæology* 4: 5–18.

SPOONER, B. (ed)

1972. *Population Growth: Anthropological Implications* Cambridge MA, MIT Press.

WRIGHT, H. E. Jr.

1977. "Environmental Change and the Origin of Agriculture in the Old and New Worlds", in *Origins of Agriculture*, edited by C. A. Reed, The Hague, Mouton.

CURRENT ARCHAEOLOGICAL EVIDENCE FOR THE BEGINNING OF AMERICAN AGRICULTURE

C. EARLE SMITH, JR.
University of Alabama

During this week of conference dedicated to the memory of V. Gordon Childe, I felt it appropriate to review the current knowledge of the beginnings of agriculture in the Americas. During Childe's productive years, the recovery of archæological plant remains was developing in the Old World in sufficient volume that interpretation became worthwhile. In part, the plant remains were providing indications of agricultural development contrary to many of the then current hypotheses. For Childe (1953), Tell Hassuna represented the earliest agricultural date although indications from Jarmo were portending even earlier agriculture. He conceived of the origin of food production as the greatest economic revolution "after the mastery of fire". While this may be true and is conceded by many authors on the origin of agriculture, the factors influencing man to undertake cultivation are still widely argued.

I am going to forego discussion of the forces which led to the taming of animals and the domestication of plants, and deal principally with the current factual, archæological evidence for the development of agriculture in the New World. I will not debate the beginning, but I do hope to indicate the progression of cultivation as the evidence now can be interpreted.

Most of the story of cultivation in the Americas can be placed in the American tropics. The recent publication of *Prehistoric Food Production in North America* (Ford 1985) clearly indicates that all of the major crops of temperate North America originated to the south, with the probable exception of sunflower (*Helianthus annuus*) (Heiser 1985). Evidence is still accumulating on the cultivation of several other temperate North American species (Asch and Asch 1985; Cowan 1985), without definitive evidence that any of these were 1. of primary importance in prehistoric diets and, 2. in the case of *Chenopodium*, whether this was definitely temperate North American and not derived from a Mexican species. Throughout the eastern deciduous forest region of North America, the archæological record indicates that a dependence on hickory nut and acorn developed in the Archaic and it remained uppermost in importance long after the introduction of maize. Contact period accounts (Harriot 1972 [1590]) suggest that both hickory nuts and acorns were of major importance in the diet at that time, although English observers, accustomed to a grain dominated diet, made much of the maize which they encountered.

The western part of temperate North America has not produced any cultivated crop of note beyond the sunflower (*Helianthus annuus*) (Heiser 1969, 1985). The Pueblo and Zuni groups have depended heavily on cultivated crops which have largely been imported from Mesoamerica. The remainder of the region was inhabited by gathering and hunting folks some of whom practiced minimal cultivation either of local plants or of Mesoamerican species (Steward 1938). The people of the Northwest Coast found resources too bountiful to experiment with cultivation.

The Mesoamerican area has long been recognized as a part of the world in which cultivation of plants developed at an early date (Vavilov 1951). The plants listed by Vavilov (1951) as originating in Southern Mexico and Central America include *Zea mays; Phaseolus vulgaris, P. lunatus, P. multiflorus, P. acutifolius* var. *latifolius; Cucurbita ficifolia, C. moschata, C. mixta; Pachyrhizus erosus, Ipomoea batatas, Capsicum annuum, C. frutescens, Gossypium hirsutum, Agave sisalana; Annona cherimola, A. reticulata, A. squamosa, A. muricata, A. purpurea, A. cinerea, A. diversifolia, A. glabra; Sapota achras (Manilkara achras), Casimiroa edulis, Calocarpum mammosum, Lucuma salicifolia, Carica papaya, Persea schiediana, P. americana, Psidium guayava, P. friedrichsthalianum, Spondias mombin, Cratægus mexicana, Diospyros ebenaster (D. digyna), Chrysophyllum cainito, Anacardium occidentale, Prunus serotina (P. capuli), Agave atrovirens, Cereus* spp., *Nopalea coccinellifera, Tigridia pavonia, Physalis æquata, Lycopersicon cerasiforme, Salvia chia, Theobroma cacao, Bixa orellana* and *Nicotiana rustica.*

In addition, Vavilov recognized that a number of species of cultivated plants originated or had a center of diversity in certain areas of South America. The list for the Andean axis includes a number of species of locally grown potatoes (*Solanum* spp.) and the following plants: *Oxalis tuberosa, Tropoeolum tuberosum, Ullucus tuberosus, Lupinus mutabilis, Chenopodium quinoa, C. canahua, Amaranthus caudatus, Lepidium meyenii, Phaseolus lunatus* var. *macrospermus, P. vulgaris, Xanthosoma sagittifolium, Canna edulis, Arracacia xanthorrhiza, Solanum muricatum, Lycopersicon esculentum, Cyphomandra betacea, Physalis peruviana, Cucurbita maxima, Capsicum baccatum, C. pubescens, C. frutescens, C. chinese, Erythroxylon coca, Bixa orellana, Gossypium barbadense, Fourcroya cubensis, Passiflora ligularis, P. quadrangularis, Lucuma obovata, Psidium guayava, Nicotiana tabacum, Anona cherimola, Inga feuillei, Bunchosa armeniaca, Matisia cordata, Caryocar amygdaliferum, Guilielma speciosa, (Bactris gassipaes), Malpighia glabra*, and *Solanum quitoense.* He separates out a Chiloe center from which come *Solanum tuberosum*, and *Fragaria chiloense.* In the eastern portion of South America, Vavilov places the ancestral area for the following: *Manihot utilissima (M. esculenta), Arachis hypogæa, Theobroma cacao, Hevea brasiliensis, Ilex paraguariensis, Eugenia uniflora, E. uvalha, E. dombeyi, Myrciaria jaboticaba, Ananas comosa, Anacardium occidentale, Feijoa sellowiana, Passiflora edulis.* A few other plants could be added to these lists, but these provide sufficient species for comment.

Since the time when Vavilov compiled the evidence which led him to list these species, examples of many of them have been found as archæological plant remains. We, thus, have better information on the possible area of origin for some of these cultivated species. For others, the record remains very much the same as it was in Vavilov's day.

Maize (*Zea mays*) is one of the best represented plants recovered from archæological sites. Both the maize cob and the kernels are particularly persistent as carbonized specimens. Both dried and carbonized archæological materials have been recovered.

The origin of maize and its diffusion through temperate North America has recently been reviewed by Galinat (1985). As he observed, the oldest known archæological maize dates to about 5,000 BC from the dry cave deposits of the Tehuacan Valley (Mangelsdorf *et al.*, 1967). While originally interpreted as wild maize and the subsequent products of human selection, it is now conceded that the most primitive form is cultivated maize evolved under human selection pressure. The Tehuacan maize record covers the period from 5000 BC to AD 1500 and illustrated a steady progression of selection for larger cobs with greater amounts of fruits. The original podcorn form becomes the present-day maize with small glumes. A number of different selection trends can be illustrated by the many maize plant parts recovered, but no teosinte was found nor is there any material which can be interpreted as transitional between wild teosinte and early cultivated forms of maize.

The evidence from other Mesoamerican archæological sites does not provide either earlier or more informative remains concerning the origin of this crop. To the north at Bat Cave, New Mexico were found maize cobs which, in the lower levels, are similar to the most primitive cobs from the Tehuacan Valley (Mangelsdorf and Smith 1949). Debate still rages over the correct age for this material, but the morphology clearly indicates its affinities to early maize elsewhere. The cobs are small, delicate and soft, the glumes partially enclose the individual grains and the cobs were apparently topped by a tassle of male florets. Inasmuch as none of the hypothetical ancestors of cultivated maize could have survived the winters at this site, it is obvious that the maize was grown nearby during the summer growing season, and it must have been an introduction from tropical Mexico.

Recoveries at other Mexican sites have often provided some evidence of the evolution of lines of maize leading to present-day varieties (Mangelsdorf *et al.*, 1967; Tolstoy *et al.*, 1977; Smith and Tolstoy 1981, among others). They may also indicate a timetable for the spread of maize in many directions, but it is wise to reflect that maize may have been introduced into a region far earlier than archæological contexts would lead us to believe. At the present time, the northward shift of maize into the temperate part of North America is being debated. Some archæologists dispute the early radiocarbon dates for several deposits and assert that the earliest evidence for maize in the Southwest (Bat Cave) is not until about 300 BC. Apparently another infiltration of germplasm from Harinoso de Ocho was penetrating further westward which would cross with the Nal Tel-Chapalote to become the later Pima-Papago race (Galinat 1985). Many more years of adjustment for day length and to shortened growing season were needed before maize could reach its northern limits in eastern North America where it is represented in sites at around AD 700 (Galinat and Gunnerson 1963).

The spread of maize southward is even less well documented with archæological remains than the northward spread. The presentday diversity of maize in Guatemala indicates that the highlands are joined to southern Mexico as the primary center of diveristy of maize. At this time, we have little archæological recovery of maize in any part of Guatemala. From Belize (Cuello), Bird (1984; Miksicek *et al.*, 1981) reports maize that may be as much as 4000 years old. It probably is Chapalote-Nal Tel which has been observed in fields in the Yucatan Peninsula in historic times. While samples were collected from several chronological levels, little variation is observed in the remains.

Lentz (1984) reports the recovery of carbonized maize remains by the Cajon Project from sites in the Sulaco River Valley, Honduras. While carbonized remains are far less informative than dried remains, enough data were gathered to indicate that the Cajon Project maize is probably also Chapalote-Nal Tel. How widely distributed this may have been in Honduras is not indicated by the current evidence.

From Costa Rican sites, a number of specimens of carbonized maize have been recovered. The earliest are from the Pavas Phase (200 BC–AD 400) and include, from Barrial de Heredia, a 10-rowed cob and, from Sitio 43-SE, 10-row and 12-row cobs as well as kernel fragments (Smith, notes). Material recovered from an El Bosque Phase level (100 BC–AD 500) from the site of Severo Ledesma was examined by Galinat and identified as Chapalote-Nal Tel-Pollo (Snarskis 1976). Recoveries from La Fabrica near Grecia were kernel fragments from the Curridabat Phase (AD 400–700) (Smith, notes). Late prehistoric maize (*c.* AD 800–1300) recovered from Aguacaliente (C-35-AC) near Cartago and from La Ceiba, Guanacaste (G-60-LC) apparently represent two races (Smith, notes). One of these is small kerneled and appears to be related to the Chapalote-Nal Tel-Pollo previously mentioned. The other is apparently an 8-row variety with large, floury kernels (the interior of the burned kernel does not exhibit the highly vesicular appearance of more sugary burned kernels). The 8-row races from Central America have been interpreted as introductions from South America (Wellhausen *et al.*, 1952; 1957).

The current evidence shows no continuity in maize cob or kernel morphological change from the north to the south as might be expected if the crop was gradually diffused overland. The lack of recoveries in much of Central America may contribute heavily to this impression. However, it cannot explain the apparent relatedness of the Costa Rican carbonized maize fragmnents to Colombian collections.

On the other hand, Galinat (1980) in reporting on maize from Chiriqui, Panama, has chosen to interpret the identification of Chapalote-Nal Tel-Pollo from Sitio Pitti (BU-17) as indicating the gradual divergence of Pollo from the Chapalote-Nal Tel complex as it has moved southeastward from Mexico-Guatemala. Should further archæological remains be recovered in Honduras and Nicaragua this interpretation may well be confirmed.

Additional maize cob fragments and kernels have been recovered from Carabali Rockshelter (SF-9) which clearly fall into two varieties of maize kernels, (Smith, notes). The smaller kernels are Pollo-like and clearly fit with the maize reported from Sitio Pitti (Galinat 1980). The others are notably wider and may represent an 8-row maize variety.

The northwestern region of South America has had little plant remains recovery from archæological sites. Only recently from sites in the Cauca Valley have maize remains become available. The earlier from the Yotoco Phase (300 BC–AD 1200) do not appear to differ dramatically from carbonized maize kernels found in Sonso Phase (AD 1200–1500) context (Smith, notes). Two size classes occur, the more nearly iso-diametric fitting into the Pollo-Nal Tel complex, and the much larger kernels probably deriving from a line of evolution leading to the modern race Clavo (Roberts *et al.*, 1957).

From adjacent Venezuela, at least two areas have produced archæological maize remains. The find at the El Tiestal site in Lara (Mangelsdorf and Sanoja 1965) consists of cob and kernel fragments identified as an early form of the race Pollo. Other maize remains from the Orinoco Valley near Parmana are currently under study by Galinat. These carbonized fragments first appear in the Corozal phase (800 BC–AD 400) and continue through the Camoruco Phase (AD 400–1500) (Smith and Roosevelt, ms.)

In Ecuador, the presence of maize plants has been suggested by phytolith analysis of archæological site fill from Real Alto and OGCH-20, Santa Elena Peninsula by *c.* 2450 BC (Pearsall 1978, 1980). This evidence has lately been reconfirmed (Pearsall and Piperno 1986). It accords well with the evidence from the coastal sites of Peru where dates of about 2500 BC have been given for the earliest maize remains (Grobman *et al.*, 1977; Grobman and Bonavia 1978). The majority of the cobs could be assigned to a line of evolution leading to the modern race, Confite Morocho. It is a matter of conjecture that the coastal maize varieties were introduced into the coastal valleys from the highlands. Relatively few recoveries of maize have been made in the highlands. One of these dated to the Chihua Phase (4300–3800 BC) of Pikimachay included maize identified by Galinat as Confite Morocho-like (MacNeish *et al.*, 1970). A lot of dried maize cobs from Guitarrero Cave in the Callejón de Huaylas came from Complex III [perhaps 7000 BP (Lynch *et al.*, 1985)] and Complex IV [perhaps around 2200 BP (Lynch *et al.*, 1985)]. This material is obviously more primitive in the lower levels. It can be assigned in part to the Confite Morocho line of evolution and in part to the Confite Puntiagudo line of evolution (Smith 1980). Relationship with Pollo and Pira of Colombia is obvious. The general impression from the archæological remains at the present time is that maize came into South America from the north through the lowlands of Colombia. Spread southward along the Andean chain and into the coastal valleys occurred before its spread eastward into the llanos and the rainforest. The morphology of maize in Panama and Costa Rica indicate either that the differentiation of Chapalote-Nal Tel into Pollo had begun before maize reached South America or that maize reached South America earlier and Chapalote-Nal Tel-Pollo diffused northward into Central America. It is possible that major introductions of maize into South America were made along the Pacific coast or the Caribbean coast via trading exchanges and that a number of different introductions were made. I believe the latter is the case, because the earliest introductions were maize which was similar to the early primitive maize from the Tehuacan Valley, but later more heavily teosintoid maize appears in the Peruvian area.

Common beans (*Phaseolus vulgaris*) present a much different pattern from maize. There is no doubt that cultivated varieties of common beans are derived from wild populations of common beans on the Pacific slopes of Mexico and Central America and from the population of wild common beans along the eastern flanks of the Andes in South America from Venezuela to Argentina. Among bean experts (Kaplan 1965) some question remains of how wild is wild, because some wild common bean populations may have been secondarily derived from cultivated common bean varieties. Also, while the self polination mechanism is supposed to be obligate, evidence suggests that gene exchange can take place between wild and cultivated common beans. However, the archæological record for common beans does not include forms illustrating the morphological modifications from wild to cultivated forms.

The earliest known common beans in archæological contexts are the beans from Complex II in Guitarrero Cave in Peru (Kaplan 1980). Reevaluation of the chronology (Lynch *et al.*, 1985) places these two varieties of common beans at about 8000 BC. Kaplan asserts that they are fully cultivated common beans which show little variation in the deposit over an 8000 year time span. Two other land races of common beans were recovered from the Guitarrero Cave fill.

Both the Tamaulipas caves and the Tehuacan Valley caves included cultivated common beans at about 5000 BC (Kaplan and MacNeish 1960; Kaplan 1967). Altogether, three varieties of cultivated common beans were recognized from the Tehuacan plant remains and three varieties of common beans were recognized from the Tamaulipas Caves. Both deposits included common bean remains at all time levels from about 5000 BC to AD 1500. Carbonized remains of common beans are found sparsely in most deposits where careful separation is practiced, but common bean remains are rarely plentiful. If they are fragmentary, they are difficult to identify positively. Electron microscopy identified common bean fragments in the Proyecto Cajón sites in Honduras at the late Classic (600–900 AD) chronological level (Lentz 1984).

Carbonized remains of probable common beans have been recovered in Costa Rica from La Ceiba, Guanacaste (800–1300 AD), from Barrial de Heredia and Sitio H43 near San José in Tumbas Phase contexts (200 BC–AD 400) (Smith, notes). In Panama, carbonized common beans have been recovered from several sites including Carabali Rockshelter (SF-9), Cueva de los Ladrones (CL-6) (Smith, notes) and Cerro Punta (BU-17). The samples from the latter site are dated about AD 200–400 (Linares and Ranere 1980). They are not distinctive enough to indicate whether they might have been introduced from the Andean region or from Mesoamerica.

Common beans have been found in many sites along the desert coast of Peru (Towle 1961). Among the early finds are pods and beans from the Cupisnique levels of Huaca Prieta (Towle 1961). More recently, excavations at La Galgada in the Tablachaca Cañón have revealed cultivated common bean pods at *c.* 2300 BC (Smith and Grieder, ms.). In general, the coastal finds of common beans have all been in chronological contexts later than those of Guitarrero Cave. Inasmuch as the known native distribution of wild common beans in South America is east of the Andes, it must be assumed that all of the cultivated coastal beans are introduced from the highlands. They generally are large-seeded races which could either have been derived from wild common beans in the South America area or they could have been introduced from Mesoamerica (Kaplan, personal communication).

Squash is the third member of the so-called classical American Indian diet. It represents at least five species in cultivation (*Cucurbita pepo, C. moschata, C. mixta, C. maxima*, and *C. ficifolia*) (Whitaker and Cutler 1965). In the Mesoamerican area, the first three are widely known and widely recovered in archæological contexts. Like common beans, the archæological material is generally cultivated although two early recoveries of *C. pepo* seed [Ocampo Caves, Tamaulipas (Whitaker *et al.*, 1957); Oaxaca, (Whitaker and Cutler 1971)] are judged to be from wild or scarcely cultivated squash based on their size. The *C. pepo* seed found in the lower level of Guila Naquitz Cave, Oaxaca, is the first of a series of obviously cultivated seeds in upper levels (Flannery 1986) and the supposition is that it was introduced into the Oaxaca Valley at the 8,000 BC chronological level. All of the other squash species are found in later levels both in Mesoamerica and in South America. *C. moschata* was recovered from the 4900–3500 BC levels in the Tehuacan Valley caves and it appears in the Huaca Prieta, Peru, deposits at about 3000 BC. It is in the La Galgada plant remains at about 2000 BC (Smith, ms.). It is obvious that the *C. pepo* races moved northward into temperate North America while *C. moschata* races moved southward into South America (Whitaker 1983).

The archæological finds of *C. mixta* from the Tehuacan Valley deposits date from before 5200 BC (Whitaker 1981). They make their way northward to appear in the southwestern United States by AD 380. They did not apparently move southward. On the other hand, *C. ficifolia*, the highland, perennial species, was recovered from Huaca Prieta, Peru, at 3000 BC while the North American archæological recovery in the Oaxaca Valley dates to AD 700. *C. maxima* is in the plant remains from La Galgada, Peru at about 2000 BC and in the Viru Valley around 1800 BC. It did not get introduced into North America prehistorically according to current evidence.

Lagenaria siceraria, the bottle gourd, is an early cultivated (or wild) plant in the Americas appearing in the Piki complex of Ayacucho between 5000 and 3800 BC (MacNeish *et al.*, 1969). It was abundant at La Galgada around 2000 BC (Smith and Grieder, ms.). Its obvious use for containers made it important and it may be supposed that a similar use for the shells of *Cucurbita pepo* and *C. moschata* led to their early cultivation.

The center of diversity for squash is southern Mexico and most of the *Cucurbita* species seem to have originated there. *C. pepo* is closely related to *C. texana* but doubt still exists that it was a direct ancestor of the cultivated forms (Heiser 1985). *C. pepo*'s early date in Oaxaca suggests that it may have originated nearby. *C. maxima* was not known prehistorically outside of South America and it must have come into cultivation there, but its predecessor is unclear. No clear archæological evidence exists for the development of cultivated forms from wild forms of squash and gourds.

The biological evidence for the movements of chili peppers (*Capsicum*) has recently been summarized by Pickersgill (1984). So far as can now be determined, like the *Cucurbita* spp., the peppers in cultivation can be assigned to five species. The most easily resolved origin of a chili species is that for the Mexican cultivate. *Capsicum annuum* has an unusual chromosomal arrangement which indicates that the cultivated forms could only have derived from wild forms in central Mexico (Mesoamerica) with the same chromosomal structure (Pickersgill 1984). Archæological evidence places chilis in the Tehuacan Valley in the El Riego Phase between 7000 and 5000 BC (Smith 1967). Unfortunately, chili peppers are not often found with intact stems which are needed for some hope of species identification. The earliest pepper remains cannot be identified to species in Mexico. In the Mexican area, carbonized chili pepper seeds are the most usual recovery. They have been found in the Basin of Mexico at Teotihuacan (Tapia 1979: 163). An intact carbonized chili stem, clearly *C. annuum*, was found in the contact period (AD 1520) level in the Nochixtlan Valley (Smith 1976). It is interesting that no chili remains could be certifiably assigned to preceramic levels in Guila Naquitz Cave in the Oaxaca Valley (Smith 1986). At later levels, they were not uncommon (Smith, ms.).

Chili seeds have not been found in the Central American archæological sites from which I have examined carbonized material. This may be a result of the recovery technique, but it is most likely absent below the Maya area.

The remainder of the *Capsicum* species in cultivation apparently originated in South America (Pickersgill 1984; Pickersgill and Smith 1981). The earliest known chili pepper was recovered from the Guitarrero Cave deposit in contexts that place it at about 7000 BC (Lynch *et al.*, 1985). It is a fruit with stem attached indicating that it is cultivated and it appears to be *C. chinense*. *Capsicum* seeds were recovered in quantity from the La Galgada site, but the seeds are unidentifiable to species (Smith and Grieder, ms.). They appear in the Huaca Prieta site during the Early Horizon (1200 BC–AD 150) and may be *C. chinense* and/or *C. frutescens* (Pickersgill 1984). Other Peruvian coastal finds are equally uninformative although associated plant species suggest that the peppers were introduced from east of the Andes.

The archæological record, thus, does little to solve the beginnings of cultivation of chili peppers, although, combined with biological evidence, the ancestral wild peppers can be speculated on. Most of the related wild species of *Capsicum* have native ranges in South America leading to the probability that all of the species have their beginnings there, but the chromosomal evidence for *C. annuum* definitively places its early cultivation in Mesoamerica.

The Mesoamerican area appears to have been the area of domestication of an assortment of beans other than common beans. Perhaps the most important of these at the present time is the ayocote (*Phaseolus coccineus*) which is present as a wild bean (6000–4500 BC) in the Tamaulipas Caves (Kaplan and MacNeish 1960). It occurs in the Tehuacan Valley deposits fairly late in a cultivated form (*c.* 2200 BC upward) (Kaplan 1964) and it was not recovered in the Oaxaca Valley (Kaplan 1986). Thus, the archæological record is short and not very informative, except that wild *P. coccineus* is not known to be in the Tamaulipas areas at the present time.

Another species of bean in cultivation in Mexico and the southwestern United States is the tepary bean (*Phaseolus acutifolius* var. *latifolius*). It appeared at about 3000 BC in the Tehuacan Valley deposits and about AD 900 in the Southwestern (Kaplan 1965). Archæological records do not indicate that this bean ever moved far outside of its natural range and it is very little used at the present time.

Phaseolus lunatus, like *P. vulgaris*, is now known as a wild plant on the Pacific slopes of Central America and the eastern Andean slopes. The seiva and Lima beans are found archæologically in both areas, the smaller seiva being the northern form and the larger Limas being the South American form. Seiva beans are known archæologically from the southwestern United States (AD 1200), the Ocampo Caves (AD 100), the Tehuacan Valley (AD 500) and as carbonized seivas, from Dzibilchaltún, Yucatan (AD 700) (Kaplan 1964). In South America, the potential earliest record is from Guitarrero Cave, Peru, in association with common beans and chili peppers at about 8000 BC. However, the context from which the big Lima bean was recovered might have been disturbed (Kaplan 1980). It is definitely in the La Galgada site at about 2000 BC (Smith and Grieder, ms.) and it appears in many of the other coastal Peruvian sites. The wild population of this species in South America is east of the Andes and it must be assumed that the Guitarrero Cave forms and the Peruvian coastal valley forms were all introduced. While the archæological record clearly indicates that the seiva beans races were undoubtedly brought into cultivation in the northern area and the big Limas were brought into cultivation in South America, no evidence is given of the original changes under cultivation. All of the finds have been of completely cultivated bean races (Kaplan, personal communication).

Probably the next most important food species in Mesoamerica was the avocado (*Persea americana*). The earliest archæological record comes from the Ajuereado Phase of the Tehuacan Valley and the cotyledon may date to about 8000 BC (Smith 1967). Upward in the archæological deposits avocado cotyledons become more numerous and an increase in size indicates that they have been brought into cultivation (Smith 1966; 1969). Carbonized avocado cotyledons in Teotihuacan excavations are good proof that avocados were being carried from more tropical regions of Mexico to supply the upland valleys where they do not grow (Tapia 1979). Cotyledons were also found in the Nochixtlan deposits (Smith 1976) and in the Oaxaca Valley (Smith, ms.). Plant remains recovered from Kaminaljuyú, Guatemala include many fragments of burned avocado cotyledons (Hatch, personal communication).

Probable carbonized avocado cotyledon fragments are in Costa Rica (Guanacaste) at AD 800–1300 (Smith, notes). It appears again in the coastal valleys of Peru where its earliest recovery is from La Galgada at about 2000 BC (Smith, ms.). It is a complete dried seed. From this period on, avocado appears in other coastal Peruvian sites with some regularity (Towle 1961). The increase in size in the avocado cotyledons from the Tehuacan Valley sites leaves no doubt that avocado was both under cultivation and selection for improved forms. Inasmuch as the wild avocado is native to the forest of the slopes above the Valley, it is probable that it was first brought into cultivation here where irrigation would be needed to help the trees survive during the six months dry season. Avocado seems not to have been introduced at an early period anywhere in the lowland tropics of Central or South America, but its appearance in Peruvian coastal valleys, along with maize, indicates contact with Mesoamerica.

Among the tree species on Vavilov's list of plants he considers as originating in Mexico and Guatemala are several that were recovered from sites in the Tehuacan Valley. These are the zapote blanco (*Casimiroa edulis*), zapote negro (*Diospyros digyna*) and ciruela (*Spondias mombin*). All are first recovered from the cave deposits dated around 5000 BC and they continue upward in the deposits. All of these tree fruits require supplemental water to survive the dry season in the valley. Generally these are the earliest records for these species. Zapote negro has been recovered from Lambityeco in the Oaxaca Valley (Houston 1983) and zapote blanco in known from the Nochixtlan Valley and Lambityeco as carbonized seeds (Smith 1976; Houston 1983). Ciruela is much better known archæologically. It was found in the Nochixtlan Valley and has been recovered from Tikal in the Peten (Smith, notes), from the Ocos area of Guatemala (Coe and Flannery 1967), from Honduras (Lentz 1984) and from the Parmana area of Venezuela where it is quite late (AD 700–1500) (Smith and Roosevelt, ms.).

The archæological evidence for these three species does not provide the evidence for the development of cultivated forms from wild forms. Zapote blanco may be from the deciduous forest on the lower slopes of Mexico's interior as is probably zapote negro. Ciruela may be from the Pacific drainage area of Mexico-Central America from seasonally deciduous woodlands. None of them appear to have been native to the Tehuacan Valley flora. Thus, their appearance in the plant remains in the caves indicates that they must have been introduced in cultivation by people who knew that they would have to have supplementary water during the long dry season. Whether this meant merely planting on the banks of natural permanent streams or whether it meant watering with an artificial irrigation system we may never know. Only the ciruela appears to have been introduced widely around the Americas or it may have been widely distributed naturally and was brought into cultivation several different times. For example, *Spondias* is represented by a wild form on the Venezuelan llanos. Is this an escape from early introduction and cultivation or is it truly native to the llanos? Perhaps someday genetics and serological studies will provide a clue.

For a number of other species considered by Vavilov to be native to Middle America, the evidence is incomplete or it suggests that Vavilov was in error. The Guava (*Psidium guayava*) is well represented in La Galgada Peru, at around 2000 BC (Smith and Grieder, ms.) and it appears early in a number of coastal Peruvian sites. It was found in the Tehuacan Valley at about 200 BC (Smith 1967). Its scanty representation in Mexico, as well as the time-frame, suggests that it was first cultivated in South America (or was it a native of the coastal river valleys of Peru?) and later introduced into Mesoamerica.

A number of fragments of seeds of species of Sapotaceae have been recovered from archæological sites around the Americas. Only when whole seeds are encountered or fragments are sufficiently large can a generic or specific identification be made. Plant remains from the coastal valleys of Peru often include lucuma seed fragments (*Pouteria lucuma*) as did the earliest levels of Guitarrero Cave (8000 BC) (Smith 1980). Whether this species is represented by small fragments of sapotaceous seeds in Panama and Costa Rica (Smith, notes) or whether these represent a local species of *Pouteria* is difficult to discover. I suspect that it is the latter. Lentz (1984) has recovered fragments of the seeds of *Calocarpum mammosum* (*mamey*) in Post Classic levels in Honduras. Sapotaceae represented in the Tehuacan Valley included *Bumelia laetivirens* (*tempixquistle*) and *Sideroxylon* cf. *tempisque* (*cosahuico*) neither of which are known archæologically elsewhere (Smith 1967). It is beginning to look as though sapotaceous fruit of most species was sufficiently popular that local species were brought into cultivation with relatively little introduction into other areas. It seems unlikely that the archæological record will be so specific in the future that the exact areas of origin of sapotaceous species will be identifiable.

A number of species listed for Middle America by Vavilov have either never been recognised among archæological fragments or their presence is uninformative. Into the latter category fall all of the species which are traditionally vegetatively cultivated. These include the *Agave* species of which many were probably in prehistoric cultivation. The archæological remains from dry Mexican caves include large quantities of maguey quids and smaller amounts of leaf fragments and spines. Cacti fall in the same category, being represented abundantly in dry caves and in open sites by carbonized seeds. How many of the species of *Opuntia, Lemaireocereus, Escontria*, and others were in cultivation we will probably never discover. It is probable that the fragments of the cacti recovered in the Peruvian coastal valleys are those species native to Peru, although it is possible that *Opuntia ficus-indica* was an early introduction.

Vavilov has proposed another group of cultivated species as originating in South America in the Andes, on the island of Chiloe and from eastern South America in the region of the Brasilian–Paraguayan border (Vavilov 1951). The Andean species include a number of the root and tuber crops which are currently the major carbohydrate resource for millions of Andean Indians. Unfortunately, the root and tuber crops are rarely found as carbonized remains. Even the dry remains of these species are difficult to identify, because they have so few distinctive morphological characteristics. From Guitarrero Cave, numerous tuber remains were recovered from all levels. Undoubted manioc from Complex IV (Smith 1980) was from a disturbed area and might have been modern. The probable oca (*Oxalis* spp.) and ullucu (*Ullucus tuberosa*) tubers could not be distinguished by starch grains and the general morphology was not distinctive enough to make positive identifications. Their presence from 8000 BC upward indicates that they were important to the occupants, but the remains give no clue to whether they might have been gathered from the wild or if they were cultivated. I suspect the latter, as they generally appeared to be larger than the usual wild species underground parts which I collected in the Callejón de Huaylas. Towle (1961) reports tuberous material (probably ullucu, oca and potato) from coastal sites while recent evidence verifies potato in the Casma Valley at 2000–1700 BC (Ugent *et al.*, 1982). Inasmuch as none of these are native in the area, it is evident that they were either traded in or were grown in the valleys as cultivated plants in irrigation agriculture.

In general, root crops are rarely found carbonized and dry sites are few in the higher elevations where most of them originated. It is unrealistic to expect much information concerning the development of cultivated root crops to be furnished by archæological remains. An exception to this may be achira (*Canna edulis*) which is currently used in the coastal valleys of Peru (Ugent 1984). It is frequently encountered as tuber or leaf material in archæological excavations including La Galgada at 2000 BC (Smith and Grieder, ms.). Achira may be native to the river valleys although it is now distributed from Colombia to Chile. The currently known archæological remains tell little about the origin and development of the crop.

Another important Andean crop is quinoa (*Chenopodium quinoa*) which is valued for its seeds. About the earliest evidence for this crop come from the Piki Phase (5500–4300 BC) in the Ayacucho Project. None were found in the Guitarrero Cave deposit in spite of expectation that they would be there. Most of the records of quinoa are from coastal Peruvian sites (Towle 1961). None of the archæological finds provide us with evidence for the beginning of cultivation of the pseudocereal.

Related to the quinoa is the species *Amaranthus caudatus* used for its seed which also seems to have come into cultivation in South America. It was first reported archæologically from Pampa Grande, Argentina, by Hunziker (1943). Another find of the species was made in the Gruta del Indio del Rincón del Atuel (Hunziker and Planchuelo 1971). Both were found mixed into a lot of quinoa with which it was apparently grown. Neither lot of archæological material tells much about the original development and introduction into cultivation.

Some of the tree crops of South America are different from those discussed earlier. In the Peruvian area, one of the early tree crops is pacay (*Inga feuillei*). It apparently was grown in the Callejón de Huaylas where it was found principally in the upper Complex IV. Pacay was popular in La Galgada around 2000 BC judging by the pod fragments found. Towle (1961) reports a number of finds in the coastal desert area. The tree may have been naturally distributed in the river fringes in prehistoric times.

Ciruela del fraile (*Bunchosia armeniaca*) also appears plentifully in the La Galgada deposits at about 2000 BC and slightly later chronologically in the Casma Valley (Ugent *et al.*, 1986). It, also, appears to have been popular in most of the coastal valleys and was perhaps native there, but it does not appear in the highland contexts.

Guava (*Psidium guayava*) has been previously mentioned, because it appears to have been introduced into Mexico from South America. It is a small, weedy tree which may have also come from the fringe forests along the coastal rivers of Peru. It is represented in the La Galgada site (*c.* 2000 BC) by two whole fruits and numerous seeds.

One of the plants sometimes found in the Peruvian coastal valleys, coca (*Erythroxylon coca*), is probably from the eastern Andean slopes. While it is used throughout the Andes today, it is not represented in highland prehistoric sites unless the possible find by the Ayacucho Project can be confirmed. This plant does not seem to have reached Mesoamerica prehistorically.

While Vavilov credits the Chiloe center with both potato and the South American strawberry (*Fragaria chiloense*) it is probable that both were brought into cultivation in the highlands. *Fragaria* has not been found archæologically and potato has already been discussed.

The plants from the eastern South American center of origin (Vavilov 1951) include several which are important crop plants today. Manioc (*Manihot esculenta*) is certainly a prominent Amazonian crop today, but there is some doubt that it was brought into cultivation in South America (Rogers and Fleming 1973). While it may have originated in Mexico, most of the archæological recoveries have been in the desert valleys of coastal Peru (Ugent *et al.*, 1986). The fragment found in Guitarrero Cave (Smith 1980) came from disturbed fill and may well represent modern use in the Callejón de Huaylas. As with other root crops, methods of preparation and discard of garbage do not promise a very extensive archæological record. It certainly was never native to the sites where it has its best archæological record dating back to about 1900 BC.

Another prominent world crop today is the peanut (*Arachis hypogæa*) which is, without doubt, a native of the eastern South American area (Krapovickas 1973). It appears in the Guitarrero Cave deposit under less than secure provenience in Complex IV. It is present in La Galgada at around 2000 BC and it appears in a number of other coastal sites (Towle 1961; Ugent *et al.*, 1986). A fragment of legume and a seed securely place the peanut in the Tehuacan Valley at 200 BC, but it apparently was not popular, because it has never been found in quantity in Mesoamerica in prehistoric deposits. We currently have no archæological evidence for peanut in the part of South America where it originated.

Vavilov (1951) credits the origin of *Theobroma cacao* to the eastern South American area. However, the archæological evidence for this species is almost non-existant [one possible identification from Guanacaste, Costa Rica (Smith notes) and several fragments from Belize (Crane, personal communication) which might be cacao]. I feel that it could have been brought into cultivation in Gulf coast Mexico or Caribbean coastal Guatemala as readily as in South America. Cacao, as with a number of other cultivated species, has related forms in the wild in many areas which may be native to the area or they may represent escapes into the wild after early introduction.

Two other crops of modern importance are credited to the eastern South American area. They are pineapple (*Ananas comosus*) and cashew (*Anacardium occidentale*). While many species of the Bromeliaceae (pineapple family) have been recovered archæologically in widely distant sites throughout the Americas as sources of fiber, pineapple has not yet been identified archæologically. In spite of its prominence throughout lowland tropical America, cashew has never been identified in archæological contexts, probably because native Americans are well aware of the dangers of poisoning if the fruits are burned.

The one group of plants that are now becoming prominent among archæological plant remains are the palms. Vavilov (1951) lists only *Bactris gassipaes* (*Gassipaes speciosa*) as originating in the Andean area of South America. To the list should be added coyol (*Acrocomia mexicana*), juana (*Sabal mayarum*), palma de vino (*Scheelia* spp.) and many others. Coyol appears archæologically about 5000 BC in the Tehuacan Valley deposits where it is dependant upon irrigation during the dry season, thus indicating that it was introduced under cultivation. It is now known archæologically from Belize (Miksicek, personal communication), Honduras (Lentz 1984), Costa Rica (Smith notes), Panama (Smith 1980 and notes) and Venezuela (Smith and Roosevelt, ms.). In Panama, when the fragments are sufficiently large to indicate whole seed shape, it is obvious that some of the palm fruit remains are from a species of *Scheelia*. Costa Rica and Panama both have identifiable archæological fragments of pejibaye palm (*Bactris gassipaes*) (Smith, notes). Before the introduction of maize in much of lowland Central and northern South America, palm fruits were apparently prominent in the diet. A palm fruit was recovered at the La Galgada site (Smith and Grieder, ms.) where it could only have been carried in from the Amazon basin. After the introduction of maize, the volume of palm remains is only lightly reduced although isotopic analysis of human skeletal material from the Parmana, Venezuela sites indicated that the inhabitants were cosuming around 80% maize at around AD 400 (van der Merwe *et al.*, 1981). I would suggest that more analysis of this nature is needed before the 80% figure be completely accepted. It may be that the ^{4}C is accumulated rather than being processed by the human body proportionally with ^{3}C foods.

The final material to be examined is the archæological record of cotton (*Gossypium hirsutum* and *G. barbadense*). The former is found in the Tehuacan Valley cave deposits at about 3000 BC (Smith 1967; Smith and Kerr 1968). In the Oaxaca area, cotton is in the upper levels of Guila Naquitz at AD 700–1500 (Smith, ms.). Because of the nature of the morphology of cotton plants, most of the archæological cotton remains cannot be identified to species. A peduncle on a boll from OC-47 in the Oaxaca Valley allowed a positive identification of this cotton as *Gossypium hirsutum* (Smith and Stephens 1971). In Costa Rica at Barrial de Heredia in the Tumbas Pavas Phase (200 BC–AD 400) several possible carbonized cotton seeds were found (Smith, notes). They cannot be identified to species and it is impossible to know if they represent *G. hirsutum* from Mexico or *G. barbadense* from Peru. In the coastal area of Peru, cotton is often encountered. It was present in abundance at La Galgada where seeds were often found splity lengthwise as if they were opened and the kernals eaten by humans (Smith and Grieder, ms.). Cotton of about the same age (Conchas Phase, 2750–1850 BC) is reported by Stephens and Moseley (1973, 1974) who found a peduncle which could be critically identified as *Gossypium barbadense*. Stephens (1975) reexamined the Huaca Prieta cotton remains and concluded that they and the Ancon remains represented *G. barbadense* intermediate in characteristics between wild and fully domesticated forms. It would seem logical to conclude that cotton was brought into cultivation in Mexico (*G. hirsutum*) and in southern Ecuador and northern Peru where wild forms of *G. barbadense* still can be collected.

Summary

Maize is probably the most extensively documented of the New World crops. It is found archæologically from Canada to southern South America. In spite of the archæological evidence, we still have no proof that maize is either derived from wild maize or from teosinte (Galinat 1971). We do have the evidence that maize had been introduced as far south as Ecuador and Peru by 2500 BC. It moved into the llanos along the Orinoco at a later date. The variability exhibited in Mexico seems to be matched in Peru. The maize remains from Central America do not yet show this variability. Some have theorized that the South American variability arose from hybridization with species of *Tripsacum*. I believe it is more logical to account for the variability by postulating a continuing series of introductions from Mesoamerica via coastal trading expeditions. This would easily account for the increasing teosintisation of Peruvian races of maize without recourse to hybridization that is highly unlikely anyway.

Common beans present a picture which could be interpreted in two directions: beans came into cultivation in Mesoamerica and in South America from different wild common bean strains or wild beans were first brought into cultivation in South America (before 8000 BC) and early cultivated beans diffusing northward, became the base for both cultivated Mesoamerican beans and for feral populations which are today truly wild on the Pacific slopes of Mexico and Central America. One of the logical reasons for the latter view is the similarity of the cultivated bean races in both North and South America in prehistoric times. We know that European settlers were responsible for widespread interchange of crop plants around the world and might have been responsible for the homogenization of bean varieties since AD 1500, but they cannot be accused of pre-1492 homogenization. All common bean archæological remains are fully cultivated forms.

The archæological records of squash do little to explain present-day distributions of the cucurbits. Biologically the center of *Cucurbita* variation is the southern edge of Mesoamerica. Archæologically, the center of variation is central Mexico where most of the species occur in the plant remains. *Moschata* squash arrives in Peru about 2000 BC, presumably from Mesoamerica, but *maxima* squash appears about at the same time level in South American contexts only. It may eventually be shown that the South American wild squash relatives are derived out of the Central American center and that *C. maxima* then was selected out of these forms. No archæological series illustrating the derivation of cultivated from wild forms is known.

The archæological remains of the cultivated chili peppers are even less informative than the evidence for the squashes. The earliest record of cultivated pepper is for probable *C. chinense* in South America at 8000 BC. The biological evidence indicates that the center of speciation is in South America and that the cultivated forms of the Mexican chili pepper arose in Mesoamerica from wild forms with the same chromosomal anomaly. For the most part, Mesoamerican archæological pepper remains are unidentifiable to species, but those calyces which have survived are those of *C. annuum*, the expected species. In South America with several species in cultivation, the situation is more complicated. The purple-flowered species and three white flowered species are not identifiable with certainty from most of the archæological remains. The latter are mostly from the coastal desert valleys where they are not native and must have come from east of the Andes.

The archæological records of species of *Phaseolus* other than common bean indicate that the scarlet runner bean or *ayocote* was in cultivation in the Tehuacan Valley by about 2200 BC but it apparently never moved out of the Mexican highlands. Tepary beans did move into the southwest United States, but never in quantity anywhere. The most archæological evidence is for seiva and Lima beans which apparently were derived from wild plants both in Mesoamerica, where the seiva form is the only one known archæologically and South America where the big Lima is the only form known archæologically. All archæological seiva and Lima beans are fully cultivated. We have no series illustrating the morphological changes from the wild beans to the cultivated beans.

Avocado is one of the few plants for which we may have a series of seeds illustrating purposeful selection. It is a tree of mesophytic forests and its early inclusion in the plant remains of the Tehuacan Valley indicates that it was either carried in from the mesophytic forest higher on the mountain slopes to the east or that its water requirements were known and satisfied. The seed series from the Tehuacan cave deposits shows an ever increasing mean size upward in the deposits. The presence of avocado in archæological deposits in the central Basin of Mexico where frost prevents its cultivation, shows that it was highly enough regarded to have been carried into the Basin. It also was traded into South America where archæological avocado seeds are in the plant remains of the archæological sites by about 2000 BC. Avocado may have been introduced into South America about the same time that maize was introduced.

The records of the fleshy fruits in cultivation in prehistoric contexts are difficult to interpret. Part of the difficulty concerns the original wild form of several species which become confused with persistent feral forms escaping from cultivation through time which may have returned progeny to the wild flora far removed from the original natural range of the species. On the other hand, it is possible that some species have abnormally wide tolerances for environmental variability and were very widely distributed. In my experience, tree species are not known to have the wide tolerance exhibited by polyploid herbaceous complexes like dandelions (*Taraxacum officinale*). I prefer to interpret the wide distribution of the trees bearing fleshy fruits as the distribution of species in cultivation for their useful fruits. Certainly, the remains of these species in dry intermontane valleys is clear evidence of early cultivation.

On occasion, the archæological remains provide indisputable evidence for long distance introduction like that indicated by remains of guava. Found early in chronological levels in the coastal desert valleys of Peru, it is represented about 200 BC by the fragment of dried fruit in the Tehuacan Valley of Mexico.

The archæological record of the seeds from sapotaceous fruits is such that it can only be interpreted that local genera and species were brought into cultivation where humans found them. With increased recovery of archæological plant remains, this interpretation may change.

Little more need be said of the archæological remains of other species of plants. The various magueys and cacti found archæologically are so similar morphologically to their wild counterparts that it is impossible to tell when they first came into cultivation. I personally believe that they may have been among the earliest cultivated plants because of the ease with which they may be propagated, but this is without proof. The difficulty with root crop preservation will probably always hinder interpretation of the few dried remains recovered. We cannot expect to have much information from carbonized remains.

The pseudocereals also, have not been given much attention in this paper as much for the difficulty in interpreting the finds that we have as for anything else. *Chenopodium* and *Amaranthus* have been discussed for South America, but the numerous finds of carbonized seeds of these genera in Mesoamerica do not tell us much. The suggestion has been raised that the *huauhzontle* of Mesoamerica may be closely related to quinoa (Wilson and Heiser 1979), but the archæological record does not provide aid. Several recoveries of dried *Amaranthus* in the Tehuacan Valley (Sauer 1967) provide evidence of cultivation and selection, but carbonized *Amaranthus* seeds tell us little. The problem in Mesoamerica is compounded, because the wild, weedy forms of these genera are encouraged and harvested by current population in the Mesoamerican area and we suspect that the same pattern has great depth in prehistory. Note also the complications introduced by the finding of carbonized *Chenopodium* and *Amaranthus* seeds in archæological sites in temperate North America.

Another group of cultivated plants not treated in the previous pages is the complex of *Canavalia* species. Found early and abundantly in the coastal valleys of Peru, they also appear in the Oaxaca Valley at about AD 900 (Sauer and Kaplan 1969). Most of the recoveries have been from desert areas where the *Canavalia* beans must have been in cultivation, but they provide no clues to areas of origin or patterns of distribution.

Altogether, the archæological plant remains from the Americas have generally supported the biological evidence on the areas of origin of cultivated American plants and have often suggested a chronological level within which the beginnings of cultivation may have taken place. The evidence is so diverse, though, that no single postulate on origin of agriculture in the New World will suffice to cover all situations.

The hypothetical places of beginning cultivation for the American crop species generally agree with theoretical centers of origin suggested by present day crop plant variability. However, we should always remember that these may well be secondary centers of diversity far from the original point of cultivation of a species. The archæological plant remains present a more definitive pattern for the beginnings of cultivation in that they definitely exclude some areas. For the greatest number of cultivated American species, we can positively say that crops did not originate where the archæological remains have been found. The single exception may be avocado; it certainly is not true for any of the annual crops like maize, beans, squash, and chili peppers. Maize would seem to have come into cultivation in what is now southern Mexico and may have had a single point of origin. Common beans, seiva and Lima beans, squashes, and chili peppers all appear to have had several areas of origin. For most of the remaining species, we must await further archæological evidence to come to any conclusions.

Chronologically, the situation is somewhat more specific. The presence of cultivated common beans and possibly cultivated pepper and Lima beans at around 8000 BC in Guitarrero Cave indicates that experiments with cultivation in South America had begun earlier to the east of the Andes. In North America, the presence of pepo squash seeds at about 8000 BC in the Oaxaca Guila Naquitz Cave indicates the beginning of experimentation with cultivation there. The presence of a number of seeds from tree fruits adapted to more mesophytic environments in the desert Tehuacan Valley indicate a knowledge of tree fruit cultivation and the water needs of all of them implying that water was artifically supplied to these species during the long dry season. Altogether, the evidence is for extensive and wide-spread knowledge of cultivation around or before 5000 BC.

The kinds of plants being brought into cultivation represent no fixed pattern. Annual cultivated crops are highly represented as they always are, but many of them seem to be derived from perennial relatives. The beans all seem to have derived from forms with perennial bases as do the squashes and the peppers. The many succulent species show no morphological changes which might be due to cultivation. A surprisingly large number of fruit trees were brought into cultivation at an early period in the Americas. In fact, a combination of palms and dicotyledenous tree crops are indicated as widespread throughout the tropical lowlands from Belize through Honduras, Costa Rica, and Panama to the llanos of Venezuela and they may have constituted the principal dietary cultivates over a long period in prehistory. In South America, in the highland, basic subsistence appears to have largely been based on root crops for which we have a very restricted archæological record. At the same time, annual seed crops were supplementing the Andean diet and have remained important despite the introduction of maize.

REFERENCES

ASCH, D. L. and N. B. ASCH

1985. "Prehistoric Plant Cultivation in West-central Illinois", in Richard I. Ford (ed), *Prehistoric Food Production in North America*, Anthropological Papers of the Museum of Anthropology no. 75, University of Michigan.

BIRD, R. McK.

1984. "South American Maize in Central America", in Doris Stone (ed), *Pre-Columbian Plant Migration*, Papers of the Peabody Museum of Archæology and Ethnology vol. 76.

CHILDE, V. G.

1969. *New Light on the Most Ancient East*, W. W. Norton, New York.

COE, M. D. and K. V. FLANNERY

1967. *Early Cultures and Human Ecology in South Coastal Guatemala*, Smithsonian Press, Washington, DC.

COWAN, C. W.

1985. "Understanding the Evolution of Plant Husbandry in Eastern North America: Lessons from Botany, Ethnography and Archæology", in Richard I. Ford, *Prehistoric Food Production in North America*, Anthrop. Pap. no. 75, Museum of Anthropology, University of Michigan.

FLANNERY, K. V. (ed)

1986. *Guila Naquitz. Archaic Foraging and Early Agriculture in Oaxaca, Mexico*, Academic Press, New York.

FORD, R. I. (ed)

1985. *Prehistoric Food Production in North America*, Anthropological Papers no. 75, Museum of Anthropology, University of Michigan.

GALINAT, W. C.

1971. "The Origin of Maize", *Annual Review of Genetics* 5: 447–478.

1980. "11. The Archæological Maize Remains from Volcan Panama —A Comparative Perspective", in O. F. Linares and A. J. Raners (eds), *Adaptive Radiations in Prehistoric Panama*, Peabody Museum Monographs no. 5.

1985. "Domestication and Diffusion of Maize", in Richard I. Ford (ed), *Prehistoric Food Production in North America*, Anthropological Papers no. 75, Museum of Anthropology, University of Michigan.

GALINAT, W. C. and J. H. GUNNERSON

1963. "Spread of Eight-rowed Maize from the Prehistoric Southwest", *Botanical Museum Leaflets*, Harvard University 20: 117–160.

GROBMAN, A. and D. BONAVIA

1978. "Pre-ceramic Maize on the North-central Coast of Peru", *Nature* 276: 386–387.

GROBMAN, A. *et al.*

1977. "Study of Pre-ceramic Maize from Huarmey, North Central Coast of Peru", *Botanical Museum Leaflets*, Harvard University 25: 221–242.

HARRIOT, T.

1590. *A Brief and True Report of the New Found Land of Virginia*, Dover, New York, (1972).

HEISER, C. B., Jr.

1985. "Some Botanical Considerations of the Early Domesticated Plants North of Mexico", in Richard I. Ford (ed), *Prehistoric Food Production in North America*, Anthropological Papers no. 75, Museum of Anthropology, University of Michigan.

1969. "The North American Sunflowers (*Helianthus*)", *Mem. Torr. Bot. Club* 22: 2–213.

HOUSTON, M. S.

1983. The Paleoethnobotany of Oaxaca, Mexico, PhD Diss., University of North Carolina.

HUNZICKER, A. T.

1943. "Granos hallados en el yacimiento arqueológico de Pampa Grande (Salta, Argentina)", *Revista Argentina de Agronomía* 10: 146–154.

HUNZICKER, A. T. and A. M. PLANCHUELO

1971. "Sobre un nuevo hallazgo de *Amaranthus caudatus* en tumbas indígenas de Argentina", *Kurtziana* 6: 63–67.

KAPLAN, L.

1965. "Archeology and Domestication in American *Phaseolus* (Beans)", *Economic Botany* 19: 358–368.

1967. "Chapter 20. Archæological *Phaseolus from Tehuacan*", in Douglas S. Byers (ed), *The Prehistory of the Tehuacan Valley. Environment and Subsistence*, University of Texas Press, Austin.

1980. "Variation in the Cultivated Beans", in Thomas F. Lynch (ed), *Guitarrero Cave. Early Man in the Andes*, Academic Press, New York.

1986. "Preceramic *Phaseolus* from Guila Naquitz", in Kent V. Flannery (ed), *Guila Naquitz. Archaic Foraging and Early Agriculture in Oaxaca, Mexico*, Academic Press, New York.

KAPLAN, L. and R. S. MACNEISH

1960. "Prehistoric Bean Remains from Caves in the Ocampo Region of Tamaulipas, Mexico", *Botanical Museum Leaflets*, Harvard University 19: 33–56.

KAPLAN, L. *et al.*

1973. "Early Cultivated Beans (*Phaseolus vulgaris*) from an Intermontance Peruvian Valley", *Science* 179: 76–77.

KRAPOVICKAS, A.

1973. "Evolution of the Genus Arachis", in Rom Moav (ed), *Agricultural Genetics. Selected topic*, National Council for Research and Development, Jerusalem.

LENTZ, D. L.

1984. A Description of the Plant Communities and Archæoethnobotany of the Lower Sulaco and Humuya River Valleys, Honduras; PhD Dissertation, University of Alabama.

LINARES, O. F. and A. J. RANERE

1980. *Adaptive Radiations in Prehistoric Panama*, Peabody Museum Monographs no. 5.

LYNCH, T. F., *et al.*

1985. "Chronology of Guitarrero Cave, Peru", *Science* 229: 864–867.

MACNEISH, R. S.

1969. *First Annual Report of the Ayacucho Archæological-Botanical Project*, Phillips Academy, Andover.

MACNEISH, R. S. *et al.*

1970. *Second Annual Report of the Ayacucho Archæological-Botanical Project*, Phillips Academy, Andover.

MANGELSDORF, P. C. and M. SANOJA O.

1965. "Early Archæological Maize from Venezuela", *Botanical Museum Leaflets*, Harvard University 21: 105–112.

MANGELSDORF, P. C. and C. E. SMITH, Jr.

1949. "New Archæological Evidence on Evolution in Maize." *Botanical Museum Leaflets*, Harvard University 13: 213–259.

MANGELSDORF, P. C., *et al.*

1967. "Bat Cave Revisited", *Botanical Museum Leaflets*, Harvard University 22: 1–31.

MANGELSDORF, P. C., *et al.*

1967. "Prehistoric Maize, Teosinte, and *Tripsacum* from Tamaulipas, Mexico", *Botanical Museum Leaflets*, Harvard University 22: 33-63.

MERWE, N. van der *et al.*

1981. "Isotopic Evidence for Prehistoric Subsistence Change at Parmana, Venezuela", *Nature* 292: 536–538.

MIKSICEK, C. H. *et al.*

1981. "Preclassic Lowland Maize from Cuello, Belize", *Nature* 289: 56–59.

PEARSALL, D. M.

1978. "Phytolith Analysis of Archæological Soils: Evidence for Maize Cultivation in Formative, Ecuador", *Science* 199: 177–178.

1980. "Analysis of an Archæological Maize Kernel Cache from Manabi Province, Ecuador", *Economic Botany* 34: 344–351.

PEARSALL, D. M. and D. R. PIPERNO

1986. Antiquity of Maize Cultivation in Ecuador: Summary and reevaluation of the Evidence, paper for the Society for American Archæology, 51st Annual Meeting Program and Abstracts, Society American Archæology, Washington, DC.

PICKERSGILL, B.

1984. "Migrations of Chili Peppers, *Capsicum* spp., in the Americas", in Doris Stone (ed), *Pre-Columbian Plant Migration*, Papers of the Peabody Museum of Archæology and Ethnology 76: 106–123.

PICKERSGILL, B. and R. T. SMITH

1981. "Adaptation to a Desert Coast: Subsistence Changes through Time in Coastal Peru", in Don Brothwell and Geoffrey Dimbleby (eds), *Environmental Aspects of Coasts and Islands*, BAR International Series 94: 89–115.

ROBERTS, L. M. *et al.*

1957. *Races of Maize in Colombia*, NAS, NRC Publication 510.

ROGERS, D. J. and H. S. FLEMING

1973. "A Monograph of *Manihot esculenta* —with an Explanation of the Taximetric Methods Used", *Economic Botany* 27: 1–113.

SAUER, J. D.

1967. "The Grain Amaranths and their Relatives: A Revised Taxonomic and Geographic Survey", *Annual Memoir of the Botanical Garden* 54: 103–137.

SAUER, J. and L. KAPLAN

1969. "Canavalia Beans in American Prehistory", *American Antiquity* 34: 417–424.

SMITH, C. E., Jr.

1966. "Archæological Evidence for Selection in Avocado", *Economic Botany* 29: 169–175.

1967. "12. Plant Remains", in D. S. Byers (ed), *The Prehistory of the Tehuacan Valley,* Environment and Subsistence, University of Texas Press, Austin.

1969. "Additional Notes on Pre-conquest Avocados in Mexico", *Economic Botany* 23: 135–140.

1976. *Modern Vegetation and Ancient Plant Remains of the Nochixtlan Valley, Oaxaca,* Vanderbilt University Publications in Anthropology no. 16.

1980. "5. Plant Remains from Guitarrero Cave", in Thomas F. Lynch (ed), *Guitarrero Cave. Early Man in the Andes*, Academic Press, New York.

1980. "6. Ancient Peruvian Highland Maize", in Thomas F. Lynch (ed), *Guitarrero Cave. Early Man in the Andes*, Academic Press, New York.

1986. "19. Preceramic Plant Remains from Guila Naquitz", in Kent V. Flannery (ed), *Guila Naquitz. Archaic Foraging and Early Agriculture in Oaxaca, Mexico,* Academic Press, New York.

n. d. Notes on Archæological Plant Remain Identifications in Costa Rica and Panama.

n. d. Pre-Conquest Plant Remains from the Oaxaca Valley, ms.

n. d. Floral Remains from La Galgada, Peru, ms.

SMITH, C. E. and T. GRIEDER

n. d. Floral Remains from La Galgada, Peru; ms.

SMITH, C. E., Jr. and T. KERR

1968. "Pre-conquest Plant Fibers from the Tehuacan Valley, Mexico", *Economic Botany* 22: 354–358.

SMITH, C. E., Jr. and A. C. ROOSEVELT

n. d. Prehistoric Plant Use in the Middle Orinoco Basin: Collectible Resources, ms.

SMITH, C. E., Jr. and S. G. STEPHENS

1971. "Critical Identification of Mexican Archæological Cotton Remains", *Economic Botany* 25: 160–168.

SMITH, C. E., Jr. and P. TOLSTOY

1981. "Vegetation and Man in the Basin of Mexico", *Economic Botany* 35: 415–433.

SNARSKIS, M. J.

1976. "Stratigraphic Excavations in the Eastern Lowlands of Costa Rica", *American Antiquity* 41: 342–353.

STEPHENS, S. G.

1975. "A Reexamination of the Cotton Remains from Huaca Prieta, North Coastal Peru", *American Antiquity* 40: 406–419.

STEPHENS, S. G. and M. E. MOSELEY

1973. "Cotton Remains from Archæological Sites in Central Coastal Peru", *Science* 180: 186–188.

1974. "Early Domesticated Cottons from Archæological Sites in Central Coastal Peru", *American Antiquity* 39: 109–122.

STEWARD, J. H.

1938. *Basin-Plateau Aboriginal Sociopolitical Groups*, U. S. Bureau of American Ethnology Bulletin no. 120.

TAPIA, E. M. de

1979. Plants and Subsistence in the Teotihuacan Valley AD 100–750; PhD Dissertation, Brandeis University.

TERRELL, E. E.

1977. *A Checklist of Names for 3,000 Vascular Plants of Economic Importance*, USDA, ARS, Agric. Handbook no. 505.

TOLSTOY, P. *et al.*

1977. "Early Sedentary Communities of the Basin of Mexico", *Journal of Field Archæology* 4: 91–106.

TOWLE, M. A.

1961. *The Ethnobotany of Pre-Columbian Peru*, Aldine, Chicago.

UGENT, D. *et al.*

1982. "Archæological Potato Tuber Remains from the Casma Valley of Peru", *Economic Botany* 36: 182–192.

1984. "New Evidence for Ancient Cultivation of *Canna edulis* in Peru", *Economic Botany* 38: 417–432.

1986. "Archæological Manioc (*Manihot*) from Coastal Peru", *Economic Botany* 40: 78–102.

VAVILOV, N. I.

1951. "The Origin, Variation, Immunity and Breeding of Cultivated Plants", *Chronica Botanica* 13: 1–364.

WELLHAUSEN, E. J. *et al.*

1952. *Races of Maize in Mexico. Their Origin, Characteristics and Distribution*, The Bussey Institution, Harvard University.

1957. *Races of Maize in Central America*, NAS, NRC Publication no. 511.

WHITAKER, T. W.

1981. "Archæological Cucurbits", *Economic Botany* 35: 460–466.

1983. "Cucurbits in Andean Prehistory", *American Antiquity* 48: 576–585.

WHITAKER, T. W. and H. C. CUTLER

1965. "Cucurbits and Cultures in the Americas", *Economic Botany* 19: 344–349.

1971. "Prehistoric Cucurbits from the Valley of Oaxaca", *Economic Botany* 25: 123–127.

WHITAKER, T. W. *et al.*

1957. "Cucurbit Materials from Three Caves near Ocampo, Tamaulipas", *American Antiquity* 22: 352–358.

WILSON, H. D. and C. B. HEISER, Jr.

1979. "The Origin and Evolutionary Relationships of 'Huauzontle' (*Chenopodium nuttalliae* Safford), Domesticated Chenopod of Mexico", *American Journal of Botany* 66: 198–206.

UNIV DUNELM

PLANT OPAL PHYTOLITHS: A PROBABLE FACTOR IN THE ORIGINS OF AGRICULTURE

IRWIN ROVNER
North Carolina State University

V. Gordon Childe is justifiably renowned for his insights into the problem of the origins of food production. In popularizing Raphael Pumpelly's "oasis theory" as the mechanism that pushed the agricultural revolution, Childe recognized that this was more than an event —it was a *process* fundamentally altering the social and economic structure of culture through the interaction of humans and their environment. Although details of Childe's revolutionary model may be obsolete, his concept of agriculture as a dynamic process continues to influence every serious investigator of the problem. For example, David Rindos reminded us in a forceful and provocative way that this process, in a Darwinian sense, is a co-evolution in which both cultigens and cultures change in response to each other.

Today we have substantially more information derived from many new analytical techniques increasing our understanding of early agriculture. One of the recent methods to be developed is plant opal phytolith analysis (Rovner 1971, 1983b). Opal phytoliths —literally, plant stones— are mineralized particles of varying shape and size produced within living plants. Silica, dissolved in ground water, is carried into the plant through its roots in the form of monosilicic acid. Eventually the silica is deposited in a mineral form identical to gemstone opal (opal A) where the water is used or lost through transpiration. When dead plant material is deposited in the ground and decays, phytoliths are left behind. Since they are entirely composed of mineral material, phytoliths do not decay in the botanical sense, but remain preserved in soil for enormous periods of time. Distinctive phytoliths have been recovered from geological deposits dating as far back as the Paleocene, some 60 million years ago. This not only far exceeds the age of earliest agriculture, but is obviously far beyond the age of the earliest archæological sites.

Phytolith analysis is particularly suited to the identification of grasses in archæological contexts, providing opportunities to study grassland paleoecology and climatic change in ways impossible only through analysis of pollen. Moreover, phytolith analysis, the best microfossil system for the study of cultivated cereals, has been used successfully in the archæological study of *Zea mays* (Pearsall 1978; Rovner 1983a; Piperno 1983; Pearsall and Piperno 1986), rice (Watanabe 1970), millet (Netolitzky 1914; Watanabe 1968), barley (Tack 1986), wheat (Rosen 1984) and others. Particularly impressive among the studies indicated above are those by Pearsall, who pushed back the date for the introduction of maize into Ecuador by several millennia; and by Piperno for the introduction of maize (and squash, an important non-grass cultigen) into Panama using phytolith data. Unfortunately recent study indicates that phytoliths will be a poor source of information on root crop agriculture, e. g. manioc (Piperno 1986).

All current indications suggest that phytolith analysis is a major source of significant archaeobotanical information. When used in conjunction with pollen and other sources of comparable data, our understanding of the history and of the process of the origins of agriculture should increase enormously. Moreover, phytolith analysis may provide a special source of unique insights. Other data systems, such as pollen and flotation (macro-remains) are residues of plants representing ***passive evidence*** of prehistoric agriculture. Phytoliths may represent a great deal more. Botanical studies of the role of silicon and silica phytoliths in the growth, strength and productivity of living plants suggest that phytoliths played a major *active* role in the selection and evolution of cultigens; i. e., in the co-evolution process, of the origins of agriculture itself.

It is a well known ethnographic observation that non-agricultural foragers and collectors use a much wider range of plant species than farmers cultivate. This is also true of the archaeobotanical evidence of the subsistence patterns of non-agricultural people. The process of early agriculture involved selecting among that wider range of wild plants for those that performed in superior fashion as cultigens. We have often heard that this narrower range of cultigens was "pre-adapted" to cultivation and "responded genetically" to the influence of human activity by becoming "more productive". Yet such vague and ambiguous terms tell us nothing about what specific pre-adapted qualities cultigens had in the wild state or which genetic factors present responded and in what way. Moreover, we are seldom, if ever, told what qualities were present —or absent— in other plants that precluded selection for cultivation or explain their failure to respond successfully to attempts by early farmers to cultivate them.

It is a rather curious fact that many of the most important sources of plant food to collectors, such as chenopods, amaranths, marsh elder (*Iva* sp.) are poor cultigens at best. Even where evidence is offered that these were cultivated (Asch and Asch 1977), they are eventually replaced by cereal grain cultigens (e. g. maize). On the other hand, wild ancestors of major cultigens may be poorly represented in the archaeobotanical remains of prehistoric collectors. For example, kernels of teosinte, the probable wild ancestor of maize, are rarely reported in pre-agricultural sites of Mexico. In the Tehuacan excavations substantial amounts of seeds of amaranth and of setaria, a grass that is not an important cultigen, were recovered (Smith 1967). Yet, there is no reported recovery of teosinte in any of the sites (MacNeish 1967; Mangelsdorf *et al.* 1967; Smith 1967). If we are to truly understand the process of the origins of agriculture, it is essential that we learn the reason wild plants important to collectors were abandoned during the period of incipient agriculture. Likewise, it is equally essential that we learn why wild plants unimportant to collectors became major cultigens.

Phytolith studies may help us to identify at least one of the specific reasons some important wild plants failed while insignificant plants became important. With the exception of setaria, the plants noted above as important wild plants are not silica-accumulators and do not produce significant quantities of opal phytoliths in leaf and stem tissue. On the other hand, all cereal cultigens are heavy silica-accumulators producing enormous quantities of phytoliths in leaf and stem tissue. Many studies now show that silica directly provides plants with qualities and benefits that aid agricultural success. Moreover, botanical studies have shown that simply accumulating large amounts of silica is not sufficient to guarantee that a plant will be a good cultigen. In many grasses rapidity of phytolith production and/or the resulting distribution pattern are more important than quantity alone. This may distinguish successful grass cultigens from unsuccessful species. Careful study of setaria may show that this grass belongs in the latter category.

Information in Tables 1, 2 and 3 provides a list of plants, mostly cultigens which are benefited by the presence of silica. While the majority of these are grasses, it is interesting to note how many non-grass cultigens appear, including plants not silica-accumulators in the usual sense. Many of these do, in fact, absorb significant amounts of silica with substantial benefits. Silica assists in plant growth, strength and nutrition; and provides resistance to diseases (blasts, rusts and fungi), to chewing and sucking insects, as well as protection from wilting due to cold or drought. As every farmer knows, or soon realizes, planting a single species intensively on a single field is an invitation to disease and insects in general, and to an epidemic of species-specific pests and insects in particular. However, silica-accumulator plants use phytoliths as a natural, pre-adapted defense against some of the most disastrous problems any cultigen and every farmer faces —insect and disease susceptibility. Many grasses are particularly well pre-adapted with silica defenses. Thus, the presence of silica may be one of the specific factors that determined the success or failure of a plant as a cultigen.

The rice plant, reputed to have the highest silica content among the grasses, has been the object of the most intensive study to date. Silica benefits rice growth and nutrition in a variety of ways. Mitsui and Takatoh (1963: 10) note that the effects of added silica in the rice plant were longer leaves, more root tillering and earlier head sprouting (flowering). In silica deficient plants lower parts of leaves became withered, brown and necrotic and were attacked by disease; heads were dark-colored and hull rice was smaller in size. Silica fertilizer added to rice fields makes rice stalks more rigid, preventing lodging (Anonymous 1964, 1965; Idris *et al.* 1975: 691); enhances leaf erectness (Yoshida *et al.* 1961: 55) and increases yield (Anonymous 1965), ripening percentage and weight of rice kernels (Seo and Ota 1983: 79).

On the other hand, silica deficiency in rice and in many other plants has readily observed deleterious effects.

> Symptoms of silica deficiency in rice plants and horsetails (*Equisetum* sp.) were surprisingly similar... necrosis... developed in the leaves...Both rice plants and horsetails also exhibited a 'weeping willow' habit of growth under Si-deficient conditions. Withering of the leaves and wilting of the plants were other sympotms frequently observed in Si-deficient plants. Yields of species of Gramineae (barley, oats, wheat, and rye) were substantially decreased when the plants were grown in the absence of silica. (Lewin and Reimann 1969: 293).

Silica benefits other grasses in similar fashion, significantly enhancing growth in the barley plant (Okawa 1937: 36; Okawa and Tanaka 1940: 718). Addition of silica increased barley yield (Toth 1939) and seed production (Lipman 1938). Okawa and Tanaka (1940: 718) report significant cold resistance in barley absent in silica-deficient test plants. Ponnaiya (1960) reports that silica level is directly correlated with drought resistance in sorghum. A minimun level of silicon is essential for sugarcane. No other minerals can substitute adequately. Without silica sugarcane simply will not mature (Ayres 1966).

Silica is an important nutrient for many plants including several not normally considered silica-accumulators. Chen and Lewin (1969: 125) cite evidence that the "...addition of silica...increased the growth rate and yield of rice, millet, barley, sunflower, beet, cucumber, corn, tomato and tobacco plants". Comhaire (1966: 13) notes that jute, beets, sunflower, maize, tobacco, pineapple, and surgarcane all benefit from added silica. According to Lanning (1972), sunflower is an effective silica accumulator with high concentrations especially in the hairs of leaves.

> Sunflowers...are definitely benefitted, *especially as regards seed production*, by the presence of Si in the culture medium (Lipman 1938: 197, emphasis mine.)

Non-accumulator *Cucurbita* are benefitted by silica.

> The cucumber plant does not show active Si uptake as does rice... But when sufficient amounts of soluble Si are available in the surrounding media, the cucumber plant can absorb large amounts of Si, and the Si contents in the leaves can reach values as high as those in graminaceous plants. In the field experiments, the cucumber plants which had been given large amounts of silicate fertilizers showed adequate growth and yield and also suffered less damage from wilt disease. (Miyake and Takahashi 1983b: 470–1).

Cucumbers and tomatoes respond to silica deficiency in similar ways. Early growth appears normal but leaves curl at the flowering stage and dry up. Pollen fertility and yield is reduced and fruit is often malformed (Miyake and Takahashi 1978: 175; 1983a: 71).

> [beet] Plants made rapid growth...[if supplied with silica] ...Plants in solution deficient in silica made very slow growth. [In the silica-deficient plants] roots became very dark...with a growth of [fungi]...When plants were transferred to silica-deficient solutions at an early stage, damping-off was common...When silica was added, practically no damping-off occurred...Wilting of the outer leaves during periods of high transpiration was one of the early symptoms of silicon deficiency...In younger plants the cotyledons turned yellow and in most cases soon died...Silicon is an indispensible chemical element for the growth of the beet plant. (Raleigh 1939: 826–7; 1945).

Silica also plays a significant role in providing one of the most important qualities that a plant can have as a successful cultigen —seed retention.

> Loss of seed from the ripening inflorescence (seed shattering) is a common characteristic of grasses and complicates the harvesting of seed in some species...A high silica content of the glumes was found to be one of several characteristics associated with high retention. It seems likely that solid silica would contribute to the stiffening of the glumes, a process...of some importance in seed retention. (Jones and Handreck 1967: 134).

It is often stated that improved seed retention was the result of genetic changes —deliberate selection of variant and mutant qualities after the wild plant(s) had been brought under direct agricultural manipulation by early farmers. Actually, many silica-accumulators may have been preadapted for seed retention— more likely a direct factor in original selection as superior cultigens than as a genetic change that occurs later.

Obviously early farmers did not knowingly select plants because of high silica content and did not add silica fertilizers to their fields to improve crop strength and yield. The more obvious and readily observed benefits occur in the areas of pre-adapted resistance to diseases and insects.

The presence of silica as a defense against various diseases has been noted in several grass cultigens. In rice, silica protects against blast infestation of the pedicel (Akai 1953), *Helminthosporium oryzae* [brown spot] (Gangopadhyay *et al.* 1975; Imaizumi and Yoshida 1958), leaf spot and fungi (Mitsui and Takatoh 1963). Volk *et al.* (1958) report that younger rice leaves are more susceptible to blast fungus than are older leaves which have increased silica levels to inhibit enzyme action used by the fungus to penetrate and infect plant tissue. Several studies have shown similar results in other cereal grasses, including defense against powdery mildew (Germar 1934), and rusts in wheat and rye (Palladin 1927). The differences in disease resistance between silica-accumulator and non-accumulator plants is made clear in the following:

> Primary leaves of barley and wheat, and cotyledons of cucumbers and morning glory were inoculated with *Erysiphe graminis, Alternaria kikichiana, Colletotrichum lagenarium* and *Cochliobolus miyabeanus*. Haloes appeared around the fungal penetration points in the leaves and cotyledons in all host-fungus combinations ...Silicon accumulation occurred in the halo area on barley and wheat leaves but not on cucumber and morning glory cotyledons...It seems that the host cell wall [of the two grasses] may show dynamic reaction against fungal attack. (Kunoh and Ishizake, 1975: 283).

On the other hand, studies of wheat and barley by Lanning (1966a, 1966b) failed to find any correlation between total amount of silica in plant varieties and defense against cold, fungal attack or insects. Lanning does suggest that rapid uptake of silica may be more important than total amount but otherwise provides no explanation for obvious differences in plants with similar silica levels. A possible explanation is provided by studies which show that silica up-take and deposition is not passive, but can be directed to a specific part of the plant when the part is more vulnerable to attack. When a plant part receives silica for its defense seems more important than how much silica it receives.

> The critical periods of vulnerability to attack differ with the different parts of the [rice] plant, thus during the early stages of growth of the plant the foliage is more exposed to fungus attack than the stems, the latter being covered by leaf sheaths. The time when the stem is really and badly exposed to attack is when it emerges through the flag leaf-bearing earhead i. e., at the time of flowering. At that time the stems are not only well exposed but are also very juicy conducting nutrients to panicle. Thus under actual field conditions we find that the stems mainly suffer the fungal attack...at the flowering stage while the foliage are attacked much earlier...It is found...that a definite correlation does exist between the rate of intake during the critical periods of vulnerability to attack and the resistance. Thus we find during the vegetative stage when the leaves are more exposed to attack, the intake of silica in leaves is greater in the resistant varieties than in the susceptible ones...Again from maturity to flowering when... the stem is exposed to attack, the intake by this part of the plant is greater in the resistant variety...it is similarly high in the panicles of the resistant varieties from the flowering stage to harvest...Thus, while the silica in the plant when considered as mere total contents at a particular date does not give a correct indication of the role of silica in the resistance of the plant to disease, the representation of the intake during the critical periods of vulnerability to fungal attack expressed as percentage of total intake gives the correct picture. (Venkatachalam 1954: 307).

The ability of plants to respond dynamically to specific areas of vulnerability or directly to areas under attack by disease organisms occurs in non-accumulator plants, sometimes rather impressively.

> French bean cultivars respond to cowpea rust infection by the production of naturally electron-opaque deposits on the surrounding mesophyll cell walls...[Various tests with chemical reagents] suggests that silicon, rather than phenolic compounds, is the primary electron-opaque component of the wall deposits formed in this non-host interaction in that, contrary to suggestion for silicon deposition in some other situations, formation of these deposits is controlled by metabolic activity... (Heath 1979: 141).

The actual mechanism of defense provided the bean plant by silica is to prevent the rust enzymes from penetrating the leaf tissue. If fluids from infected leaves are injected into healthy leaves, the dynamic response which sends silica to the infected area is suppressed leaving the leaf "defenseless".

> Extracts of bean rust-infected French bean leaves, when injected into uninoculated French bean tissue increased the frequency of haustoria subsequently produced by... (the cowpea rust fungus); in correlation, the increase of silicon-containing...deposits at infection sites decreased. The effect of these extracts on haustorium formation was influenced by the age of the leaves...in general, both extracts and exudates were less effective... (in older) leaves...The ability of the (fungus)...to invade its host species successfully is initially determined by its ability to suppress the deposition of silicon-containing deposits... (Heath 1981: 141).

Older leaves are less vulnerable to attack because they normally contain higher levels of silica than in young leaves even in plants that are not considered silica-accumulators. Such evidence suggests that the ability of any plant to accumulate silica rapidly is a powerful defense against disease, supporting Lanning's suggestion noted above. The bean is not the only non-accumulator to have silica defenses. In a study by Miyake and Takahashi (1983a), powdery mildew attacked all silicon deficient cucumber plants, but was absent on all silicon supplied plants.

Silica also aids the plant by preventing the effects of other toxic chemicals and minerals. It is a "potent factor" in controlling the toxic effect of MgO on tobacco (MacIntire *et al.* 1925) and decreases iron content in the rice shoot protecting plants against iron toxicity (Tanaka and Park 1966: 27). Mn toxicity in oats, barley, rye, wheat, rice, and ryegrass produced necrotic leaf spots in most cases. Addition of silica prevented the symptoms from appearing by causing the Mn to diffuse throughout the plant preventing toxic concentration levels to build up in the leaves (Vlamis and Williams 1967: 139; Williams and Vlamis 1957: 408).

High silica levels as a plant defense against insect predation has been well established. Amos (1952) notes the use of wood timbers high in silica for shipbuilding as a defense against the marine teredine ship borer. Several studies of insect attacks on rice indicate silica is a defense against a host of insects including stem borer (Anonymous 1964; Comhaire 1966; Djamin and Pathak 1967; Nakano *et al.* 1961), red mites (Mitsui and Takatoh 1963; Tanaka and Park 1966), brown planthopper (Yoshihara *et al.* 1979), leaf roller (Hanifa *et al.* 1974) and against the grazing slug, ***Agriolimax reticulus*** (Wadhan and Parry 1981).

> Among the insect pests of rice plants, the Asiatic rice borer is one of the most destructive throughout Asia ...A highly significant negative correlation was recorded between silica content of the stem and susceptibility to the rice borer. High silica content in the plant seemed to interfere with feeding and boring of the larvae and could cause defacing of their mandibles. The use of varieties with high silica content is a...practical and economical method of reducing rice borer infestation...(Djamin and Pathak 1967: 347).

Nakano *et al.* (1961: 26-27) investigated the causes for substantially greater rice borer destruction in one growing region just 5 km from a neighboring area free from severe levels of infestation. No differences were found in the nature of the insects, however, the silica supplying power of the paddy soils to the rice plant in the problem zone was substantially lower than in the other zone.

Hanifa *et al.* (1974) report that leaf roller resistance in rice depends more on the location and distribution of silica than on total quantity. Continuous epidermal phytoliths, long chains of phytoliths in the intercostal zone of the leaf, etc., provides a more effective defense when compared with varieties containing more silica, but with different distribution patterns. According to Yoshihara *et al.* (1979), the presence of dissolved silica in plant fluids is an effective defense against a sucking insect, the brown leafhopper. Thus, silica defends against the full range of chewing, boring and sucking insects.

Silica defense against insects has been noted in other cereal grains. Varieties of sorghum resistant to cinchbug absorb silica more rapidly than susceptible sorghum varieties (Lanning and Linko 1961: 465) and silica provides wheat and rye with resistance to plant lice (Palladin 1927).

Silica content in plants is also a critical variable in the coevolutionary relationship between human culture and domesticated animals. The effect of silica on animals (see Table 4) is more indirect compared to plants, resulting primarily from eating various plants which contain phytoliths. Unlike the effects of silica in plants, which are beneficial in many ways, the effects of silica in the food of animals is mostly negative, sometimes toxic. In fact, the characteristic of silica which is advantageous to a plant is often the same factor that causes harm to the animal that eats it. There can be little doubt that many of the activities of pastoralists in finding suitable food and water for their herds are based on the effects imposed by high silica content of forage and feeds being used. For the cereal grain farmer who also maintains cattle herds, the best choice of grass cultigen varieties for successful defense against insect and disease infestation may also be the worst choice of fodder and silage for cattle.

One of the most serious problems in cattle caused by plant silica is urolithiasis ("water-belly"), blockage of the urinary tract by the buildup of silica calculi, more or less like "kidney stones".

> Calculi composed mainly of silica form in ruminants ...death due to obstruction of the urethra by large calculi is a serious problem in some areas...The graminaceous species that constitute the main diet of afflicted animals contain high levels of silica (up to 6%)...Calculus formation is largely abolished by increasing the intake of water...[and] by providing supplements containing 15–25% sodium chloride. The effect was to increase urine output, lower the urinary concentration of silicic acid and prevent deposition of silica in the bladder...Obstruction and rupture of bladder or urethra is fatal...within one or two days. (Bailey 1981: 219–20).

In North America, obstructive urolithiasis most often affects range calves in winter...while among sheep in Australia obstruction occurs most frequently in the hottest months... This suggests that calculus formation may be related to changes in water intake since winter temperatures in North America cause reduction in water intake whereas water is likely to be in short supply in Australia in summer... Calculus formation can be avoided by removing susceptible animals from problem areas. (Bailey 1981: 221).

In the northern Great Plains of North America, indigenous grasses are the main diet of cattle affected by calculi. These grasses contain low levels of silica and high levels of protein at the start of the growing season. Silica increases and protein decreases as the season continues...(Bailey 1981: 225).

Forman and Sauer (1962) report identical effects of plant silica, water and salt intake on the frequency of urolithiasis in sheep. Clearly whenever such factors occur in nature, pastoralists must adjust their strategy for determining the seasonal patterns and cycles of movement of their herds to new foraging areas accordingly.

Bezeau *et al.* (1966) compared available silica in the plants of two different areas with respect to different frequency of urolithiasis in cattle. In the area of low incidence, forbs and shrubs, i. e., non-grass plants low in silica, comprised 40% of the flora, while in the area high in urolithiasis, forbs and shrubs comprised only 10%. Cattle of the first region were observed eating a larger percentage of low silica plants. Interestingly, the only year when no urolithiasis was reported in calves in the second region a severe drought forced the cattle to eat a higher percentage of available low silica weeds and shrubs. The authors also note that sheep fed more low silica silage drink less water. Parker (1957) reported similar reduction of urolithiasis when cattle were fed mixed feed containing a high percentage of low silica clover (legume). Thus, the amount of non-grass, low silica plants available to cattle and sheep will directly affect both health and water consumption —an obviously important consideration for the success of pastoralism especially in semi-arid and desert ecologies.

Ecological changes in prairie grass species which significantly affect the amount of silica ingested by cattle can have devastating consequences.

> Medusahead [*Elymus caput-medusae* L.] is invading vast acreages [in Washington State] formerly dominated by perennial grass species and more recently by downy brome grass, *Bromus tectorum* L. The latter, although not a desirable range grass, provides considerable early spring forage... [The] problem is three-fold: first, medusahead suppresses desirable vegetation due to its competitive ability; second, it is unpalatable to livestock at all stages of growth; and third, the dead vegetation decomposes slowly, thereby forming a persisting dense duff layer on the soil surface. In addition as the plant matures, it develops long barbed awns which cause mechanical injury to the eyes, noses, and mouths of grazing animals. (Bovey *et al.* 1961: 307).

Studies of ash content indicated that silica is 13.3% of dry weight in Medusahead compared to only 4.4% in downy brome. "The high silica content of Medusahead is the basis for its harshness and may partially explain its unattractivenss to livestock" (Bovey *et al.* 1961: 311). Ecological changes such as these occuring in the past should cause cultural changes in the archæological record. However, such a critical change in grass dominance simply would not appear in the pollen record. In this instance opal phytolith analysis is probably the only reliable and effective source of such profound and culturally critical paleoecological information.

Brazle *et al.* (1979) found the phytoliths in the epidermis of big bluestem, *Andropogon gerardi*, and little bluestem, *A. scoparius*, resisted proper digestion by cattle rumen microorganisms. Higher levels of silica also reduces the digestability of cultivated corn and sorghum (Harbers and Thouvenelle 1980) and other grasses (Harbers *et al.* 1981). Gupta and Pradhan (1975) found a direct correlation between digestibility of a plant and its silica content. In seven grass species tested, each unit of increase in silica content reduced digestibility by 1.4 units; while in legumes, one unit increase of silica depressed digestibility by only 0.6 units —less than half the degree of negative response to the same amount of silica. The negative effects of silica on plant digestibility is not linear; digestibility is reduced faster as the rate of silica increases in cattle feed. Understanding the relationship of plant silica to digestibility of cattle feed is vital to successful livestock management.

> Silica...is an integral part of the matrix of plant cell walls and it may similarly reduce the accessibility of the cell wall carbohydrates to attack by digestive microorganisms...In fact there is so little undertanding of silica in plant feeds that it is overlooked in chemical analyses aimed to assessing the nutritive value of feeds...It is sometimes alleged that a grass which is normally palatable becomes unapalatable on certain soils. Variation in palatability may in part be due to differences in the silica content of the plant. When this is high, the hairs or other trichomes on the leaves are heavily silicified and since the leaf surfaces thereby become rough to the touch...the grasses would probably be less palatable to the grazing animal. (Jones and Handreck 1967: 138).

Probably the most noteworthy aspect of the relationship between plant silica and animal grazing is the evidence that plants can and do respond actively to grazing attack. Just as was noted above in the examples of metabolic, dynamic responses to disease attack, plants (e. g. grasses) respond similarly to grazing. Studies by McNaughton *et al.*, (1985) of the relationship between African perennial grasses and grazing activity indicate: first, where grazing intensity increases, the amount of silica taken up by the grasses increases; second, chlorophyll levels increase apparently to encourage faster regrowth of the plant; third, silica is supplied differentially to vital or vulnerable areas of the plant such as the leaf sheath from which regrowth of the leaf develops and to the roots as a defense against destruction of regeneration areas threatened by animal grazing; and, fourth, as grazing activity increases, silica is supplied more rapidly to the regenerating leaf growth as an earlier defense against further grazing. Whereas urolithiasis is an indirect, passive result afflicting only the grazing animal, these are direct responses by plants as active defenses against the negative effects of animal grazing activity. In simple terms, it is a clear case of a coevolutionary relationship.

> We believe the high values of silica concentrations that we found are a consequence of a long period of intense coevolution between the herbivore fauna and the grasses that support them. In the field, silica concentration increases in more intensively grazed vegetation...The agreement of these patterns suggest that natural selection for silica accumulation is related to grazing exposure... Patterns of grass silicification in contemporary grasslands with an abundant herbivore fauna may assist in interpretation of the fossil record, provide insight into coevolutionary processes, and contribute to an understanding of species diversity in present grazing ecosystems... (McNaughton *et al.* 1985: 532–3).

Certainly the same can be said for the study of the origin and development of pastoral systems and animal husbandry.

Finally, plant silica has significant effects on the most important component of cultural systems —humans. Baker (1961a) reports that airborne wheat dust from handling large quantities of wheat in storehouses and silos contains large amounts of opal phytoliths. These are small enough to be inhaled into the lungs where sharp edges, spines and points of the fragments can cause conjunctival, dermal, naso-pharyngeal and broncho-pulmonary disease symptoms noted in grain handlers. Baker (1961b) reports a similar situation in the airborne dust of a sugarcane processing factory. DeNee *et al.* (1975) conducted a lung biopsy of a 56-year-old male who suffered from *Bronchiolitis obliterans* after being exposed to burning jute fibers. Silica particles found in the lung tissue were similar to phytoliths identified in jute fibers. The authors concluded that phytoliths were the direct cause of the disease resulting in death. The possible toxic effects arising from consistent use of slash-and-burn agriculture in areas with flora high in silica is worthy of investigation.

Recent public health studies have shown that silica in human plant foods can present substantial dangers to life and health.

> Foxtail millet, *Setaria italica* (L.) Beauv., is a major food grain in Asia. It is the most important millet species in N. China. Lin Xuan, in the Henan Province of northern China, is an area with a very high incidence of oesophageal cancer. Recent work has indicated that siliceous plant structures originating from *S. italica*, may be involved in the ætiology of oesophageal cancer in the Lin Xuan area. The inhabitantss of the area produce a cake from the bracts of this millet by roasting the bracts and mixing them with the pulp of persimmon fruit. This cake forms part of the staple diet of the inhabitants...the inflorescence bracts and the caryopsis of *S. italica* have both been shown to contain silicon. In the bracts, high concentrations of silicon [are present]...Siliceous hairs from the inflorescence bracts of the Mediterranean grass *Phalaris minor* Rets. have been strongly implicated as the causal agents of oesophageal cancer in N. E. Iran. These hairs have been shown to contaminate the wheat flour of the region. (Hodson *et al.* 1982: 843).

The emerging evidence clearly shows that plant silica is a major factor that influenced coevolutionary developments connected with the origins of agriculture. It may have been one of the important genetic characteristics that pre-adapted certain wild plants for selection and success as cultigens. The absence of significant levels of silicia, slow rates of phytolith production, inferior patterns or distribution or lack of dynamic responses in insect/disease vulnerable or invaded areas, etc. may serve to explain why other plants failed. Plant silica and phytoliths are also major factors in the success or failure of pastoral systems and animal husbandry and it apparently also causes serious problems directly affecting human health and safety.

Plant silica is certainly not the only factor important to the origin and development of agricultural systems, but for research in archæology it may prove to be the most important. Opal phytoliths are both a cause and an effect —an important contributor to agricultural processes in earlier times and its own preserved fossil evidence of those processes available for analysis today.

Table 1

PLANTS IN WHICH SILICA BENEFITS HEALTH, GROWTH, AND/OR PRODUCTIVITY

REFERENCE	FAMILY TAXON	COMMON NAME
Anonymous, 1964	Gramineae	rice
Anonymous, 1965	Gramineae	rice
Ayers, 1966	Gramineae	sugarcane
Chan and Lewin, 1969	Equisetaceae	scouring rush
Comhaire, 1966	several	
Fox, *et al.*, 1967	Gramineae	sugarcane
Idris, *et al.*, 1975	Gramineae	rice
Jones and Handreck, 1967	Gramineae	
Lanning, 1972	Compositae	sunflower
Lewin and Reimann, 1969	several	
Lipman, 1938	several,	barley, sunflower
Mitsui & Takatoh, 1963	Gramineae	rice
Miyake & Takahashi, 1978	Solanaceae	tomato
Miyake & Takahashi, 1983a	Cucurbitaceae	cucumber
Miyake & Takahashi, 1983b	Cucurbitaceae	cucumber
Nakagawa, *et al.*, 1954	Gramineae	rice
Okawa & Tanaka, 1940	Gramineae	barley, rice
Okawa, 1936	Gramineae	barley
Okawa, 1937	Gramineae	barley
Ponnaiya, 1960	Gramineae	sorghum
Raleigh, 1939		beet
Raleigh, 1945		beet
Seo & Ota, 1983	Gramineae	rice
Takahashi, 1961	several	
Toth, 1939	Gramineae	barley
Yoshida, *et al.*, 1961	Gramineae	rice
Yoshida, *et al.*, 1959	Gramineae	rice

Table 2

PLANTS IN WHICH SILICA PROVIDES DISEASE RESISTANCE

REFERENCE	FAMILY	PLANT	DISEASE
Akai, 1938	Gramineae	rice	blast fungus
Akai, 1953	Gramineae	rice	blast, spores
Gangopadhyay, *et al.*, 1975	Gramineae	rice	brown spot
Germar, 1934	Gramineae		mildew
Heath, 1979	Leguminosae	pintobean	cowpea rust
Heath, 1981	Leguminosae	pintobean	bean rust
Imaizumi & Yoshida, 1958	Gramineae	rice	blast, brownspot
Jones & Handreck, 1967	Gramineae		several
Kunoh & Ishizake, 1975	several		fungi
Lanning & Linko, 1961	Gramineae	sorghum	smut, milo
Lanning, 1966a	Gramineae	barley	mildew, smut
Lanning, 1966b	Gramineae	wheat	rust, smut
Macintire, *et al.*, 1925	Solanaceae	tobacco	MgO toxicity
Mitsui & Takaton, 1963	Gramineae	rice	necrosis, fungi
Miyake & Takahashi, 1983a	Cucurbitaceae	cucumber	powdery mildew
Miyake & Takahashi, 1983b	Cucurbitaceae	cucumber	wilt disease
Okawa & Tanaka, 1940	Gramineae	barley & rice	black spot
Padallin, 1972*	Gramineae	wheat & rye	rust
Raleigh, 1945		beet	fungi
Tanaka & Park, 1966	Gramineae	rice	iron toxicity
Vankatachalam, 1954	Gramineae	rice	blast
Vlamis & Williams, 1967	Gramineae	cereals	Mn toxicity
Volk, *et al.*, 1958	Gramineae	rice	blast fungus
Williams & Vlamis, 1957	Gramineae	barley	Mn toxicity

Table 3

PLANTS IN WHICH SILICA PROVIDES INSECT RESISTANCE

REFERENCE	FAMILY	PLANT	INSECT
Amos, 1952	several	wood	marine borer
Anonymous, 1954	Gramineae	rice	stem borer
Comhaire, 1966	several		stem borer
Djamin & Pathak, 1967	Gramineae	rice	rice borer
Hanifa, *et al.*, 1974	Gramineae	rice	leaf roller
Jones & Handreck, 1967	Gramineae	several	
Lanning & Linko, 1961	Gramineae	sorghum	chinch bugs
Lanning, 1966a	Gramineae	barley	greenbugs
Lanning, 1966b	Gramineae	wheat	Hessian fly
Mitsuit & Takatoh, 1963	Gramineae	rice	red mite
Nakano, *et al.*, 1961	Gramineae	rice	stem borer
Palladin, 1927	Gramineae	wheat, rye	plant lice
Tanaka & Park, 1966	Gramineae	rice	mites
Volk, *et al.*, 1958	Gramineae	rice	unspecified
Whadam & Parry, 1981	Gramineae	rice	slugs
Yoshihara, *et al.*, 1979	Gramineae	rice	brown planthopper

Table 4

STUDIES ON THE EFFECTS OF PLANT SILICA INGESTED BY ANIMALS

REFERENCE	ANIMALS	CONDITION
Bailey, 1981	cattle, sheep	urolithiasis
Baker, 1961	humans	pulmonary
Baker, 1961	humans	pulmonary
Baker, *et al.*, 1959	sheep	tooth wear
Bezeau, *et al.*, 1966	cattle, sheep	urolithiasis
Bovey, *et al.*, 1961	cattle	
Brazle, *et al.*, 1979	cattle	
DeNee, *et al.*, 1975	humans	pulmonary
Forman and Sauer, 1962	sheep	urolithiasis
Gupta and Pradhan, 1975	in vitro tests	
Habers & Thouvenelle, 1980	cattle	rumen tests
Habers, *et al.*, 1981	steers	
Hodson, *et al.*, 1982	humans	esophageal cancer
Jones & Handreck, 1967	cattle	
McNaughton, *et al.*, 1985	African herbivores	
Parker, 1957	cattle	urolithiasis
Parry & Hodson, 1982	humans	esophageal cancer

REFERENCES

AKAI, S.

1938. "On the Ash Figures of Leaves of the Rice Plant Grown under a Combination-practice of Several Effective Measures for the Control of Blast Disease", *Annals of the Phytopathological Society of Japan* 7: 173–92.

1953. "Helminthosporium Blight of Rice Plants. VII. The Relation of Silicic Acid Supply to the Outbreak of Helminthosporium Blight or Blast Disease in Rice Plants", *Annals of the Phytopathological Society of Japan* 17: 109–112.

AMOS, G. L.

1952. "Silica in Timbers", *Bulletin of the Commonwealth Scientific and Industrial Research Organization* 267, Australia: 1–59.

ANONYMOUS

1964. "Effects of Forms of Silicon on Growth, Yield", *Annual Report: The International Rice Research Institute, Los Banos, The Phillipines*: 99–102.

1965. "Silica Fertilization in Low Land Soils", *Annual Report: The International Rice Research Institute, Los Banos, The Phillipines*: 216–218.

ASCH, D.L. and N.B. ASCH

1977. "Chenopods as Cultigen: a Re-evaluation of some Prehistoric Collections from Eastern United States", *Midcontinent Journal of Archæology* 2 (1): 3–46.

AYERS, A. S.

1966. "Calcium Silicate Slag as a Growth Stimulant for Sugar Cane on Low Silicon Soils", *Soil Science* 101: 216–27.

BAKER, G.

1961. "Opal Phytoliths from Sugar Cane, San Fernando, Phillipine Islands", *Memoirs of the Queensland Museum* 14 (1): 1–12.

BEZEAU, L.M., A. JOHNSTON, and S. SMOLIAK

1966. "Silica and Protein Content of Mixed Prairie and Fescue Grassland Vegetation and its Relationship to the Incidence of Silica Urolithiasis", *Canadian Journal of Plant Science* 46: 625–631.

BOVEY, R.W., D. Le TOURNEAU, and L.C. ERICKSON

1961. "The Chemical Composition of Medusahead and Downy Brome", *Weeds* 9: 307–311.

BRAZLE, F.K., L.H. HARBERS, and C.E. OWENSBY

1979. "Structural Inhibitors of Big and Little Bluestem Digestion Observed by Scanning Electron Microscopy", *Journal of Animal Science* 48 (6): 1457–63.

CHEN, C. and J. LEWIN

1969. "Silicon as a Nutrient Element for *Equisetum arvense*", *Canadian Journal of Botany* 47: 125–131.

COMHAIRE, M.

1966. "The Role of Silica for Plants", *Agri Digest* 7: 9–19.

DE NEE, P.B., J.L. ABRAHAM, and A.H. GELDERMAN

1975. "*Bronchiolitis obliterans* and Silica in Human Lung Following Exposure to Burning Jute: a Scanning Electron Microscope Study", *American Journal of Pathology* 78 (1): 7A–8A.

DJAMIN, A. and M.D. PATHAK

1967. "Role of Silica in Resistance to Asiatic Rice Borer, *Child suppressalis* (Walker), in Rice Varieties", *Journal of Economic Entomology* 60 (2): 347–51.

FORMAN, S.A. and F. SAUER

1962. "Some Changes in the Urine of Sheep Fed a Hay High in Silica", *Canadian Journal of Animal Science* 42: 9–17.

FOX, R.L., J.A. SILVA, O.R. YOUNGE, D.L. PLUCKNETT, and G.D. SHERMAN

1967. "Soil and Plant Silicon and Silicate Response by Sugar Cane", *Soil Science Society of America, Proceedings* 31: 775–779.

GANGOPADHYAY, S. and S.B. CHATTOPADHYAY

1975. "Total Silica and Brown Spot Disease Development of Rice under Varying Levels of Nitrogen", *Current Science* 44 (3): 92–94.

GERMAR, B.

1934. "Über einige Wirkungen der Kieselsaure in Getreidepflanzen, insbesondere auf deren Resistenz geüber Mehltau", *Zeitschrift der Pflanzenernaehrung Dungung Bodenkunde* 35: 102–115.

GUPTA, P.C. and K. PRADHAN

1975. "A Note of the Effect of Silica on in Vitro Dry Matter Digestibility of Legume and Non-legume Forages", *Indian Journal of Animal Science* 45 (7): 497–8.

HANIFA, A.M., T.R. SUBRAMANIAN, and B.W.X. PONNAIYA

1974. "Role of Silica in Resistance to the Leaf Roller, *Cnaphalocrocis medinalis guenee*, in Rice", *Indian Journal of Experimental Biology* 12: 463–465.

HARBERS, L.H., L.K. MCNALLY, and W.H. SMITH

1981. "Digestibility of Three Grass Hays by the Horse and Scanning Electron Microscopy of Undigested Leaf Remnants", *Journal of Animal Sciences* 53 (5): 1671–7.

HARBERS, L.H. and M.L. THOUVENELLE

1980. "Digestion of Corn and Sorghum Silage Observed by Scanning Electron Microscopy", *Journal of Animal Science* 50 (3): 514–526.

HEATH, M.C.

1979. "Partial Characterization of the Electron Opaque Deposits Formed on Non-host Plant, French Bean, after Cowpea Rust Infection", *Physiological Plant Pathology* 15 (2): 141–148.

HODSON, M.J. and D.W. PARRY

1982. "The Ultrastructural and Analytical Microscopy of Silicon Deposition in the Aleurone Layer of the Caryopsis of *Setaria italica* (L.) Beauv.", *Annals of Botany* 50 (2): 221–8.

IDRIS, M., M. HOSSAIN, and F.A. CHOUDHURY

1975. "The Effect of Silicon on Lodging of Rice in Presence of Added Nitrogen", *Plant and Soil* 43 (3): 691–695.

IMAIZUMI, K. and S. YOSHIDA

1958. "Edaphological Studies on Silicon Supplying Power of Paddy Fields", *Bulletin of the National Institute of Agricultural Sciences, Series B.* 8: 261–304.

JONES, L.H.P. and K.A. HANDRECK

1967. "Silica in Soils, Plants, and Animals", *Advances in Agronomy* 19: 107–149.

KUNOH, H. and H. ISHIZAKI

1975. "Silicon Levels near Penetration Sites of Fungi on Wheat, Barley, Cucumber and Morning Glory Leaves", *Physiological Plant Pathology* 5: 283–287.

LANNING, F.C.

1966a. "Pattern of Distribution: Relation of Silicon in Wheat to Disease and Pest Resistance", *Journal of Agricultural and Food Chemistry* 14 (4): 350–352.

1966b. "Barley Silica: Relation of Silicon in Barley to Disease, Cold and Pest Resistance", *Journal of Agricultural and Food Chemistry* 14 (6): 636–638.

1972. "Ash, SiO_2, CaO, Fe_2O_3, and Al_2O_3 in Helianthus", *Hortscience* 7 (2): 184–186.

LANNING, F.C. and Y.Y. LINKO

1961. "Role of Silica in Plants: Absorption and Deposition of Silica by Four Varieties of Sorghum". *Journal of Agricultural and Food Chemistry* 9 (6): 463–465.

LEWIN, J. and B. REIMANN

1969. "Silicon and Plant Growth", *Annual Review of Plant Physiology* 20: 289–304.

LIPMAN, C.B.

1938. "Importance of Silicon, Aluminum and Chlorine for Higher Plants", *Soil Science* 45: 189–98.

MACINTYRE, W.H., W.H. SHAW, and J.B. YOUNG

1925. "The Role of Silica in Counteracting Magnesium-induced Toxicity", *Soil Science* 19 (5): 331–341.

MACNEISH, R.S.

1967a. "Introduction", in *The Prehistory of the Tehuacan Valley, V.I. Environment and Subsistence*, D.S. Byers (ed), University of Texas Press, Austin.

MANGELSDORF, P.C., R.S. MACNEISH, and W.C. GALINAT

1967. "Prehistoric Wild and Cultivated Maize", in *The Prehistory of the Tehuacan Valley, V.I Environment and Subsistence*, D.S. Byers (ed), University of Texas Press, Austin.

MCNAUGHTON, S.J., J.L. TARRANTS, M.M. MCNAUGHTON, and R.H. DAVIS

1985. "Silica as Defense against Herbivory and a Growth Promotor in African Grasses", *Ecology* 66 (2): 528–535.

MITSUI, S. and H. TAKATOH

1963. "Nutritional Study of Silicon in Graminaceous Crops (Part I)", *Soil Science and Plant Nutrition* 9 (2): 7–11.

MIYAKE, Y. and E. TAKAHASHI

1978. "Silicon Deficiency of Tomato Plant", *Soil Science and Plant Nutrition* 24 (2): 175–189.

1983a. "Effects of Silicon on the Growth of Solution-cultured Cucumber Plant", *Soil Science and Plant Nutrition* 29 (1): 71–83.

1983b. "Effect of Silicon on the Growth of Cucumber Plant in Soil Culture", *Soil Science and Plant Nutrition* 29 (4): 463–71.

NAKAGAWA, M., A. KITAMOTO, and I. TANGE

1954. "Effect of Fused Magnesium Phosphate on Paddy Growth on 'Akiochi' Soil with Particular Reference to Phosphorous and Silicon", *Soil and Plant Food* 1: 27–28.

NAKANO, K., G. ABE, N. TAKETA, and C. HIRANO

1961. "Silica as an Insect Resistance Component of Host Plant, Found in the Relation between the Rice Stem Borer and Rice Plant", *Japanese Journal of Applied Entomology and Zoology* 5: 17–27.

NETOLITZKY, F.

1914. "Die Hirse aus Antiken Funden", *Sitzbuch der Keisliche Akademie Wissenshaft der Mathematisch-Naturwissenshaft* 123 (6): 725–59.

OKAWA, K.

1936. "Investigation on the Physiological Action of Silic Acid for Plants", *Journal of the Science of Soil and Manure* vol. X, no. 1, March: 95–110.

1937. "Investigation on the Physiological Action of Silic Acid for Plants (VI)", *Journal of the Science of Soil and Manure* vol. XI no. 1, February: 23–36.

OKAWA, K. and R. TANAKA

1940. "The Relation between Silicic Acid and Sulphuric Acid on Plants", *Journal of the Science of Soil and Manure* vol. XIV, no. 11, Japan: 703–718.

PALLADIN, V.I.

1927. *Plant Physiology*, Blakiston, Philadelphia.

PEARSALL, D.M.

1978. "Phytolith Analysis of Archæological Soils, Evidence for Maize Cultivation in Formative Ecuador", *Science* 199 (4325): 177–78.

PEARSALL, D.M. and D.R. PIPERNO

1986. "Antiquity of Maize Cultivation in Ecuador: Summary and Reevaluation of the Evidence", paper presented at the 51st Annual Meeting of the Society for American Archæology, New Orleans.

PIPERNO, D.R.

1984. "A Comparison and Differentiation of Phytoliths from Maize (*Zea mays L.*) and Wild Grasses: Use of Morphological Criteria", *American Antiquity* vol. 49, no. 2: 361–83.

1986. "A Survey of Phytolith Production and Taxonomy in Non-graminaceous Plants: Implications for Paleoecological Reconstructions", in *Plant Opal Phytolith Analysis in Archæology and Paleoecology, Occasional Papers #1 of the Phytolitharien*, I. Rovner (ed), Raleigh.

PONNAIYA, B.W.X.

1960. "Silica Deposition in Sorghum Roots and its Possible Roles", *Madras Agricultural Journal* 47: 31.

RALEIGH, G.J.

1939. "Evidence for the Essentiality of Silicon for Growth of the Beet Plant", *Plant Physiology* 14: 823–828.

1945. "Silicon in Plant Growth", *Soil Science* 60: 133–5.

ROSEN, A. M.

1984. "Phytoliths of Cereals from two Negev Chalcolithic Sites: Identification and Edaphic Implications", Ms. submitted to *Journal of Archæological Sciences.*

ROVNER, I.

1971. "Potential of Opal Phytoliths for Use in Paleoecological Reconstruction", *Quaternary Research* 1 (3): 345–359.

1983a. "Applications of Phytolith Analysis in Cultural Resource Management, Site Survey, Testing and Significance", Paper presented at the 48th Annual Meeting, Society for American Archæology.

1983b. "Plant Opal Phytolith Analysis: Major Advances in Archæobotanical Research", in *Advances in Archæological Method and Theory*, edited by M. Schiffer, Academic Press, New York.

SMITH, C.E., Jr.

1967. "Plant Remains", in *The Prehistory of the Tehuacan Valley, V.I Environment and Subsistence*, D.S. Byers (ed), University of Texas Press, Austin.

SEO, W.S. and Y. OTA

1983. "Role of the Hull in the Ripening of the Rice Plant. VII. Effects of Supplying of Silica and Potassium during Reproductive Growth State on the Form and Function of Hulls", *Japanese Journal of Crop Science* 52: 79.

TACK, M.

1986. "Phytolith Analysis as Corroborative Physical Evidence", in *Plant Opal Phytolith Analysis in Archæology and Paleoecology, Occasional Papers #1 of the Phytolitharien*, I. Rovner (ed), Raleigh.

TAKAHASHI, E.

1961. "Effect of Silicon on the Growth of Barley, Tomato, Radish, Green Onion, Chinese Cabbage, and their Nutrients Uptake", *Journal of the Science of Soil and Manure* 32: 623–6.

TANAKA, A. and Y.D. PARK

1966. "Significance of the Absorption and Distribution of Silica in the Growth of the Rice Plant", *Soil Science and Plant Nutrition* 12 (5): 191–6.

TOTH, S.J.

1939. "The Stimulating Effects of Silicates on Plant Yields in Relation to Anion Displacement", *Soil Science* 47 (2): 123–41.

VENKATACHALAM, S.

1954. "The Intake of Silica by the Rice Plant with Reference to Blast Disease", *The Madras Agricultural Journal* 41 (9): 304–10.

VLAMIS, J. and D.E. WILLIAMS

1967. "Manganese and Silicon Interaction in the Gramineae", *Plant and Soil* 27 (1): 131–140.

VOLK, R.J., R.P. KAHN, and R.L. WEINTRAUB

1958. "Silicon Content of the Rice Plant as a Factor Influencing its Resistance to Infection by Blast Fungus, *Piricularia oryzae*", *Phytopathology* 48: 179–184.

WADHAM, M.P. and D.W. PARRY

1981. "The Silicon Content of *Oryza sativa L.* and its Effect on the Grazing Behavior of *Agriolar reticulatus muller*", *Annals of Botany,* 48: 399–402.

WATANABE, N.

1968. "Spodographic Evidence of Rice from Prehistoric Japan", *Journal of the Faculty of Science of the University of Tokyo* 3 (3): 217–235.

1970. "A Spodographic Analysis of Millet from Prehistoric Japan", *Journal of the Faculty of Science, University of Tokyo, Sec. V,* 3 (5): 357–379.

WILLIAMS, D.E. and J. VLAMIS

1957. "The Effect of Silicon on Yield and Manganese-54 Uptake and Distribution in the Leaves of Barley Plants Grown in Culture Solutions", *Plant Physiology* 32: 404–9.

YOSHIDA, S., S.A. NAVASERO, and E.A. RAMIREZ

1961. "Effects of Silica and Nitrogen Supply on some Leaf Characters of the Rice Plant", *Plant and Soil* 31: 48–56.

YOSHIDA, S., Y. OHNISHI, and K. KITAGISHI

1959. "Role of Silicon in Rice Nutrition", *Soil and Plant Food* 5 (3): 127–133.

YOSHIHARA, T., K. SOGAWA, M.D. PATHAK, B.O. JULILAND, and S. SAKAMURE

1979. "Soluble Silicic Acid as a Sucking Inhibitory Substance in Rice against Brown Plant Hopper (Delphacidae, Homoptera)", *Entomologia Experimentalis et Applicanta* 26 (3): 314–322.

RACIAL SYSTEMATICS AND THE EVOLUTION OF MEXICAN MAIZE

BRUCE F. BENZ
University of Wisconsin–Madison

Taxonomic investigations of Mexican maize have a long history. Beginning with Sahagún in the early sixteenth century (see Dibble and Anderson 1963) and ending with Anderson (1946), early attempts at the classification of maize races grown in the environmentally diverse Mexican landscape focused more on recognition of morphological similarities than on patterns of evolutionary descent (Bonafous 1836; Kuleshov 1929). Mangelsdorf and Reeve's (1939) tripartite hypothesis was undoubtedly the impetus for the first attempt (Wellhausen *et al.* 1951; 1952) to reconstruct the phylogeny of Mexican maize. Renewed interest in teosinte (*Zea mays* subsp. ***parviglumis*** Iltis and Doebley) as the ancestor of maize (Iltis 1971, 1983; Iltis and Doebley 1980, 1984; Doebley and Iltis 1980; Doebley 1983; Doebley *et al.* 1984; see also Galinat 1970, 1983) has once again stimulated inquiry to trace the origin and evolution of maize's racial diversity through taxonomic research based on that culturally revered and botanically marvelous artifact, the maize ear. Based on the new theoretical insights gained through comparative morphology (Iltis 1983), taxonomic and phylogenetic pursuits have begun to unlock the mysteries of the origin and racial diversification of Mexican maize.

In the following discussion, a comparison of maize ear and teosinte tassel raceme morphology and anatomy summarizes the changes that occurred in the transformation of the teosinte tassel into the maize ear. Following this a newly derived taxonomic scheme for Mexican maize races is summarized. This scheme is then analyzed using the cladistic method to derive a phylogenetic tree for Mexican maize. The taxonomic and phylogenetic patterns of maize evolution are then reviewed by comparing them to known relationships between indigenous Mexican linguistic groups. The final section notes the disparity between predictions that domesticated *Zea* arose in the Balsas River watershed and the lack of evidence in the archæological record for human use of teosinte.

The Initial Attempts to Reconstruct Racial Phylogeny

The first attempt at phylogenetic classification of Mexican maize was conducted under a theoretical umbrella devised by Mangelsdorf and Reeve's presumption that maize's ancestor must have had covered grains that would pop (Mangelsdorf and Reeves 1939: 232). Logically then their candidate for the predecessor of maize was a pod-popcorn. Believing that the ancestor of maize was a wild pod-popcorn, Wellhausen *et al.* (1952) reasoned that primitive Mexican maize races could be identified as those possessing seeds that would pop, those whose glumes were large relative to seed size, and, of course, those with small ears. Though perhaps inevitable, it was nevertheless unfortunate that they chose to rely upon such genetically plastic characteristics as kernel size and endosperm texture because it is exactly these characteristics that would be expected to change with human selection. As an indication that their choice of characters to identify primitive races was inadequate, the geographic distribution of primitive maize races they reconstructed could not be explained without postulating that two or more races of wild maize, both of which are now extinct, had at one time existed in Mexico; one at low elevations, the second at high elevations (Wellhausen *et al.* 1952: Figure 10; Manglesdorf *et al.* 1967; Mangelsdorf 1974).

This earliest attempt at deciphering the phylogeny of Mexican maize, though based upon a magisterial effort in gathering and describing the vast diversity of Mexican maize, also suffered from an overzealous appeal to hybridization in an effort to explain the origin of over 70% of the then recognized races. The emphasis placed on hybridization to explain the diversity of Mexican races is perhaps understandable when it is recalled that during the 1930's and 1940's enormous yield gains due to hybrid vigor obtained from the famous double-cross hybrid were widely touted, and when hybridization was even used to explain such biological mysteries as the origin of the flowering plants (Anderson 1934). To be sure, hybridization may have been important in the diversification of maize races, whether between maize and teosinte, or between races of maize, but the great diversity of Mexican cultures and microenvironments together suggest instead that racial diversification has resulted primarily by way of geographic isolation.

A New Framework for Phylogenetic Reconstruction in Cultivated Zea

Phylogenetic reconstruction from natural classification by comparative morphology is a method dating back to Cuvier and Darwin (Mayr 1982). The method, simply stated, compares the various taxa whose relationships are established by overall morphological similarity, and, based on a few key characters selected for their relevance in describing phylogenetic change, designates those which possess the greatest number of derived characteristics as advanced and those with the greatest number of generalized characters as primitive. As already pointed above, this method was in fact used by Wellhausen *et al.* in their recognition of four classes of Mexican maize races: Ancient Indigenous, Precolumbian Exotics, Prehistoric Mestizos and Modern Incipient. For example, Wellhausen *et al.* (1952) recognized Chalqueño and numerous other races as highly specialized because their ears are large, their glumes are short relative to the length of the grains, and because their grains do not pop. Employing the same rationale, Chapalote was recognized as primitive because of its small ears and grains, its long glumes relative to the size of the grain, and its poppable grains. The method is sound but the determination of primitive versus advanced is entirely dependent upon the correct recognition of the ancestor.

Reevaluation of racial relationships and construction of a phylogenetic scheme for Mexican maize applies the same method described above, though it differs dramatically from previous investigations in recognizing teosinte as the ancestor of maize (Benz 1986). Moreover, we do not subscribe to what has been called the Orthodox Teosinte Hypothesis (OTH; Galinat 1983 and references therein, see also Iltis n.d.), which posits the maize ear as being derived from the female spike of teosinte, but subscribes instead to the theory that the maize ear originated as a result of feminization of the central raceme of the male tassel terminating each of the primary lateral branches of teosinte, the Catastrophic Sexual Transmutation Theory (CSTT; Iltis 1983).

Instead of subjecting anthropological readers to a discussion of the morphological and anatomical intricacies which point out why the maize ear could have only a single antecedent (see Iltis 1983, n.d.; Gould 1984; Benz 1986), the similarities and differences between the maize ear and teosinte tassel raceme are only briefly summarized here. Their similarities are: 1) both the maize ear and teosinte tassel (especially the central raceme) are borne terminally on primary lateral branches, 2) both the tassel of teosinte and ear of maize bear a pair of spikelets per individual rachis segment, and 3) in each rachis segment of the maize ear and teosinte tassel raceme there is an evascular region above the point of attachment of the spikelet pair which contains a lignified hypodermis (a spikelet gap, which in the maize ear is the so-called cupule while in the teosinte tassel it is recognized simply as a hypodermis). Attributes which distinguish the two inflorescences are: 1) the maize ear bears spikelets on many sides of the inflorescence axis (polystichous) while the tassel raceme bears spikelet pairs on only two sides (distichous), 2) each spikelet in the maize ear has only a single fertile floret (the terminal or second floret), which is female, while each teosinte tassel spikelet contains two fertile florets, which are both male, and 3) in the maize ear, presumably because it is female, the lignified evascular region above each spikelet pair is enlarged, folded into the rachis and variously shaped (the cupule), while in the tassel raceme, this lignified evascular region is thin, very narrow and hardly noticeable (see Iltis 1983; Benz 1986).

Recent morphological and anatomical examination of a feminized teosinte tassel indicates that the traits which distinguish the teosinte tassel (distichy, two male florets per spikelet and lack of a well-developed cupule) and the maize ear (polystichy, one female floret per spikelet and well-developed cupule) are, for the most part, secondary sexual characteristics (Iltis 1983; Gould 1984; Benz 1986). In effect, the well-developed cupule and single floret per spikelet are demonstrably due simply to the production of female flowers in the male tassel. For example, in the feminized teosinte tassel specimens examined (Iltis and Doebley no. 7bb, [km 20.2 on the road from Texcoco to Los Reyes, Mexico] WIS), feminization of an otherwise male tassel raceme has produced a flattened, cup-shaped rachid with a thickened hypodermis, in effect making the triangular teosinte tassel rachis all at once cupulate. This suggests that feminity by itself explains how the cupule originated in the maize ear.

The loss of fertility in the lower of the two florets in each spikelet is somewhat more difficult to describe but seemingly as simply explained as the origin of the cupule. It is widely recognized that loss of floret fertility in the spikelets of genera belonging to the Andropogoneae proceedes from the base to the apex. In the tribe of the Grass Family to which *Zea* and *Tripsacum* belong, it is always the lower floret that loses its sexual parts while it is the terminal floret that maintains its sexual function. The mechanism promoting loss of the lower floret's fertility in the maize ear is presumably related to apical dominance; in other words, the uppermost flower is developmentally dominant and usurps all nutrients made available to the spikelet, probably because the developing seed requires prolonged sustenance long after anthesis. Thus in the maize ear the loss of fertility in the lower floret fits a pattern, one shared by its close relatives (*e. g.* teosinte, and *Tripsacum*), and one associated with femininity.

Polystichy in the maize ear is somewhat more difficult to explain. Femininity by itself does not induce polystichy. Polystichy is believed to have occurred in the fashion described by Collins (1919; *cf.* Iltis 1983; Galinat 1985); that is by condensation, yoking and twisting of the rachids because of man's tendency for the collection and eventual cultivation of rigid rachis phenotypes and selection for larger, more productive ears. At the present time, human selection for a rigid rachis would seem to be the simplest explanation of the single most recognizable difference between maize and teosinte, namely, polystichy.

In sum, positional and structural homologies between the maize ear and teosinte tassel argue convincingly that the maize ear evolved by Catastrophic Sexual Transmutation (Iltis 1983, Gould 1984). Morphological and anatomical examination of a feminized teosinte tassel (Benz 1986) demonstrates that two of three major morphological differences between the teosinte tassel and the maize ear (one fertile floret per spikelet and a well-developed cupule in the latter) are due simply to femininity. Though not a direct result of femininity *per se*, polystichy was probably a secondary consequence of human harvesting and selection on the seed-bearing tassel. The adaptive radiation of maize leading to its eventual diversification into regional forms thus resulted from the formation of the eminently dispersable maize ear.

Taxonomy of Mexican Maize

In recognizing teosinte as the ancestor, and specifically identifying the teosinte male tassel raceme as the precursor of the maize ear, classification of Mexican maize and designation of the primitive races is significantly different from those of previous workers (Benz 1986). Though focusing on ear morphology in much the same way as Wellhausen *et al.*, a more exacting description of a wider variety of morphological characters have been employed which are believed to be more conservative under human selection than kernel shape, size and texture. Thus emphasis has been placed on interracial comparisons based upon features of the female inflorescence which, for one reason or another are not visible to the cultivator, and so, are not directly acted upon and modified in the continuing process of selection for new and more productive forms. Therefore, because taxonomic and subsequent phylogenetic analyses focus on what are believed to be conservative characters, the classification derived is expected to reflect patterns of evolutionary descent from teosinte which can, in turn, be used to infer where maize originated, the direction and pathways of its adaptive radiation, and to hypothesize how and why such races diversified.

Taxonomic studies summarized here have focused on the maize races more or less as they were described by Wellhausen *et al.* (1952). In exceptional circumstances new races have been recognized where the previous investigations were insufficiently detailed to detect subtle differences and where maize from previously unexplored areas became available for study. Unfortunately, not all of the Mexican races described by previous workers could be examined during the course of our these studies. While the majority of races described by previous authors are also recognized in our analyses, there are notable differences between the two (Table 1). Methods of analysis are described in detail elsewhere (Benz 1986). The discussion which follows summarizes the recently established taxonomic relationships which provide the basis for outlining maize's descent from teosinte and its subsequent racial diversification.

Thirty Mexican races have been classified into two racial complexes. These racial complexes, though distinguished principally on the basis of ear morphology also exhibit strict elevational allopatry; the Mexican Pyramidal Ear Complex is found primarily above 1800 meters, while the Mexican Narrow Ear Complex occurs exclusively below 1800 meters. These complexes are named following Anderson's (1946) recognition of the characters that distinguish races grown at high elevations, namely, those with conical (hence the name Pyramidal) or abruptly tapering ears that possess a distinctly enlarged butt with irregular rows of grain; from the races grown at mid-to-low-elevations, these exhibiting cylindrical or fusiform ears with straight rows of grain. Each complex is further subdivided into two or more distinct groups of races, recognized here as racial alliances. Each of these alliances is believed to represent monophyletic group, that is, a number of races that descended from a single common ancestor.

Two alliances are recognized within the Mexican Pyramidal Complex: the Central Highlands Alliance, which is composed of five races found only in the high plateau of central Mexico; and the Trans-Sierra Madre Alliance, representatives of which are scattered from the Chiapas Central Mesa to the sierra adjoining Volcán Colima in Jalisco. The Mexican Narrow Ear Complex is composed of three alliances and a group which contains four races of as yet unaffiliated taxonomic status: the Balsas-West Mexico Alliance, which is, as the name implies, composed of races found principally at mid-elevations in Balsas River Drainage, the Jaliscan Plains, and in coastal western Mexico; the Isthmian Alliance consists of races existing primarily at low elevations from at least as far north as Sinaloa, along the "Costa Chica" of Guerrero to the coast of southern Chiapas, across the Isthmus of Tehuantepec, and into and across the northern Yucatan Peninsula; and the Mixe Alliance, composed of races occurring in the San Juan River drainage of northeastern Oaxaca; the remaining four races, as yet "unaffiliated", occur at low elevations along both the Pacific and Gulf coasts.

Before proceeding to discuss the evolution of racial diversity in Mexican maize it is appropiate to describe the phylogenetic patterns evident from the study of their comparative morphology.

Mexican Maize Phylogeny

Each of the five alliances is believed to comprise a monophyletic group because, 1) all races in each alliance tend to share the same derived character states, 2) all are morphologically quite similar, and 3) all races in each alliance are found in close geographic proximity to one another and in similar ecological habitats. Morphological characters suitable for constructing a phylogeny of all 30 Mexican races are too few to resolve a tree with so many branches. Hence, the racial alliances described above are believed to be a more appropiate unit of analysis for reconstructing the path of evolutionary descent of maize from teosinte.

Phylogenetic inquiry must rely upon the relatively few morphological attributes that describe the transformation of the teosinte tassel into the maize ear. Such characters are those which would also be unaffected by human selection either because they are invisible to the cultivator, or because they can be overcome by selection for some other attribute. Certainly quantitative changes such as the dramatic increase in ear size describe one of many general trends in maize race evolution (*cf.* Mangelsdorf *et al.* 1967). However, size alone cannot tell the entire story because it is also affected by certain environmental factors such as those related to elevation (*e. g.* temperature, Benz unpub.) and thus does not provide a clear picture of which alliances are primitive and which are specialized. Rather, recounting maize's descent from teosinte focuses upon attributes that both differ between alliances and also describe how the alliances differ from the teosinte tassel raceme. The similarities and differences between maize and teosinte mentioned above provide one such suite of characters.

The method of choice in reconstructing the phylogeny of Mexican maize employs one of the many formal cladistic methods (see Wiley 1981). The associated algorithms, such as Wagner Parsimony or Camin-Sokal Parsimony, are programatic means of resolving phylogenetic trees from a data matrix comprised of characters believed to represent the changes experienced by the members of the lineage in the course of their evolution. Thus changes in a character, either by loss or gain, are used to "map" the phylogenetic paths over which the lineage has diversified (see Felsenstein 1983, 1985 and references therein).

Morphological changes describe the origin of the maize ear from a teosinte tassel, and distinguish racial alliances include: 1) the change from exclusively distichous inflorescences to predominantly polystichous inflorescences, 2) the loss of dissimilar spikelet pedicel lengths, 3) the loss of the sulcus (a narrow furrow separating the ranks of the inflorescence), 4) the loss of soft, flexible glumes, 5) the loss of male flower parts and the gain of female parts, and 6) the loss of a flexible rachis. These character state changes were then coded for each alliance, the teosinte tassel as the assumed ancestor is used as the outgroup to root the tree, and a tree was constructed. The method of establishing the branching pattern, namely Wagner Parsimony, constructed a tree with the fewest number of character state changes (see Fig. 1; Table 2; Felsenstein 1983).

The tree depicted in Figure 1 indicates that races belonging to the Balsas-West Mexico Alliance exhibit the fewest character state changes when compared to the teosinte tassel raceme; in other words, races belonging to this alliance retain the largest suite of presumably primitive character states. The primitive character states retained are: 1) ears are occasionally distichous, 2) pedicels of distinctly different lengths subtend the spikelet pair, 3) ears possess well-developed sulci and 4) soft glumes, 5) vestigial stamens are present in the fertile floret, and 6) all possess a more or less flexible rachis. At the other end of the tree, races belonging to the Central Highlands Alliance exhibit none of the primitive character states and because all character states are of the derived condition, they are together believed to be the most recent and highly specialized in Mexico.

From this it should be immediately evident that the most specialized races are those occurring at high elevations (*i. e.* above 1800 meters). Less specialized races, though certainly not the most primitive, are those found at lower elevations along the coast (Isthmian Alliance), and those found at intermediate elevations along the Gulf Coastal Plain (Mixe Alliance). The most primitive races, those belonging to the Balsas West Mexico Alliance, are found at intermediate elevations in the Balsas River drainage and in Jalisco, and along the coast of Nayarit.

Where, How and When (?) Maize Races Diversified

Factors responsible for the evolution of Mexican maize races include such well-known processes as dispersal and geographic isolation, human and natural selection, and mutation and hybridization (see Hernández and Alanís 1970). At one level, mutation, dispersal, selection, isolation, and extinction are processes that all operate to give rise to distinct organisms. At the same time, these same processes operate in a slightly different way in the evolution of maize because the sole agent responsible for maize's survival, reproduction, dispersal, geographic isolation, and extinction, is man. Because humans figure so prominently in the continued existence of maize, we should expect that prehistoric societies are ultimately responsible for the existing racial diversity and that taxonomic relationships among races and even alliances correspond in one way or another to patterns of specific cultural evolution (*sensu* Sahlins 1960). Significantly, the same processes responsible for the racial divergence of Mexican maize are responsible for the specific evolution of cultural groups though they are recognized in the different disciplines by different terms. Just as isolation and extinction are processes that affect cultures in the same ways they affect biological populations, hybridization and introgression between biological organisms are processes analogous to what cultural anthropologists see happening as a result of culture contact, and mutation analogous to innovation or invention. Because both maize and certain ethnic groups are believed to have evolved *in situ* in central Mexico, we should not be surprised, indeed we should expect to find corresponding patterns of relatedness among maize races and the ethnic groups who have fostered their survival and facilitated their diversification from some remote common ancestor. This model is adapted from an analogous one termed the "Genetic Model" by cultural anthropologists and linguists. The Genetic Model has been specifically adopted and successfully applied to such Mesoamerican groups as the Maya (Vogt and Ruz 1968) and the Mixtec-Zapotec (Marcus 1983). Thus its adoption here, in a somewhat modified form, should be acceptable to most anthropologists and ethnobotanists.

One measure of relatedness among societies believed to have evolved from a common ancestor is determined on the basis of linguistic similarity. Numerous methods are employed by linguists to determine how closely two languages/language groups are related (Swadesh 1967; Longacre 1967; Rensch 1973; Kaufman 1974). Because it is beyond the scope of this paper to discuss the specifics of each method, it can be simply noted that relationships among Mesoamerican languages are very well-studied. Hence comparison of racial taxonomy and phylogeny of Mexican maize with relations between Mesoamerican languages should enhance our understanding of how and where maize races evolved because while the two methods of estimating "group" relatedness are more or less independent they both attempt to describe the phylogenetic basis of diversification.

Perhaps the simplest example of this approach is demonstrated by the relationships between maize races presently found in the Mexican Meseta Central. Palomero Toluqueño, Arrocillo, Cónico, Chalqueño, and Cacahuacintle are all closely related one to another and without doubt comprise a monophyletic group (see also McClintock *et al.* 1981; Doebley *et al.* 1985). However, these races also exhibit considerable morphological diversity, especially Cónico, Chalqueño and Cacahuacintle, perhaps attesting to their having been exposed to a wide variety of selection regimes. It is widely recognized that the Valleys of Mexico and Puebla were occupied at various times by numerous ethnic groups representing perhaps the whole of Mesoamerica (see Carrasco 1972) over the course of its ten or more millennia of human occupation. Hence the morphological diversity evident in the races found in this region is probably due in part to its multiethnic cultural history.

Evidence in the form of archæological maize from Terminal Formative deposits at Teotihuacan provide corroborative evidence for the demonstrably specialized character and suggests a relatively recent origin of the races in the Central Highlands Alliance (McClung 1977; Benz *et al.* n. d.). While it is suspected that this alliance might be present at some earlier period in the Valley of Mexico, it has not yet been determined which race(s) in this alliance evolved first.

A second example comes from the races that together constitute the Isthmian Alliance. In this alliance, all but three of the races are cultivated by peoples who speak Zapotecan or Mixtecan languages. More specifically, the race Bolita, cultivated exclusively by Zapotec speakers in the Valley of Oaxaca, is most closely related to Chatino Maizón, a race grown by the Chatino of southern Oaxaca. The Chatino language is very closely related to Zapotecan [in fact Swadesh (1967) separates the two on the basis of lexiscostatistics by only 20 minimum centuries]. Further, Bolita is also very similar to Zapalote, the latter race grown by the Zapotec and Huave who both reside along the southern coast of the Isthmus of Tehuantepec. Rensch (1973) notes that the Zapotec and the Huave have been in contact for a considerable period of time. Similarly, Conejo, a race cultivated by the Amuzgo of Guerrero and Oaxaca, is closely related to Chatino Maizón and Zapalote. Conejo is more than likely also grown by the Mixtecs along the Pacific coast to the southeast. Though the Amuzgo language is quite distinct from both Mixtec and Zapotec, it shares a large number of innovations (shared-derived glosses) with a number of these languages suggesting the Amuzgo have maintained contact with a variety of cultural groups such as Otopamean, Chinantec and Popolocan. Whether Conejo's similarity to Nal Tel, Chatino Maizón and Zapalote is due to cultural contact or because they are cultivated in similar environments is, as yet, still unclear.

The early origin of Nal Tel might indeed partially explain the overall similarity among the low elevation races in southern and eastern Mexico, *i. e.* Isthmian Alliance members. If the presence of "Nal Tel" in the Tehuacán assemblage (Mangelsdorf *et al.* 1967) is any indication of the antiquity of this alliance, there is good reason to expect that the antecedent of this racial alliance experienced widespread adaptive radiation early, perhaps by the fourth millennium BC. Only during the subsequent Late Formative Period and later cultural regionalization so evident in the archæological record did it then give rise to local races such as Zapalote, Nal Tel and Conejo.

A third example comes from the great morphological similarity of two races now recognized as members of the Balsas-West Mexico Alliance, namely, Olotillo and Tabloncillo. Though cultivation in similar habitats was offered by Wellhausen *et al.* (1952:40) to explain their similarity, a more parsimonious explanation (Benz 1986) attributes it to relatively recent dispersal and minor divergence following isolation. Olotillo is today the predominant race cultivated by the traditional farmers in the villages with Chiapanec place names located in the Central Depression of Chiapas (see Figure 2; Navarrete 1966). Wellhausen *et al.* (1952) reported collecting Olotillo from the vicinity of Coyuca de Catalán in Guerrero. Tabloncillo is presently cultivated in the mid-elevation plains of Jalisco in the vicinity of Volcán Colima, with occasional collections coming from the vicinity of Teloloapan, Guerrero. Carrasco (1972) hypothesized that the Chiapanec, who at the time of the conquest were residing in the Grijalva River Basin above Chiapa de Corzo, had migrated there from somewhere near the Valley of Mexico at some undetermined date during the Classic Period. Morphological similarity considered together with the present-day ranges of these two races argue strongly that the Chiapanec migrated to their conquest period location from somewhere in the Balsas River Valley, presumably from its middle to lower reaches, perhaps in the vicinity of Coyuca de Catalán. Though the exact date of the hypothesized migration is still conjecture, the identification of Olotillo-like maize from the archæological record in this region of Chiapas should provide the temporal as well as the cultural context of their move.

Unfortunately we have only indirect evidence for the prehistoric existence of the Balsas-West Mexico Alliance. Understanding the evidence requires a short digression on our recent reexamination of the earliest archæological maize from San Marcos Cave in the Valley of Tehuacán (Benz and Iltis n. d.). We have compared twenty-two of the earliest specimens (from Zones F and E; Coxcatlán Phase) from San Marcos Cave to 30 extant Mexican maize races and a single race of primitive popcorn from Argentina. Results of the comparison indicate that the earliest archæological maize is essentially identical to the primitive popcorn from Argentina. While demonstrating that the most ancient archæological maize from Mexico is fully domesticated maize, the results also indicate that the closest and still extant Mexican relatives of this ancient maize races found today in western Mexico, viz., Reventador and Chapalote. Reventador is most closely related to races assigned to the primitive Balsas-West Mexico Alliance (Table 1). Although the distant relationship between the Tehuacán maize and Reventador does not unequivocally demonstrate that maize evolved in the Balsas River Valley, it does suggest considerable antiquity for the Balsas-West Mexico Alliance and supports cladistic hypotheses presented above that the phylogenetically most primitive and ancient races exist today in western Mexico; hence that maize evolved somewhere in the Balsas River drainage or plains of western Mexico. Furthermore, this corroborates the findings of Doebley *et al.* (1984; see also Iltis and Doebley 1984) who, based upon the analysis of isoenzyme variation of selected maize races and all taxa of teosinte, designated the Balsas River Valley as the region where maize originated.

The Most Ancient Maize

That the most ancient maize from Tehuacán was fully domesticated cannot now be denied (*cf.* Benz 1986; Mangelsdorf 1986). For that reason we can only speculate on the stages leading up to the fully developed polystichous ear with its well-developed cupules. The earliest archæological maize from Tehuacan clearly indicates that the precursor of the maize ear was a distichous raceme in which two spikelets associated with each segment of the rachis were borne upon pedicels of distinctly different lengths. The cupules of these ancient specimens are thin, quite broad, very shallow, and in some cases clearly reminiscent of the cupule present in the rachid of the feminized teosinte tassel discussed earlier. The glumes of these early specimens are quite soft, and though they undoubtedly partially covered the grain, they were so not because of some hoped for expression of the tunicate locus (*cf.* Galinat 1983) but because they are slightly modified tassel glumes that covered a seed approximately the size of the relatively tiny seeds (caryopses) of Chalco teosinte (see Benz and Iltis n. d.).

Implied in this discussion of the Tehuacán maize is the belief that these representatives of fully domesticated maize evolved at the hand of man, from the discovery, gathering and eventually cultivation of the seeds obtained from the initially fragile and finally rigid feminized tassel of teosinte. But the evidence for human use of teosinte prior to the first appearance of maize is incredibly sparse. With the exception of a few pollen grains from a cave in Oaxaca (Schoenwetter 1974), and a few fruitcases and a handfull of pollen grains from the Archaic Period deposits at Zohapilco in the Valley of Mexico (Niederberger 1976, 1979), all other associations of humans and teosinte clearly postdate the initial appearance of maize at Tehuacán. This negative evidence cannot be assumed to cast poor light on the evolution of maize from teosinte because we have yet to seek such documentation from the region of Mexico where the closest wild relative of maize thrives (Iltis and Doebley 1984). The evidence to date indicates that archæological documentation of the exact nature of the relationship between maize, teosinte and man must be sought in the region where primitive maize races and Mexican annual teosinte are sympatric!

Table 1: Comparison of Two Existing Classifications of Mexican Maize Races

Wellhausen *et al.* (1951, 1952)

ANCIENT INDIGENOUS: Arrocillo Amarillo, Chapalote, Nal-Tel, Palomero Toluqueño.

PRECOLUMBIAN EXOTICS: Cacahuacintle, Harinoso de Ocho, Maize Dulce*, Olotón.

PREHISTORIC MESTIZOS: Cónico, Comiteco*, Jala, Olotillo, Pepitilla, Reventador, Tabloncillo, Tehua*, Tepecintle, Tuxpeño, Vandeño, Zapalote Chico, Zapalote Grande.

MODERN INCIPIENT: Celaya, Chalqueño, Cónico Norteño*, Bolita.

POORLY DEFINED: Conejo, Mushito, Mountain Yellow, Zamorano*, Maíz Blando de Sonora*, Onaveno*, Dulcillo*.

* Race not studied by Benz (1986).

Benz (1986)

MEXICAN NARROW EAR COMPLEX:

BALSAS-WEST MEXICO ALLIANCE: Harinoso de Ocho, Tabloncillo, Reventador, Jala, Maíz Ancho†, Pepitilla, Olotillo.

ISTHMIAN ALLIANCE: Conejo, Bolita, Zapalote, Chatino Maizón†, Nal Tel, Dzit Bacal, Chapalote.

MIXE ALLIANCE: Chiapaneco†, Mixeño†.

UNAFFILIATED: Tuxpeño, Vandeño, Tepecintle, Celaya.

MEXICAN PYRAMIDAL COMPLEX:

CENTRAL HIGHLANDS ALLIANCE: Palomero Toluqueño, Arrocillo, Cónico, Chalqueño, Cacahuacintle.

TRANS-SIERRA MADRE ALLIANCE: Serrano Tapalpa (=Mountain Yellow), Elotes Occidentales, Mixteco†, Serrano Mixe†, Olotón.

† Newly described race.

Table 2: Characters Used in Construction of Phylogenetic Tree

TAXA	CHARACTERS*					
(ALLIANCE)	1	2	3	4	5	6
MEXICAN HIGHLANDS	1	1	1	1	1	1
TRANS-SIERRA MADRE	1	1	0	1	1	1
MIXE	1	0	0	0	1	1
ISTHMIAN	1	0	0	0	1	0
BALSAS-WEST MEXICAN	0	0	0	0	0	0

* 0 = primitive character state, 1 = derived character state.

Characters:
1. Number of ranks or ear: few ranks —0, many ranks —1.
2. Difference in pedicel lengths: > 0.25 mm —0, < 0.25 mm —1.
3. Sulcus: present —0, absent —1.
4. Glumes: soft —0, indurate —1.
5. Vestigial male parts in fertile floret: present —0, absent —1.
6. Rachis: flexible —0, stiff —1.

Acknowledgements

This research was supported by NSF grant DEB 80-22772 and by the Pioneer HiBred International Inc. Edgar Anderson Fellowship, both awarded to H. Iltis. This research was conducted under the supervision of H. Iltis. Travel support to participate in the Childe Colloquium was provided by the University of Wisconsin–Madison Graduate School. Thanks are due to K. Laitner Benz who participated in the fieldwork and read and greatly improved the clarity of the prose presented here.

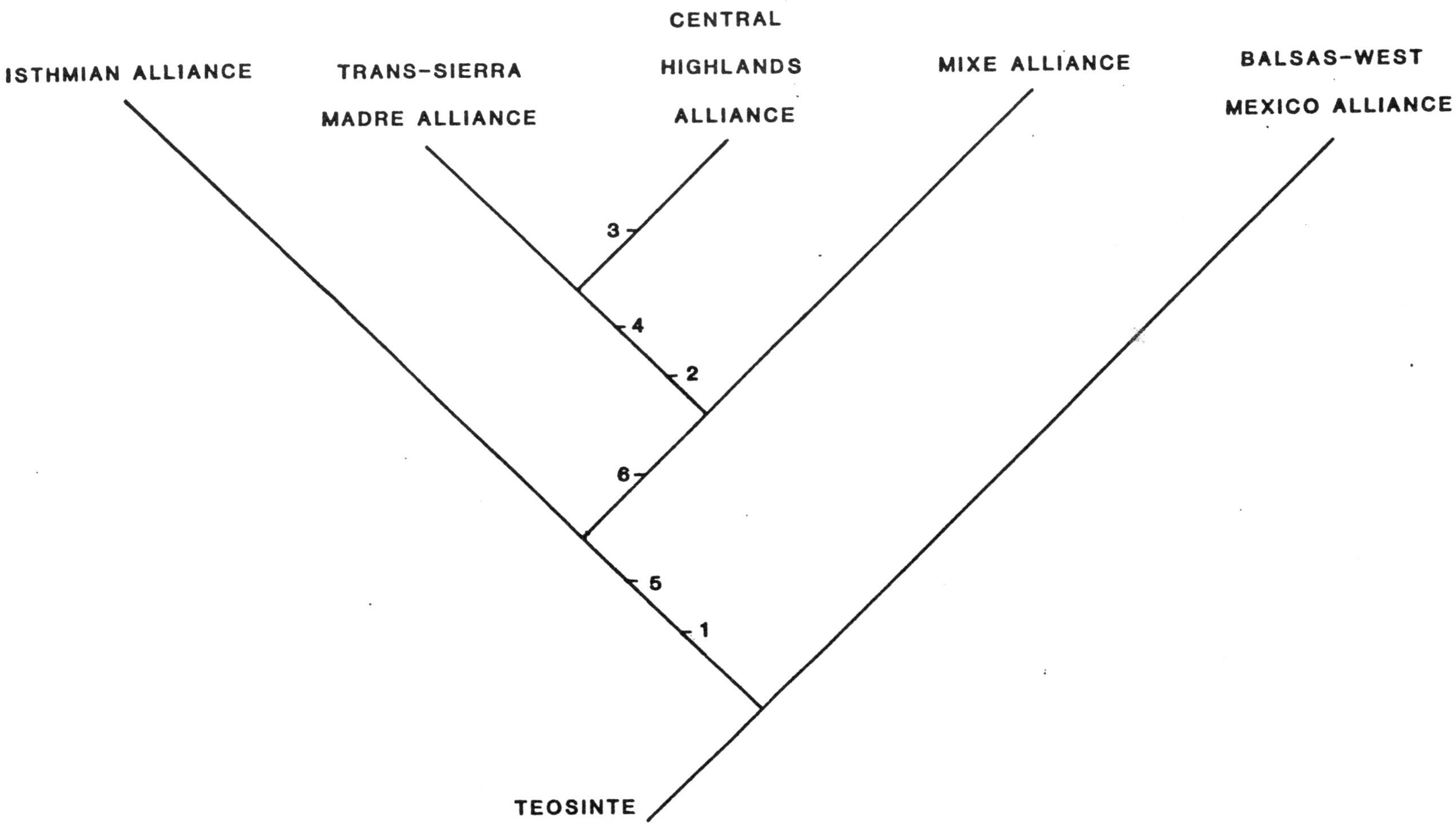
CENTRAL
HIGHLANDS
ALLIANCE
ISTHMIAN ALLIANCE
TRANS-SIERRA
MADRE ALLIANCE
MIXE ALLIANCE
BALSAS-WEST
MEXICO ALLIANCE
3
4
2
6
5
1
TEOSINTE

Figure 1.

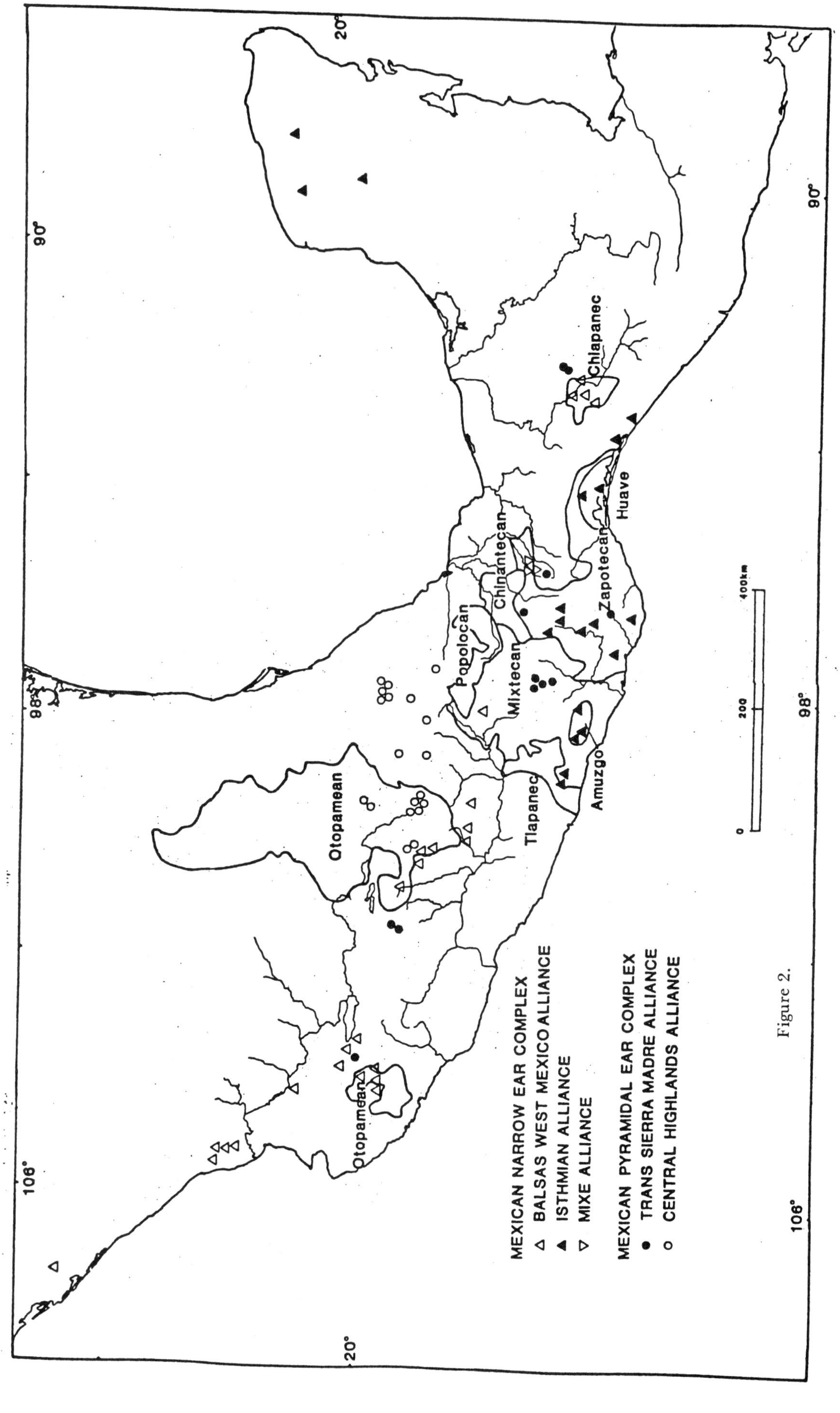
MEXICAN NARROW EAR COMPLEX
BALSAS WEST MEXICO ALLIANCE
ISTHMIAN ALLIANCE
MIXE ALLIANCE
MEXICAN PYRAMIDAL EAR COMPLEX
TRANS SIERRA MADRE ALLIANCE
CENTRAL HIGHLANDS ALLIANCE
Otopamean
Otopamean
Tlapanec
Amuzgo
Mixtecan
Popolocan
Chinantecan
Zapotecan
Huave
Chiapanec
20°
20°
90°
90°
98°
98°
106°
106°
0
200
400km

Figure 2.

REFERENCES

ANDERSON, E.

1934. "Origin of the Angiosperms", *Nature* 133: 462.

1946. "Maize in Mexico. A Preliminary Survey", *Annals of the Missouri Botanical Garden* 33: 147–247.

BENZ, B.F.

1986. Taxonomy and the Evolution of Mexican Maize. Unpublished PhD dissertation, University of Wisconsin, Madison.

BENZ, B.F. and H.H. ILTIS

n. d. a.Studies in Archæological Maize. II: The Identity of the "Wild" Maize from San Marcos Cave, Tehuacán Valley, unpublished manuscript.

BENZ, B.F., E. McCLUNG DE TAPIA, and H.H. ILTIS

n. d. b.Studies in Archæological Maize. The Origin of the Mexican Pyramidal Racial Complex. unpublished manuscript.

BONAFOUS, M.

1836. *Histoire Naturelle, Agricole et Economique*, Paris

CARRASCO P., P.

1979. *Los Otomíes. Cultura e historia prehispánicas de los pueblos de habla Otomiana*, Edición facsimilar de la de 1950. Biblioteca Enciclopédica del Estado de México, México.

COLLINS, G.N.

1919. "Structure of the Maize Ear as Indicated in *Zea-Euchlaena* Hybrids", *Journal of Agricultural Research* 17: 127–135.

DIBBLE, C.W. and A.J.O. ANDERSON

1963. *English Translation of Fray Bernardino de Sahagún's General History of the Things of New Spain,* Book 11. Monographs of the School of American Research and the Museum of New Mexico No. 14, Part XII, Santa Fe, N.M.

DOEBLEY, J.F.

1983. "The Maize and Teosinte Male Inflorescence: A Numerical Taxonomic Study", *Annals of the Missouri Botanical Garden* 70: 32–70.

DOEBLEY, J.F. and H.H. ILTIS

1980. "Taxonomy of *Zea* I: Subgeneric Classification with Key to Taxa", *American Journal of Botany* 67: 986–993.

DOEBLEY, J.F., M.M. GOODMAN, and C.W. STUBER

1984. "Isozyme Variation in *Zea* (Gramineae)", *Systematic Botany* 9: 203–218.

1985. "Isozyme Variation in the Races of Maize from Mexico", *American Journal of Botany* 72: 629–639.

FELSENSTEIN, J.

1983. "Parsimony in Systematics", *Annual Review of Ecology and Systematics* 14: 313–333.

1985. "Confidence Limits on Phylogenies: an Approach using the Bootstrap", *Evolution* 39: 783–791.

FLANNERY, K.V. (ed)

1986. *Guila Naquitz*, Academic Press, N.Y.

GALINAT, W.C.

1970. *The Cupule and its Role in the Origin and Evolution of Maize*, Massachusets Agricultural Experiment Station Bulletin 585.

1983. "The Origin of Maize as Shown by Key Morphological Traits of its Ancestor, Teosinte", *Maydica* 28: 121–138.

GOULD, S.J.

1984. "A Short Way to Corn", *Natural History* 93: 12–20.

HERNANDEZ X., E. and G. ALANIS F.

1970. "Estudio morfológico de cinco nuevas razas de maíz de la Sierra Madre Occidental de México: implicaciones filogenéticas y fitogeográficas", *Agrociencia* 5: 3–30.

ILTIS, H.H.

1971. "The Maize Mystique —A Reappraisal of the Origin of Corn", Mimeo manuscript abstract of a lecture given at the Corn Conference University of Illinois, Urbana in 1969, and Iowa State University, Ames in 1970. (Reissued as Contributions to the University of Wisconsin Herbarium No. 5, April 1985, 4 pp.).

1980. "Taxonomy of *Zea* (Gramineae). II: Subspecific Categories in the *Zea mays* Complex and a Generic Synopsis", *American Journal of Botany* 67: 994–1004.

1983. "From Teosinte to Maize: The Catastrophic Sexual Transmutation", *Science* 222: 886–894.

n.d. The CSTT and its Critics. unpublished manuscript.

ILTIS, H.H. and J.F. DOEBLEY

1984. "*Zea* —a Biosystematical Odyssey", in W. Grant (ed), *Plant Biosystematics*, Academic Press, Canada.

KAUFMAN, T.

1974. "Meso-American Indian Languages", *Encyclopaedia Britannica* 11: 956–963.

KULESHOV, N.

1929. "The Geographical Distribution of the Varietal Diversity of Maize in the World", *Bulletin of Applied Genetics and Plant Breeding* 20: 506–510.

LONGACRE, R.

1967. "Systemic Comparison and Reconstruction", in R. Wauchope and N. McQuown (eds): *Handbook of Middle American Indians*, vol. 5. University of Texas Press, Austin.

MANGELSDORF, P.C.

1974. *Corn, its Origin, Evolution and Improvement*, Belknap Press, Harvard University, Cambridge.

1986. "The Origin of Maize", *Scientific American* August.

MANGELSDORF, P.C. and R.G. REEVES

1939. *The Origin of Indian Corn and its Relatives*, Texas Agricultural Experiment Station Bulletin no. 574.

MANGELSDORF, P.C., R.S. MACNEISH, and W.C. GALINAT

1967. "Prehistoric Wild and Cultivated Maize", in D. Byers (ed): *The Prehistory of the Tehuacan Valley,* vol. 1, University of Texas Press, Austin.

MARCUS, J.

1983. "The Genetic Model and the Linguistic Divergence of the Otomangueans", in K.V. Flannery and J. Marcus (eds): *The Cloud People*, Academic Press, New York.

MAYR, E.

1982. *The Growth of Biological Thought*, The Belknap Press, Cambridge.

McCLINTOCK, B., T.A. KATO Y. and A. BLUMENSHEIN

1981. *Chromosome Constitution of Races of Maize*, Colegio de Postgraduados, Chapingo.

McCLUNG DE TAPIA, E.

1977. "Recientes estudios paleoetnobotánicos en Teotihuacan", *Anales de Antropología* vol. 14: 49–61.

MIRANDA, C.S.

1966. Discusión sobre el origen y la evolución del maíz, Memorias del Segundo Congreso Nacional de Fitogenética, Sociedad Mexicana de Fitogenética, A.C., Monterrey.

NAVARRETE, C.

1966. *The Chiapanec History and Culture*, Papers of the New World Archæological Foundation 21, Salt Lake City, Utah.

NIEDERBERGER, C.

1976. *Zohapilco. Cinco milenios de ocupación humana en un sitio lacustre de la cuenca de México*, INAH Colección Científica No. 30.

1979. "Early Sedentary Economy in the Basin of Mexico", *Science* 203: 131–142.

RENSCH, C.

1973. "Otomanguean Isoglosses", in T.A. Sebeok (ed): *Current Trends in Linguistics* vol. 11, Mouton, The Hague, Paris.

SAHLINS, M.D.

1960. "Evolution: Specific and General", in Sahlins, M. and E. Service (eds): *Evolution and Culture*, University of Michigan Press, Ann Arbor.

SCHOENWETTER, J.

1974. "Pollen Records of Guila Naquitz Cave", *American Antiquity* 39: 292–303.

SWADESH, M.

1967. "Lexicostatistic Classification", in R. Wauchope and N. McQuown (eds): *Handbook of Middle American Indians* vol. 5, University of Texas Press, Austin.

VOGT, E.Z. and A. RUZ L. (eds)

1964. *Desarrollo cultural de los Mayas*, Universidad Nacional Autónoma de México, Seminario de Cultura Maya, México.

WELLHAUSEN, E.J., L.M. ROBERTS, and E. HERNANDEZ X. in collaboration with P.C. MANGELSDORF

1952. *Races of Maize in Mexico*, The Bussey Institution of Harvard University, Cambridge.

WILEY, E.O.

1981. *Phylogentics: The Theory and Practice of Phylogenetic Systematics*, Wiley, N.Y.

THE DOMESTICATION OF *PHASEOLUS*: A SUPPLEMENTARY CROP IN PREHISTORY

LAWRENCE KAPLAN and LUCILLE N. KAPLAN
University of Massachusetts

Introduction

In 1936 V. Gordon Childe spoke of "the first revolution that transformed the human economy and gave man control over his own food supply", the Neolithic Revolution. He emphasized the major crops, wheat, rice, barley, millet, maize, yams, and sweet potatoes and saw the natural potential of the wheat and barley crops as central to the western heritage: they are nutritious, storable, and yield a high return over the expenditure of seed and labor. The result he pointed out was leisure at times other than those of the peak labor seasons. Rice growers had no such respite.

As the plant remains uncovered by excavation were studied by specialists such as Helbaek, and as the work of Vavilov became known to archæologists, the attention of prehistorians was inevitably attracted to the minor crops. A body of data was accumulated on which to test major divergences in the cultural patterns that distinguished whole geographic regions. A synergism between archæology and botany was generated which was to continue for the next few decades.

Inevitably the supplementary and minor crops came to receive the attention that would prove necessary for a functional approach to understanding differences in patterns of domestication. It became possible to look at clusters of domesticates to assess the effects of coevolution in crop complexes. In the Near East the domestication and early history of certain food crop legumes were reconstructed on a model of coevolution with the major grain crop. Using the Near Eastern model we would look, in the New World, for coevolution with maize in order to discern the early history of *Phaseolus* domestication. Because of the close association of these crops in the diet and agriculture of indigenous America it has been tacitly assumed that maize and beans were jointly domesticated. Such an assumption is probably not justified. In a large measure it was Carl O. Sauer (1969) in a series of 1952 lectures who brought the attention of botanists to the distinct differences between the processes of agricultural beginnings in the Old and New Worlds. Among the objectives of this paper is an examination of the evidence for joint or independent domestication of maize and beans.

Prehistorians should know that botanical science now defines the genus *Phaseolus* (previously regarded as including 150 species of worldwide distribution) in such a way that only New World species remain in the genus (see Delgado, this volume). Old World species such as the mung, adsuki, and rice beans now are classified as part of the genus *Vigna* along with the cow peas. Some confusion has thus been eliminated, or should have been.

Seeds constitute the principle source of evidence for the presence of beans in prehistoric sites. These may be well-preserved when dessicated or carbonized. In the uncarbonized state the seeds and other plant parts decompose rapidly under humid conditions. Pods are found in great abundance in some sites but vegetative parts, stems or vines, leaves, and roots are only rarely encountered. These pods and vegetative tissues decay rapidly under humid conditions and are of such a fragile structure when carbonized that they are never reported or recognized in that condition.

The pollen of legumes, perhaps that of *Phaseolus*, has been reported from sediments in the United States Southwest and in coprolites but the identification of this type of pollen may be uncertain. In any event, because beans are insect pollinated we do not expect to find their pollen as an especially useful source of prehistoric information. Biogenic silicates (opal phytoliths) in *Phaseolus* are not known to be distinctive. On the molecular level, the occurence of legumes in the diet of humans and other animals may be evidenced by specific ranges in the stable isotopes of nitrogen to be found in their bones (DeNiro and Hastorf 1985).

The archæological recovery and study of legume seeds and pods, and to a lesser extent, vegetative remains, in the New World, has solved some botanical questions: one hundred years ago DeCandolle (1964) was uncertain as to the origin of the common bean, *Phaseolus vulgaris* L., but certain that the Lima and seiva beans, *P. lunatus* L., were American. In the late 19th century Wittmak resolved the Western-Eastern Hemispheric uncertainly over the origin of the common bean on the basis of archæological finds. In fact, Volney Jones (1952) later discovered Wittmak's identification to be incorrect: Wittmak had actually examined remains of the tepary, *P. acutifolius* Gray. var. *latifolius* Freeman. The identity of the latter, awaited recognition by Freeman (1912) almost 40 years later. *P. coccineus* L., (*ayecote* in Nahuatl; *botil* in Tzeltal; scarlet runner bean in English) was well-known botanically at the time of DeCandolle but is not mentioned by him. *P. polyanthus* Greenm., cultivated to a limited extent in the southern Mexican highlands was not yet known to scientific botany.

Within the New World studies in recent decades have further refined distributional and temporal knowledge to the extent that fairly sophisticated questions regarding the origin and evolution of *Phaseolus* and New World grain-legume agriculture can be asked. This paper will briefly point to the important differences between the process of grain-legume domestication agriculture in the Near East and in the grain and seed dependent areas of the New World. A detailed discussion of the independent origins and movements of maize and beans will follow. It is hoped that this discussion will help to lay the basis for further, effective, investigation by botanists and prehistorians into the domestication of *Phaseolus* beans in the Americas.

Agricultural Origins and Legume Domestication: Old World

Agricultural origins in the Old World are best documented in the Middle East and Southeastern Europe. In that broad geographic region preservation of organic remains in archæological contexts is enhanced by aridity, by carbonization, or by waterlogging and deposition in peat.

The evolution of agriculture in Eurasia, as evidenced by the archæological and historic record of the Middle East probably differs from that of the Americas most perceptibly in its close association with the domestication of livestock. It is certain too, that the association of small grain cereals with indigenous legumes produced a context for domestication that was strikingly different from that which obtained for New World pulses.

Zohary and Hopf (1973) review pulse domestication in the Old World (see Table 1). Their evidence, archæological, genetic, and ecological is drawn from the Middle East and Europe and they conclude that the pea (*Pisum sativum* L.) and lentil (*Lens culinaris* Medik.) are indigenous to the Middle East. The broad bean (*Vicia faba* L.), bitter vetch (*Vicia ervilia* L.) and the chick pea (*Cicer arietinum* L.) appear in the archæological record of the Middle East later than the pea and lentil and thus do not have as clear an association with the beginnings of cereal grain agriculture as do pea and lentil. These two domesticates are thought to be derived from species indigenous to the Near East and thereby are sympatric with their ancestors and both occur in the same region as the better sites for archæological recovery. The seeds of cultivated pea cannot be separated confidently from wild pea in the archæological record until size and hilum length increase over time and until the smooth seeded character is established. In the lentil, only size increase is used as a criterion for cultivation (Zohary and Hopf 1973). No mention is made of indehiscence in pods by Zohary and Hopf (1973) or by Renfrew (1973) as indicators of cultivation, no doubt because post harvest threshing and winnowing tend to eliminate these structures prior to storage.

Harvest methods employed for New World *Phaseolus* rely less on threshing and winnowing. If these practices are carried out, small bundles of bush beans, or bunches of dry pods may be beaten with sticks or hand-rubbed in or around the habitation. The greater likelihood of pods to become part of the preserved achaeological record have allowed pod morphology and dehiscence to be used along with seed size as indicators of domestication in New World *Phaseolus* (Kaplan and MacNeish 1960; Kaplan 1965).

Table 1. The Archæological Record of Legumes Domesticated in the Near East
(data from Zohary and Hopf 1973)

Period	Field peas	Lentils
Neolithic 8–7500 BC Wild wheat and barley	absent	small wild seeds present
Neolithic 7–6000 BC		
Cult. einkorn emmer, barley	seeds present testa smooth	2.5–3.0 mm dia.
Neolithic 6200–5000 BC Cereal cutlivation well-established	seeds abundant testa smooth	abundant, up to 4.2 mm

Zohary and Hopf (1973) point out that there is a general long term trend toward increase in seed size in peas and include measurements of lentils to show the same process. Unlike the seed-by-seed sorting associated with the cultivation of New World *Phaseolus*, Old World storage and seed saving offers less opportunity for selection. The long term increase in seed size recorded Zohary and Hopf (1973), if it is a real trend and not a sampling phenomenon, might be explained by harvest practices and post harvest practices in which winnowing would repeatedly eliminate smaller lighter weight seed.

In contrast with the record for Old World pulses archæological evidence for a long term increase in seed size in *Phaseolus* under cultivation is not discernable in the New World archæological record (Kaplan 1981). The close correlation in time and process of Near Eastern pulse domestication, especially that of the lentil, and that of cereal grains seems to be essential for a long term selective pressure on the genetic system responsible for size in the legumes. There is no parallel in the domestication of *Phaseolus* and maize.

Agricultural Origins and Legume Domestication: New World

Wild legumes grow as vines on hundreds of plant species including some populations of teosinte in Oaxaca, in Guatemala (H. G. Wilkes, personal communication) and no doubt elsewhere. Such observations, however, tell us little about any association between teosinte as the progenitor of maize (Galinat 1983; Iltis 1983) and wild ancestral *Phaseolus* beans. Such associations appear in no archæological context. There exists no archæological evidence that maize and *Phaseolus* beans were domesticated jointly or as part at the same process in the evolution of agriculture. If such evidence is to appear, it will have to be within the phytogeographic range of those species and populations of teosinte that have some relationship to ancestral maize, which must be somewhere in Mesoamerica-Central America. Andean domesticated *P. vulgaris* and the large-seeded *P. lunatus* group were not domesticated within the range of teosinte. Therefore that group, and without doubt, all *Phaseolus* beans, can be excluded from the Near Eastern model for harvest, post harvest winnowing, and selection for large size and indehiscent pods.

In the Valley of Oaxaca at the prehistoric site of Guila Naquitz (Kaplan 1981, 1986) a small-seeded, gathered, wild legume, probably a species of the genus *Phaseolus*, is present in the record beginning with a date of 10,000 years before present time and continues up to the time of the Spanish conquest. The pods of this gathered wild species never loose their tightly twisted character and there is no increase in seed size. In a cave in Tamaulipas in northwest Mexico (Kaplan and MacNeish 1960) tightly twisted, dehiscent pods of what is probably a wild scarlet runner bean (*ayecote*), were collected over a period of thousands of years. It is evident that the mere activity of gathering wild bean pods over a long period of time, bringing them into the cave and extracting the beans, does not in itself guarantee the development of indehiscence or an increase in seed size, i. e., domestication, in *Phaseolus*.

A possible selective mechanism for seed size increase during the gathering, preagricultural, period is discussed in the section on Andean *Phaseolus* in this paper.

Despite the absence of evidence for the coevolution of maize and beans in the earliest stages of domestication, the long and widespread association of these two crops in indigenous American agriculture is well recognized. The degree to which they are associated in the archæological record is examined in the major geographic–cultural regions of the New World where these crops appear early in the prehistoric record.

In the Americas, archæological excavation has been most productive of vegetal remains in Mesoamerica, Andean South America and the Greater Southwest (northwestern Mexico and southwestern United States). Of these three broad areas, the latter two have the more recent increments of archæological evidence and the more recent investigations into early periods before the introduction of pottery. Especially when compared with maize the occurrence of beans in prehistoric contexts can seldom be described as abundant, "sparse and fragmentary" is usually more descriptive. Even in contemporary indigenous agriculture beans are usually best understood in their cultivation, dietary role and economy in relation to the major crop, maize. The same is true in prehistoric contexts, especially because archæologists are primarily interested in subsistence systems rather than evidence relating to a particular crop. Thus, if maize agriculture can be demonstrated to have been carried on at a particular site at a particular time, that to the archæologist is most of the story. Whether or not there is evidence for beans or whether the suspected bean samples are sent to a specialist for identification depends on whether the existence of diversified agriculture is a question that needs to be answered by those directing the excavation.

Agricultural Origins and Legume Domestication: The Greater Southwest

In the United States Southwest recent revisions of the dates for the introduction of agriculture have been undertaken. In this region the impact of direct dating of those specimens that are the objects of interest, crop remains for example, rather than associated organic material has both threatened and supported the previously accepted dates. Berry (1985) reviews and criticises the orthodox view that maize was introduced into the Arizona–New Mexico–Colorado area 4000–5000 years before present time. He cites the continued use of erroneous contextual dates at Bat Cave as a primary reason for the idea that maize (and presumably beans) had over 4,000 years in which to evolve under domestication in the Greater Southwest. He speaks instead of a later introduction, 500–750 BC, by farmers moving north from Mesoamerica with the consequent triggering of a high rate of human cultural activity. Common beans, *P. vulgaris*, first appear in the Southwest in well dated contexts at Tularosa Cave by 300 BC (Kaplan 1956) along with maize and *Cucurbita pepo*. Accepting Berry's evidence for a late introduction of maize would support the simultaneous entry of maize and common beans into the Southwest.

Minnis (1985) sees Jemez Cave in New Mexico as presenting the earliest (700 BC) well documented record of maize in the Southwest and accepts the date of about 300 BC for the earliest common beans. He makes a case however for the introduction of maize by the second millennium BC noting that archæologically detectable remains might not be produced until long after the introduction of the crop as a supplement to hunting and gathering. Simmons (1986) has obtained evidence in support of Minnis position. Direct radiocarbon dates were obtained from maize kernels, 1000 BC and *Cucurbita pepo* seeds, 950 BC and 270 BC, taken from Archaic period sites in the San Juan Basin which forms part of the northern sector of the United States Southwest. Artifacts that are associated with maize agriculture do not occur in conjunction with these finds and do not begin to appear until the close of the Archaic period. Beans were not detected in these San Juan Basin sites.

These recent interpretations of the plant remains differ with respect to the dates for the beginnings of agriculture in the U. S. Southwest but neither offers clear stratigraphic evidence in support of the simultaneous entry of maize, beans, and squash into that region. The archeometric evidence supports the Minnis position for the entry of common beans into the Southwest several hundred to over one thousand years later than maize.

Inasmuch as common beans begin to appear in the archæological record of the Tehuacan Valley in central Mexico 6,000–7,000 years ago (Kaplan 1967, 1981), it is clear that beans were available in areas contiguous to the U. S. Southwest long prior to their adoption there. One must conclude that the availability of domesticated beans, and similarly other crops, does not guarantee their adoption until other conditions such as population pressure, sedentism, dependence on trade and the like made that adoption favorable for the human population in question.

Agricultural Origins and Legume Domestication: Andean South America

Wild growing populations of both *P. vulgaris* and *P. lunatus* are sporadically distributed north to south from west central Mexico through Central America and into South America along the eastern Andean slopes into northwestern Argentina (Berglund–Brucher and Brucher 1976). These ranges have been mapped (Kaplan 1965, 1981) in relation to the distribution of archæological sites from which *Phaseolus* remains have been recovered. The mapped distribution shows that in Mesoamerica major sites with well preserved *Phaseolus* remains are located within the range of wild growing relatives of the common bean. In contrast, Andean highland and coastal sites from which *Phaseolus* remains have been recovered are well outside of the range of wild growing populations.

The genus *Phaseolus* in Andean prehistory is represented by two species *P. vulgaris* L., the common bean or *poroto*, and *P. lunatus* L., the cultivar group "Big Lima" or *pallares*. The scarlet runner bean *P. coccineus* L. of the Mesoamerican highlands has not been reliably identified among prehistoric South American plant remains. The tepary, *P. acutifolius* A. Gray is not currently cultivated south of Guatemala and has never been reported from prehistoric sites south of the Tehuacan Valley.

The typical *pallar* of Andean South America has a broad pod and a correspondingly wide seed in comparison with the narrower pod and narrower seed of the Mesoamerican small lima or seiva bean. Other botanical differences exist but those to be encountered among archæological remains are seed and pod characters because these parts constitute almost all of the archæological sample. On the basis of these characters alone it appears that the Mesoamerican and South American *P. lunatus* groups were domesticated independently from distinct gene pools. Another possibility should not be overlooked: that centers of domestication did not exist independently in these two regions but that differentiation resulting from natural and aritifical selection took place in Mesoamerica following their introduction. Studies of the distribution of *P. lunatus* phaseolin types (noted below for the common bean) may enlighten the matter.

Morphological differences between Mesoamerican and South American *P. vulgaris* are not obvious. At the present time the archæological evidence suggests that common bean domestication is older in Andean America that in Mesoamerica (Kaplan 1981). It may be that *P. vulgaris* was independently domesticated in these two regions or that the cultivars and certain wild growing relatives of Mesoamerica are derived from South America (Kaplan 1981). Recent electrophoretic studies (Gepts and Bliss 1985) of the bean storage protein, phaseolin, support the hypothesis of the separate origins of Andean and Mesoamerican common beans.

The distribution of archæological sites with well preserved plant remains in the Andes does not coincide with the distribution of wild *P. vulgaris* (Kaplan 1981) or with wild relatives of *P. lunatus* which are to be found east of the Andean highlands. The phytogeography of the genus *Phaseolus* rules out any likelihood of Andean cultivars being derived from the western slopes and points strongly to eastern slope derivation. Major vegetation change in human prehistory and in historic times has certainly taken place on the western Andean slopes. It is nor likely, however, that the ranges of the common bean and lima wild relatives ever extended to the western Andean slopes or Pacific coast.

Andean South America: The Highlands

Open sites in the high Andes are exposed to humid conditions and any plant remains present are likely to have been carbonized, some caves or rocks shelters are sufficiently sheltered and have produced well-preserved non-carbonized remains. On the desert coast of Peru and Chile archæological sites often yield large quantities of very well preserved remains in the form of stored crops, discarded trash, tapestries, human and animal coprolites, masticated quids, and pollen.

P. vulgaris at Guitarrero Cave, Department of Ancash, in the north central Peruvian Andes (Kaplan *et al.* 1973; Kaplan 1980) is represented by four fully domesticated races. Three of these are also present at Pichasca 4700 ± 80 BP (Kaplan 1971) a highland Andean site in northern Chile. Two of these types at Guitarrero Cave come from a highly reliable context (Unit 146, Complex IIe) dated at 7680 ± 280 BP, or 5730 BC. *P. lunatus* at this site may be about the same age or somewhat older. These remains are identical with those reported earlier by Kaplan, Lynch and Smith (1973). Dates, obtained by accelerator mass spectrometry, of approximately 10,000 years before present time for the oldest levels of Guitarrero Cave containing beans and pods have been reported by Lynch (1985) confirm the early dates for the presence of common beans and limas.

Beans at the early levels of this cave are accompained by a variety of plant species including some which must have contributed to the diet of those humans who camped there. The food species identified by Smith (1980) include tuberous remains of the Basellaceae (possibly *ullucu*), *Oxalis* spp. (*oca*), *Cucurbita* spp. squash remains and *Capsicum* (cf. *chinense*) were present in early levels. Maize, and possibly peanut and manioc, are identified from later levels. Of the manioc, Smith (1980) notes that it may have been traded from an elevation lower than that of the Callejón de Huaylas where the Cave is located. *P. vulgaris, P. lunatus*, and *Capsicum* sp. (cf. *chinense*) remains at this site are all domesticated and have ancestral types east of the Andes. Thus, by 7,000–10,000 years ago common and lima beans had already undergone domesticative changes east of the Andes in the Lithic Period during which hunting and gathering were the primary means of subsistence and millennia before the introduction of ceramics.

The Andean Highlands: Implications of Early Dates, and Selection for Seed Size Increase in the Lithic Period

The early, but confirmed, date for domesticated common and lima beans in the Andean Highlands has perplexed some archæologists. Geneticists (in conversation) have suggested that human selection over a long period of time in the Lithic period might accomplish the increase in seed size over the wild type. The senior author has long been skeptical over claims for the effectiveness of human selection for additive traits in polygenic system where gene effects are small and especially where self-pollination predominates. If we assume that a phenotypic condition, increase in seed size over the wild type, is the result of conscious human selection, then, we must assume that a human lineage has decided upon a goal in the distant future and has unswervingly pursued that goal. There is no empirical basis for such an assumption.

More likely than long term, purposeful directional selection for large seed size by nonsedentary, nonagricultural humans in the Lithic Period, is a selection mechanism which might operate without human awareness. Bogue and Kaplan (1984) proposed the following mechanism. Studies of seed size in tropical legumes by Janzen (1969) indicated that seed weevil (bruchid) predation might exercise a size depressing influence on the seeds of wild bean populations if the insects selected larger seeds for oviposition. The eggs laid on the seed surface produce larvae that burrow into the seed, pupate, and later emerge as adults. This predatory behavior generally results in the destruction of the seed so that if larger seeds are selected for oviposition and are thereby more likely to be destroyed than are smaller seeds, the result for the population is selection against large seed size. There are, at the same time, advantages such as the early vigorous growth of the plant can be supported by large seed size. It is possible that some mechanism unknown to us may have been used to protect the seed against bruchid weevils. For example, approximately 3 hours at a temperature of 60°C. destroys all phases of the bruchid like cycle without damaging seed viability (Metcalfe and Flint 1962: 936–938). Such exposure might occur if beans were stored near a cave hearth. Some of those seeds would have been inadvertantly returned to the soil and later germinated and provided part of the next harvest. The repetition of this process would have established a population of wild beans protected from the size depressing effects of weevil predation and exposed to the forces of natural selection favoring large seed size. It is very unlikely that the archæological record will ever provide clear evidence for this or any other selective mechanism for seed size increase. An experimental test of the feasibility of a seed–insect–human relationship such as this was undertaken.

Tests with two bruchid species native to tropical America showed that larger bean seeds were more likely to be destroyed than were smaller seeds (Bogue and Kaplan 1984). The oviposition behavior of *Zabrotes subfascilatus* Boh. showed a strong preference for cultivars with large seed size; for land races over related, smaller-seeded wild types; for the larger seeds within a cultivar, land race, or wild population, *Acanthoscelides obtectus* Say. eggs, unlike those of *Z. subfasciatus*, are not laid directly on the surface of the seeds but are deposited among them. The oviposition behavior of *Acanthoscelides obtectus* shows no correlation with seed size. The larvæ, however, which are mobile among the stored seeds show an absolute preference for the cultivars when presented with a choice.

The Andean Region: The Pacific Coast

Bonavia (1982: 317–322) has reviewed in detail the reports of several authors on evidence for the appearance of *P. lunatus* on the South Coast of Peru. On this section of the coast he disputes or casts doubt on frequently cited dates, especially from Chilca, which would place *P. lunatus* in the preceramic period as early as 5,000–6,000 years before present time. No doubt some reports of preceramic *P. lunatus* in coastal sites are spurious. However, at Huarmey a site of the "Final Preceramic Period" on the North-Central Coast of Peru *P. lunatus* pod valves are present in the late preceramic period (Kaplan 1982: 181).

In the Chicama valley on the North Coast of Peru, it is possible to confirm the presence of securely dated pallares, seeds and pods, identified by the author, in the preceramic midden of Huaca Prieta (Bird and Hyslop 1985: 233–234). Samples for radiocarbon dating from Huaca Prieta were processed in the original ^{14}C laboratory at the University of Chicago in 1949, 1950, 1951 marking the beginning of the application of this dating method to South American prehistoric materials. Because of the significance of this site, its radiometry has been restudied repeatedly. The most recent dating indicates the occurrence of the earliest lima bean remains between 3080 BC and 2890 BC (Bird and Hyslop 1985: 53). Maize is not found among the 20,000 specimens of preceramic plant remains at Huaca Prieta, but some other domesticates in addition to *P. lunatus* are present: *Canavalia plagiosperma* by 4257 ± 250 years before present, *Capsicum* spp., *Gossypium barbadense, Cucurbita* spp. Of this group of cultigens, lima beans and peppers have origins east of the Andes, *Canavalia* probably entered the Peruvian Coast from coastal Ecuador and the sources of the cotton and squashes are not known.

The preceramic period at Huaca Prieta ends at about 1200 BC or earlier (Bird and Hyslop 1985: 53). Sometime after the introduction of ceramics to the settlement near the midden, common beans appear in the record along with the earlier limas and canavalias both of which continue into later periods.

The Andean Region: Relationships with the Intermediate Zone

Bird (1984 :49) estimates that maize entered highland Peru from the Intermediate Area, a cultural-geographic continuum from Chiapas, Mexico southward into northwestern South America, sometime before 1,000 BC. Piperno (1985) finds maize phytoliths to be present in Panama by about 5,000 years before present time. Thus it is possible that crop plants from Mesoamerica entered the northern Andean region by way of the Intermediate Zone by the time that pottery had begun to appear in Ecuador. Because *P. vulgaris* does not appear in the early Valdivian Periods or sites along the Orinoco and Amazon in Venezuela (Roosevelt 1980) along with maize, it may be that the common bean did not accompany maize through the intermediate zone. There may be some evidence, however, that common beans from Mesoamerica did enter the Northern Andean region at a later prehistoric date.

Phaseolus remains from archæological sites in northwestern South America are only now beginning to become available. A collection of plant remains recovered from the northern Andean highland site of Huacaloma in the Cajamarca Valley by Onuki and others is currently under study. Two hundred fifty carbonized seeds and seed fragments of *P. vulgaris* from Huacaloma have been identified beginning with a small number (8) of seeds and fragments between 1,000 BC and 500 BC (Formative) associated with a culture in which are found evidences for early ceramics, deer hunting and other hunting and gathering activities. Most of the *P. vulgaris* seeds and fragments are present in the period at about 500 BC to 400 BC during a time of transition from deer hunting to llama domestication and other cultural changes. These carbonized remains vary considerably in their dimensions but their maxima, minima and median dimensions are

Table 2. Median Seed Dimensions of Prehistoric Phaseolus vulgaris

	Lenght	Width	Thickness
	mm	mm	mm
Andean Highland 1			
range of medians	10–13	6.7–8.3	4.3–6.9
average of medians	11	7.6	5.6
Andean Coast 2	9.3–19.1	8.0–10.0	5.3–8.0
	15.0	8.4	6.9
Mesoamerica 3	8–14	6–9.2	5–6.4
	11.5	7.2	5.2
Huacaloma (size reduced slightly by carbonization)	5.5–10.7 7.4	3.2–6.9 4.9	(omitted-mostly single cotyledons)

1 Two sites, 7 land races. 2 Five sites, 10 land races.
3 Three sites, 6 land races.

considerably smaller in size than in typical of the familiar prehistoric races from coastal Peru (see table 2).

These seeds are carbonized in such a way that they have lost their seed coats and have not been "puffed" so that they are a bit reduced in relative size when compared with the noncarbonized seeds in table 2. Nevertheless, they do not resemble morphologically any other Andean common beans that the author has examined. Could they represent an influx from the Intermediate Area, or Mesoamerica? Carbonization of this relatively large sample, unfortunately, obviates the possibility of studies on the storage protein, phaseolin, of the type noted above which might help to resolve their affinities, Andean or Mesoamerican.

The Andean Region: Highland–Coastal Agricultural Connections of Maize and Beans

Smith (1980) reviews the dates for maize in South America and finds reliable dates of 2500–3000 BC in coastal Ecuador, 4300–2800 BC (pollen) in the Ayacucho Valley (Central Highlands), and at Guitarrero Cave (Northern Highlands) somewhere between 6,000–4,000 BC. Recently obtained dates from Guitarrero Cave (Lynch 1985) indicate that maize is considerably more recent, 2,000–3,000 years old at that site. Bird (1984) finds that maize was introduced simultaneously to a number of Peruvian coastal sites after 850 BC and suggest that reports of the presence of maize (macrofossils) on the coast of Peru prior to 1500 BC are based on evidence from disturbed sites. The Huarmey pollen dates disagree and indicate an earlier presence for maize on the coast.

Weir and Bonavia (1985) at Huarmey on the North Central Coast found maize pollen in coprolites at 3,000 BC. Earlier (Grobman and Bonavia 1978) maize macrofossil remains, ^{14}C dated at $3,750 \pm 110$ years before present time had been reported at the same site.

Maize on the coast of Ecuador clearly predates maize macrofossil evidence in the Andean highlands but is not likely to have been the source of that grain for the Peruvian coast especially in view of findings by Grobman and Bonavia (1978). These authors identify maize cobs and tassels from Los Gavilanes, a preceramic coastal site in the district of Huarmey. The remains are carefully described and identified as predominately Proto-Confite Morocho which is related to the Confite Morocho race which presently occurs in the Andean Highlands. These authors believe that the early coastal race is derived from the Andean Highlands.

C. Earle Smith, Jr. is convinced that the crucial remains which will clearly place maize earlier in the Andean Highlands than on the Coast are still to be excavated in the Highlands (personal communication).

It is clear that the date of 6800–6200 BC, Guitarrero Cave, Unit 120, Complex IId, for limas and 5730 BC for common beans (Kaplan 1980) antedate any coastal or highland sites. *P. lunatus* seeds found in Early Nasca sites in the Ica Valley are identical with those found in Guitarrero Cave. By implication the entire array of coastal *P. lunatus* domesticates may be regarded as direct highland derivations from earlier sources on the eastern Andean slopes.

Further supporting the highland derivation of coastal *P. lunatus* is the absence of this species from the prehistory of coastal Ecuador. Here at Real Alto carbonized *Canavalia* seeds were found at levels bearing radiocarbon dates of 5495 ± 200 years, and 4900 ± 170 years before present time and later. Although fragmentary, these seeds were definitively identified by means of scanning electron microscopy (Damp *et al.* 1981). This region may well be the source of *Canavalia* beans for Huaca Prieta and other coastal Peruvian sites. However, with regard to *P. lunatus*, its absence from Real Alto provides useful evidence suggesting that the legume fraction of the prehistoric diet in this area was not beans of the genus *Phaseolus* and that the *P. lunatus* of coastal Peru did not derive from southwestern Ecuador.

The foregoing review of maize and bean dates and occurrences in highland and coastal Peru and Ecuador leads to certain conclusions: *Phaseolus* beans entered coastal Peru independently of maize; common beans are archæologically detectable in the coastal economy sometime in the Late Preceramic Period or after the beginning of the ceramic period; lima beans were present on the coast in the preceramic period and in some settlements antedated the use of maize. *Canavalia (plagiosperma)* predated *Phaseolus* beans on the South American Coast and probably had their earliest use in southwestern Ecuador. There is no evidence that maize and lima beans or maize and common beans traveled together as they were introduced into preceramic or early ceramic period sites. Lima beans and common beans too, appear to have been introduced or adopted into the economies of highland and coastal settlements independently of one another. The two *Phaseolus* species undoubtedly were introduced from the highlands to the coast.

Mesoamerica

Since the most recent review of common beans in the archæological record of Mesoamerica (Kaplan 1981) little new material has come to light. Attention, however, should be called to finds of maize and beans in the Mayan lowlands of Belize, Miksicek *et al.* (1981) summarize the maize finds at Cuello in levels varying from 2000 BC to AD 250. It may be assumed that *P. vulgaris* beans are present in this period. Poor preservation detracts from the probability that they will be found. At Lamanai in Belize, Pendergast (personal communication n. d.) recovered common beans (identified by the senior author) dating to about 2,000 years ago. These are small black beans, domesticated, and resemble black seeded vining beans that are today commonly cultivated in the Valley of Oaxaca. Although the dates are not significant in terms of the overall record for bean cultivation in the Americas, the record of these sites with regard to *Phaseolus* is interesting in two respects. First they constitute the first humid coastal lowland record north of South America (southwestern Ecuador), and second, *P. lunatus* is not present.

Several major stratified sites in Mesoamerica (and the U.S. Southwest) (Kaplan 1981) are similar with respect to their "bean profiles". Common beans first appear in preceramic levels in very low numbers: 4,000–6,000 years ago in northeastern Mexico at Ocampo in Tamaulipas; 5,000–7,000 years ago in the Tehuacan Valley of Central Mexico. Common beans (and maize) increase as components of the food plant remains in highland Mesoamerican sites sometime during the AD 100–700 period, approximating the time span from Protoclassic through Late Classic periods (Adams 1977: 332–333) in the Puebla–Oaxaca Valleys, (Kaplan 1981 and unpublished) in Tamaulipas (Kaplan and MacNeish 1960) and in Durango (Brooks *et al.* 1962). This is a time when there is evidence for explosive population growth and the founding of urban centers as well as pemanent villages. By this period in the arid highland valleys agriculture was carried on in fields watered by seasonal rains, and various forms of canal and well irrigation on hillside terraces and valley floors (Flannery and Marcus 1983: 323–339).

This time span of AD 100–700 in Mesoamerica may mark the beginning of common bean cultivation as a field crop with the bush types gaining in importance relative to vining types. Forces of population pressure, shortage of humid bottom lands and the concommitant growth of markets may have combined to press upon earlier forms of subsistence maize–bean–squash horticulture. In later times, conquest and the exacting of tribute which became marked in the late Classic and Post-classic times no doubt affected agricultural production by necessitating the production of short season harvest of storable, transportable, crops like maize and bush beans. Whole, dry bush bean plants are present in Valley of Oaxaca plant remains by about 1,000 years ago (unpublished) and in Tamaulipas by 800 years ago (Kaplan and MacNeish 1960). While it is doubtful that an increase in the production of bush beans can ever be documented by prehistoric bean remains themselves, at least we know that the selection of dwarf growing forms had been made and that these were available for field agriculture by 1,000 years ago. Studies of stable isotope ratios of nitrogen in human skeletal remains might resolve the question of the degree to which dietary protein was derived from legume sources rather than hunted animal species (DeNiro and Hastorf 1985).

P. lunatus in Mesoamerica appears relatively late in the archæological record, probably not before 1,300–1,400 years ago (Kaplan 1965). Both the flat seiva and the spherical so-called "carib" types are present in Mesoamerica at a single site in Durango by AD 1,300 (Brooks *et al.* 1962). There are no remains of *P. lunatus* reported as yet in the archæological literature of humid lowland Mesoamerica. The separation of Mesoamerican and Andean limas is long recognized on morphological grounds and should be examined by protein electrophoresis methods which are proving successful in the study of common beans (Gepts and Bliss 1985).

P. coccineus L., the runner bean, appears as a domesticate in the Tehuacan Valley securely dated by 2,200 years ago (Kaplan 1967) and in Durango by 1,300 years ago (Brooks *et al.* 1962). This species is not known in South American prehistory and probably did not enter the Southwest until historic times.

P. acutifolius Gray var. *latifolius* Freeman, the tepary, was cultivated by 5,000 years before present time in the Tehuacan Valley (Kaplan 1967) and by 1,000–1,200 years before present time in the Hohokam region of the Southwest (Kaplan 1956).

Summary

In his later writing, V. Gordon Childe (1958) observed that the Neolithic "revolution" must really be the result of ongoing and long term processes and should not be understood as a single event. Our present view of domestication is certainly in harmony with this position. Indeed the end result of the long term process may even mask the process itself. The present association of maize and beans in indigenous American agriculture and diet is the well adapted result of separate evolutionary processes and independent diffusion. Our understanding of prehistory in the New World and of the æ process of domestication in *Phaseolus* may benefit by greater awareness of these separate pathways.

Because of the linkage of *Phaseolus* and *Zea* in historic indigenous American agriculture the early prehistoric record of these crops is examined for evidence of an association that might reveal a possible role for coevolution in the early history of these crops. Of particular interest in this context are the period of transition in *Phaseolus* from the wild to the domestic form, especially the change in seed size, and the early introduction of these crops into new regions.

How would a coevolutionary process involving a cereal grain and food legume be recognized in the archæological record? A comparative approach is adopted in an effort to approach this question: the prehistory of Old World–Near Eastern cereals and food legumes is briefly examined. A gradual increase in legume seed size over the millennia of early agricultural (Neolithic) history appears to have taken place as part of a codomesticative process.

No consistent record of early joint occurrences is evident for *Phaseolus* and *Zea* (table 3). Furthermore, the transition from small to large seed size is not preserved in the New World archæological record. For these reason as well as the absence of *Zea* ancestors in South America where major branches of both the common and lima bean groups were domesticated, coevolution as a process in the early millennia of maize and bean evolution is rejected.

Table 3. Zea and Phaseolus in Early Prehistoric Contexts.
Sites are discussed in the text. Pv = common bean, Pl= lima bean.

Dates: thousands of years before present time				
Greater Southwest	Mesoamerica	Andean America		
		Highlands	Coast	Ecuador
	1.0(GN)	·	·	·
2.3(TulCv)	Pv	2–3.0(GT)	·	·
Zea–Pv	·	Zea	·	·
	·	·	·	·
2.7(Jemez)	·	·	·	·
Zea	·	·	·	·
	·	·	3.7(Huarmey)	
3.0(SJuan)	·	·	Zea	·
Zea	4.3–7.0	4.75(Pich)	·	4.5–5(Valdivia)
	(Tmps)Zea	Pv	·	Zea
	(Teh)Zea,Pv?	·	·	
TulCv=Tularosa Cave		5–6(Aya)	5 (Huaca Prieta)	
		Zea pollen	Pl	
Jemez=Jemez Cave		·	5 (Huarmey)	
SJuan=San Juan basin		·	Zea pollen	
GN=Guila Naquitz		8-10(GT)		
Tmps=Tamaulipas		Pv, Pl	GT=Guitarrero cave	
Teh=Tehuacan			Pich=Pichasca	
			Aya=Ayacucho	

Dates for the early appearance of *Phaseolus* are critically reviewed as they pertain to the major regions of indigenous agriculture in the Americas: the Greater Southwest; Andean America; Mesoamerica; the Intermediate Zone. In the Greater Southwest common beans are present by 2300 BP. Other *Phaseolus* species are later. In the Andean highlands secure dates of at least 8000 BP, long before the introduction of pottery, have been obtained for common and lima beans. On the Andean coast lima beans are present by 5000 years BP. Common beans are present at least by 2500 BP and probably will be shown to be earlier but there seems to be no connection between common bean and lima bean introduction. The prior appearance of these species in the Andean Highlands and their absence from early sites on the coast of Ecuador and along the Orinoco and Amazon Rivers in Venezuela confirms the view that common and lima beans reached the coast of Peru directly from the highlands. It is likely that these species were originally domesticated during the Lithic period on the eastern slopes of the Andes and were long established in Andean highland valleys before they were adopted into coastal agriculture.

In Mesoamerica, the early dates of 6–7000 BP are based on small samples, but are probably valid. The pattern of the distribution of these remains in Mesoamerican sites, however, suggests that *Phaseolus* beans may not have been an important part of the diet until the Late Pre-Classic and Classic Periods AD 100–700. At this time rapid changes in population growth, settlement patterns, land use and other areas of human activity were widespread in Mesoamerica.

This time period corresponds to the fruition of what V. Gordon Childe called the Urban Revolution: archaic and early states were moving to city–state level societies. Earlier theocratic proto-urban centers of the Formative Period came to be replaced by fully urban centers (see Gándara and Lumbreras in this volume). Increases in sedentarism, population size and density, specialization of craft production and exchange networks were accompanied by the intensification of agricultural production. It is to be expected that new strategies to increase land under cultivation and technological changes in water control should also involve changes in crop plant varieties grown under these new circumstances. Because such changes may have required the production of crops having short harvest season, in contrast with the extended season characteristic of vining beans, bush beans may have become more important at this time. Their remains have been found in Oaxaca, Mexico, by 1000 years BP.

In attempting to understand the evolution of *Phaseolus* under domestication, a variety of approaches must be used. As one approach to this problem the archæological record of plant utilization provides the perspectives of time depth and the relationship to human cultures. To be useful, however, the limits of the archæological method must be understood and the data resulting from the botanical analysis of archæological materials must be evaluated critically. We know too that the prehistoric record is far from complete: each of the species discussed in this paper in its prehistoric context had undergone major changes before our earliest records of its existence.

Note: A full treatment of the cultivated *Phaseolus* species can be found in: *Genetic Resources, Domestication, and Evolution of Phaseolus Beans*, edited by Paul A. Gepts, Martinus Nijhoff and Dr. W. Junk Publishers (in preparation). Many of the ideas discussed here were developed by the authors in the process of preparing a chapter on archæological *Phaseolus* for that book.

REFERENCES

ADAMS, R.E.W.

1977. *Prehistoric Mesoamerica*, Little, Brown and Co., Boston.

BERGLUND-BRUCHER, O., and H. BRUCHER

1976. "The South American Wild Bean *(Phaseolus aborigineus* Burk.) as Ancestor of the Common Bean", *Economic Botany* 30: 257–272.

BERRY, M.S.

1985. "The Age of Maize in the Greater Southwest: A Critical review", (Original Publication 1982, University of Utah Press), edited by Ford, Richard I., *Prehistoric Food Production in North America*, Anthropological Papers, Museum of Anthropology University of Michigan no. 75, Ann Arbor: 279–307.

BIRD, R. McK.

1984. "South American Maize in Central America?" in Doris Stone, *Pre-Columbian Plant Migration*, Papers of the Peabody Museum of Archæology and Ethnology, vol. 76, Harvard University, Cambridge: 39–65.

BIRD, J. B., and J. HYSLOP

1985. *The Preceramic Excavations at the Huaca Prieta, Chicama Valley, Peru*, vol. 62: Part I, Anthropological Papers of the American Museum of Natural History.

BOGUE, G., and L. KAPLAN

1984. "Insect Predation and the Origin of Large Seed Size in *Phaseolus vulgaris* L.", *Annual Report of the Bean Improvement Cooperative* 27: 104–105.

BONAVIA, D.

1982. *Los Gavilanes*, Editorial Ausonia, Talleres Gráficos S. A., Lima, Perú.

BROOKS, R.H., L. KAPLAN, H.C. CUTLER, and T.H. WHITAKER

1962. "Plant Material from a Cave on the Río Zape, Durango, Mexico", *American Antiquity* 27: 356–369.

CHILDE, V.G.

1936. *Man Makes Himself*, Watts and Co., London.

1956. *The Prehistory of European Society*, Penguin Books Ltd., Baltimore.

DAMP, J., D.M. PEARSALL, and L. KAPLAN

1981. "Beans for Valdivia", *Science* 212: 811–812.

DeCANDOLLE, A.

1964. *Origin of Cultivated Plants*, Second edition, originally published 1886, Reprint, Second printing, Hafner, New York.

DeNIRO, M.J., and C.H. HASTORF

1985. "Alteration of $^{15}N/^{14}N$ and $^{13}C/^{12}C$ Ratios of Plant Matter During the Initial Stages of Diagenesis: Studies Utilizing Archæological Specimens from Peru", *Geochimica et Cosmochimica Acta*, vol. 49: 97–115.

DRENNAN, R. D., and K.V. FLANNERY

1983. "The Growth of Site Hierarchies in the Valley of Oaxaca": Part II, in Flannery, K. V. and Joyce Marcus (eds), *The Cloud People, Divergent Evolution of the Zapotec and Mixtec Civilizations*, Academic Press, New York.

FLANNERY, K.V., and J. MARCUS

1983. *The Cloud People, Divergent Evolution of the Zapotec and Mixtec Civilizations*, Academic Press, New York.

FREEMAN, G.F.

1912. "Southwestern Beans and Teparies", *Ariz. Agr. Exp. Sta. Bulletin 68.*

GALINAT, W.G.

1983. "The Origin of Maize as Shown by Key Morphological Traits of its Ancestor, Teosinte", *Maydica* XXVIII: 121–138.

GEPTS, P., and F.A. BLISS

1985. "Usefulness of Phaseolin as an Evolutionary Marker", *Annual Report of the Bean Improvement Cooperative*, vol. 28: 60–61.

GORMAN, C.

1971. "The Hoabinhian and After: Subsistence Patterns in Southeast Asia during the Late Pleistocene and Early Recent Periods", *World Archæology* 2 (3): 300–320.

GROBMAN, A., and D. BONAVIA

1978. "Pre-Ceramic Maize on the North-Central Coast of Peru", *Nature* 276: 386–387.

ILTIS, H.H.

1983. "From Teosinte to Maize: The Catastrophic Sexual Transmutation", *Science* 222: 886–894.

JANZEN, D.H.

1969. "Seed-Eaters Versus Seed Size, Number, Toxicity and Dispersal", *Evolution* 23: 1–27.

JONES, V. H.

1952. "Material from the Hemmenway Archæological Expedition (1887-88) as a Factor in Establishing the American Origin of the Garden Bean", in *Indian Tribes of Aboriginal America*, vol. III, Proceedings of the 29th Congress of Americanists, edited by Sol Tax, University of Chicago Press, Chicago: 177–184.

KAPLAN, L.

1956. "The Cultivated Beans of the Prehistoric Southwest", *Annals of the Missouri Botanical Garden* 43: 189–251.

1965. "Archæology and Domestication in American *Phaseolus* (Beans)", *Economic Botany* 19 (4): 358–368.

1967. "Archæological *Phaseolus* from Tehuacan", in *The Prehistory of the Tehuacan Valley*, vol. 1 Environment and Subsistence, edited by Douglas S. Byers, University of Texas Press, Austin: 201–212.

1971. "Identificación de porotos arqueológicos de Pichasca", in M. Rivera, *Nuevos enfoques de la teoría arqueológica aplicada al Norte Chico*, Actas del VI Congreso de Arqueología Chilena: 308–9.

1980. "Variation in the Cultivated Beans, in Guitarrero Cave", Chapter 7, *Early Man in the Andes*, edited by T. F. Lynch, Academic Press, Inc., New York: 145–148.

1981. "What is the Origin of the Common Bean?", *Economic Botany* 35 (2): 240–253.

1982. "Pallar (*Phaseolus lunatus*)", in Duccio Bonavia, *Los Gavilanes*, Editorial Ausonia, Talleres Gráficos S. A. Lima, Perú: 181–182.

1986. "Preceramic *Phaseolus* from Guila Naquitz", in *Guila Naquitz: Early Foraging and Early Agriculture in Oaxaca, Mexico*, edited by Kent V. Flannery, Academic Press, New York: 281–284.

KAPLAN, L., T. F. LYNCH, and C. E. SMITH, Jr.

1973. "Early Cultivated Beans (*Phaseolus vulgaris*) from an Intermontane Peruvian Valley", *Science* 179: 76–77.

KAPLAN, L. and R. S. MACNEISH

1960. "Prehistoric Bean Remains from Caves in the Ocampo Region of Tamaulipas, Mexico", *Botanical Museum Leaflets*, Harvard University 19: 33–56.

LUMBRERAS, L. G.

1974. *The Peoples and Cultures of Ancient Peru*, Smithsonian Institution Press, Washington, translated by Betty Jane Meggers.

LYNCH, T. F.

1980. *Guitarrero Cave: Early Man in the Andes*, Academic Press, Inc., New York.

LYNCH, T. F., R. GILLESPIE, J. A. J. GOWLETT, and R. E. M. HEDGES

1985. "Chronology of Guitarrero Cave, Peru", *Science* 229: 864–867.

METCALFE, C. L., and W. P. FLINT

1962. *Destructive and Useful Insects, their Habits and Control*, 4th edition, McGraw-Hill Book Co., New York.

MINNIS, P. E.

1985. "Domesticating People and Plants in the Greater Southwest", in *Prehistoric Food Production in North America*, edited by Richard I. Ford, Anthropological Papers, Museum of Anthropology University of Michigan no. 75, Ann Arbor: 309–339.

PIPERNO, D.

1985. "Phytolith Taphonomy and Distributions in Archæological Sediments from Panama", *Journal of Archæological Science* 12 :247-267.

RENFREW, J.M.

1973. *Paleoethnobotany, the Prehistoric Food Plants of the Near East and Europe*, Columbia University Press, New York.

ROOSEVELT, A.C.

1980. *Parmana: Prehistoric Maize and Manioc Subsistence along the Amazon and Orinoco*, Academic Press, New York.

SAUER, C.O.

1969. *Agricultural Origins and Dispersals*, M.I.T. Press, Cambridge (original edition, 1952, American Geographical Society).

SIMMONS, A.H.

1986. "New Evidence for the Early Use of Cultigens in the American Southwest", *American Antiquity* 51 (1): 73–89.

SMITH, C.E. Jr.

1980. "Plant Remains from Guitarrero Cave", in *Guitarrero Cave, Early Man in the Andes*, edited by Thomas F. Lynch, Academic Press, New York: 87-119.

WEIR, G. H. and D. BONAVIA.

1985. "Coprolitos y dieta del precerámico tardío de la costa peruana", *Boletín Instituto Francés de Estudios Andinos* XIV (1–2): 85–140.

YEN, D.E.

1977. "Sunda and Sahul", in Allen, J., J. Golson, and R. Jones (eds), *Prehistoric Studies in Southwest Asia, Melanesia, and Australia*, Academic Press, New York: 567–599.

ZOHARY, D., and M. HOPF

1973. "Domestication of Pulses in the Old World", *Science* 182: 887–894.

ANOTHER APPROACH TO THE DOMESTICATION OF *PHASEOLUS*

ALFONSO DELGADO SALINAS
Universidad Nacional Autónoma de México

Verdcourt (1970) redefined *Phaseolus* to include a group of approximately fifty closely allied New World species morphologically similar to the common bean, *Phaseolus vulgaris* L. Additionally, significant contribution to the understanding of the genus were those of Maréchal *et al.* (1978) and Lackey (1983), which accepted and extended Verdcourt's work giving a more concise definition of *Phaseolus*. Recently, Delgado (1985) completed a systematic revision of the species of the genus for Mexico and Central America.

The genus *Phaseolus*, therefore, is a New World group, mostly centered in Mexico and Central America. Man has unquestionably been a factor in the distribution of at least a few species of *Phaseolus*. Wild species like *P. maculatus* (former *P. metcalfei*) show evidence of human manipulation and also intervention in its distribution. What I take to be the pre-historic range of the genus is shown in Figure 1. This distribution map was made considering species in Mexico and Central America, and *P. augustii* and *P. pachyrhyzoides* for South America. Only three species of *Phaseolus* show truly wide distributions (from Mexico to South America and often also the West Indies); these are the cultivated species *P. coccineus, P. lunatus* and *P. vulgaris*, and their wild and feral counterparts. This coincidental distribution clearly indicates human intervention during the past history of these species.

The genus *Phaseolus* has its center of diversity, and probably its center of origin in Mexico, where more than ninety per cent of the species occur. This great species diversity decreases drastically in number of South Guatemala. In South America, the few native species occur mainly in the humid Cordillera Oriental of the Andes, from southern Ecuador, Peru, Bolivia and northern Argentina.

Wild beans are a food resource that have been available for a long time and could not have been overlooked by gathering people. It would seem, from a tentative projection of the present situation backward to the time of incipient domestication, that the wild stands of beans in North America would have been very attractive to the food-collecting cultures. Thus, some of these wild beans could have been (and probably were) carried by them from place to place.

Phaseolus vulgaris L.

The origin of the common bean has been the object of much debate for several years. More than thirty years ago Burkart and Brucher (1953) proposed a wild bean from South America as the ancestor of the cultivated *P. vulgaris*. The wild bean was described by Burkart (1952) as *P. aborigineus*. Subsequently, Miranda Colín (1967) and then Gentry (1969) considered that the wild form of *P. vulgaris* occurs in Mesoamerica and that the South American form (*P. aborigineus*) was possibly a weedy form. Much has been written since that date, and taking into consideration the archeological evidence another hypothesis was proposed. Each geographical form gave rise to different cultivars of the common bean and therefore, independent domestications were established (Heiser 1979; Evans 1976; Kaplan 1981). The prevailing opinion therefore proposes a multiple domestication (Gepts, in press).

Delgado (1985) treated both geographical forms as varieties of *P. vulgaris*; thus *P. vulgaris* var. *mexicanus* Delgado encompasses the Mexican and Central American taxa (fig. 2), and *P. vulgaris* var. *aborigineus* (Burkart) Baudet comprises the beans from South America.

In Table 1 a comparison between *P. vulgaris* var. *mexicanus* and *P. vulgaris* var. *aborigineus* is presented. The characters concerned have been generally selected and represent those used in crop plant domestication. Most of the information used was extracted from Vanderborght (1983).

Corky stems have been reported for var. *mexicanus* by Gentry (1969) from Nayarit, Jalisco and Michoacan, and lignescent stems have been found on *Stevens 15633* from Nicaragua. The latter herbarium specimen shows that a certain degree of perennialism can be found on var. *mexicanus*.

The pods of var. *aborigineus* show a modification of the dehiscence mechanism only found on the cultivated plants. Berglund-Brucher and Brucher (1976) commented with respect to the pod dehiscence that this was "...less developed than in other wild growing Phaseolinae".

Seed number per pod in general has decreased following domestication in *P. vulgaris* (Evans 1976). The number of seeds per pod as well as the increment in seed weight and hilum lenght in var. *aborigineus* may reflect some degree of domestication. Kaplan (1981) on the other hand, suggested that seed size increase does not have to follow selection by humans "... but may have involved simple practices by preagricultural, gathering folks anywhere that wild *P. vulgaris* was utilized as a food resource". Also greater seedling size is observed in cultivars and domesticated plants than on wild ones.

The above information shows that var. *aborigineus* has attributes of a domesticated or semidomesticated plant, and that it is either an early escape or else was introduced as a wild or incipient cultivar by man from the northern hemisphere. The latter was considered as a possible explanation for the extensive distribution of the wild form of *P. vulgaris* by Heiser (1979).

Remains of botanical material have been uncovered in several archeological sites in Mexico and Peru. Most of these sites have yielded cultivated and in few occasions (Kaplan 1981) wild beans, most if not all of which has been critically examined by Dr. L. Kaplan. Archeological evidence shows that Andean Indians had domesticated beans by 7,600 BP. In Tehuacan, Mexico domesticated common beans were found dating to 6,000 BP. Therefore bean domestication evidently took place at some time preceding 7,000 BP. Also, this domestication took place independently in both regions.

Most recently, Gepts (in press) using the major seed storage protein, phaseolin, as an evolutionary marker together with seed size data found different phaseolin types for the North American and the South American wild beans. Also, he suggests that at least three domestication regions can be hypothesized in the origin of *Phaseolus vulgaris.*

1. In Mexico and Central America, where "S" phaseolin-type wild beans gave rise to "S" phaseolin-type cultivars (small-seeded).

2. In Colombia, where "B" phaseolin-type wild beans gave rise to "B" phaseolin-type cultivars (small-seeded).

3. In the southern Andes, where "T" phaseolin-type wild beans gave rise to "T" phaseolin-type cultivars.

The above data from electrophoretic techniques agree in many ways with what morphology has shown with respect to "wild" populations. That is, it confirms the existence of var. *mexicanus* in Mexico and Central America, and also what Berglund-

Brucher (1976) mentioned with respect to quite different populations of var. *aborigineus* from Colombia.

Phaseolus coccineus

Three are the archeological sites that have yielded beans of *P. coccineus* in Mexico. Remains of domesticated *P. coccineus* were discovered in Río Zape, Sonora by 1300 BP and in Tehuacan, Puebla by 2,200 BP (Kaplan 1967).

Kaplan and MacNeish (1960) in the Ocampo Cave site in Tamaulipas identified one almost complete pod valvæ and a fragment of a seed as a wild *P. coccineus* in occupational layers dated 7500–9000 BP. They based their hypothesis "by the extreme age of the remains and their occurrence long before the practices of ceramic cooking and agriculture, and by the apparent lack of selection for pod characteristics found in modern cultivated varieties".

Kaplan and MacNeish (1960) recognized that the anatomical features of the pod valvæ corresponded more closely to a dehiscent wild pod. However, dehiscent pods are not restricted to wild forms but are also present in domesticated *P. coccineus*. Moreover, the total length and width of the recovered valves (greater than 11.0 cm long and 1.5 or more cm wide) are diagnostic traits only found in modern cultivars. The largest pod of a wild *P. coccineus* was only 8.0 cm long and 1.3 cm wide (Delgado, in press).

Therefore, the pod fragments provide diagnostic traits for recognition of incipient domestication already in existence around 7,500–9,000 years BP. Archeological evidence appears to support the idea of an earlier domestication of beans, at least for *P. coccineus*.

Figure 1. Distribution of the *Phaseolus* genera.

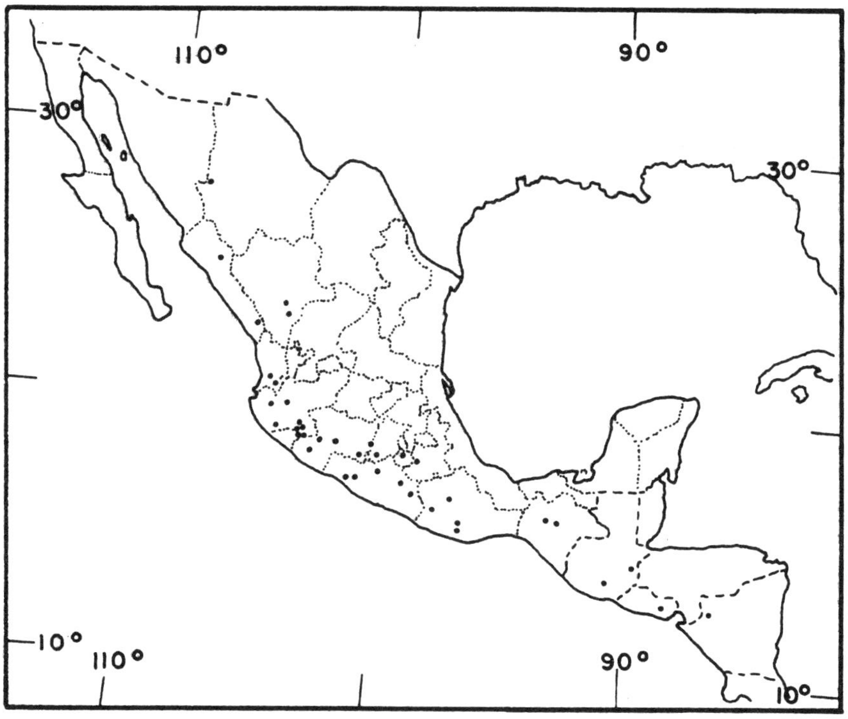

Figure 2. Distribution of *Phaseolus vulgaris* var. *mexicanus* A. Delgado

TABLE 1. Comparison between *P. vulgaris* var. *mexicanus* Delgado and *P. vulgaris* var. *aborigineus* (Burk.) Baudet.

Characters	var. *mexicanus*	var. *aborigineus*
Habit	Annual, occasionally some degree of perennialism	Annual
Stems	Herbaceous, some plants develop corky (Gentry, 1969) to lignescent ones (*Stevens 15633*, Nicaragua).	Herbaceous
Pod dehiscence	Elastically to tardily	Different degree of dehiscence presence of fiberless pods.
No. seeds per pod	$\bar{x}$ = 5.8	$\bar{x}$ = 3.3
Weight 100 seeds	$\bar{x}$ = 6.2 gr	$\bar{x}$ = 13.9 gr
Hilum length	*Ca.* 1.0 mm	*Ca.* 1.5 mm
Seedling eophyyll length	$\bar{x}$ = 4.0 cm	$\bar{x}$ = 5.6 cm
Seedling eophyll width	$\bar{x}$ = 2.9 cm	$\bar{x}$ = 4.3 cm

REFERENCES

BERGLUND-BRUCKER, O., and H. BRUCHER

1976. "The South American Wild Bean (*Phaseolus aborigineus* Burk.) as Ancestor of the Common Bean", *Economy Botany* 30: 257–272.

BURKART, A.

1952. *Las leguminosas argentinas, silvestres y cultivadas*, edited by Z. Acme, Buenos Aires.

BURKART, A., and H. BRUCKER

1953. "*Phaseolus aborigineus* Burkart, die mutmasslich andine Stammform der Kulturbohne", *Der Züchter* 23: 65–72.

DELGADO S., A.

1985. Systematics of the Genus *Phaseolus* "Leguminosae" in North and Central America, Ph. D. Dissertation, University of Texas, Austin.

in press "Variation, Taxonomy, Domestication, and Germplasm Potentialities in *Phaseolus Coccineus*", in P. Gepts (ed), *Genetic Resources, Domestication and Evolution in Phaseolus*, Martenus Nijhoff, Netherlands.

EVANS, A.M.

1976. "Beans", in N. W. Simmonds (ed), *Evaluation of Crop Plants*, Longman, London.

GENTRY, H.S.

1969. "Origin of the Common Bean", *Economic Botany* 23: 55–69.

GEPTS, P.

in press "Phaseolin as an Evolutionary Marker", in P. Gepts (ed), *Genetic Resources, Domestication and Evolution in Phaseolus*, Martenus Nijhoff, Netherlands.

HEISER, CH.B.

1979. "Origin of Some Cultivated New World Plants", *Annual Review of Ecology and Systematics* 10: 309–326.

KAPLAN, L., and R. S. MACNEISH

1960. "Prehistoric Bean Remains from Caves in the Ocampo Region of Tamaulipas, Mexico", *Botanical Museum Leaflets of Harvard University* 19: 33–56.

KAPLAN, L.

1967. "Archæological *Phaseolus* from Tehuacan. Chapt. 10", in D. S. Byers (ed), vol. I, *Environment and Subsistence. Prehistory of the Tehuacan Valley*, University of Texas Press.

1981. "What is the Origin of the Common Bean?", *Economic Botany* 35: 240–254.

LACKEY, J. A.

1983. "A Review of Generic Concepts in American Phaseolinae (Fabaceae, Faboideae)", *Iselya* 2: 21–64.

MARECHAL, R., J. MASCHERPA, and F. STAINIER

1978. "Etude taxonomique d'un groupe complexe d'especies des genres *Phaseolus* et *Vigna* (Papilionaceae) sur la base de doneés morphologiques et polliniques, traitées par l'analyce informatique", *Boissiera* 28: 1–273.

MIRANDA COLIN, S.

1967. "Origen de *Phaseolus vulgaris* L. (Frijol común)", *Agrociencia* 1: 99–104.

VANDERBORGHT, T.

1983. "Evaluation of *Phaseolus vulgaris* Wild Types and Weedy Forms", *Plant Genetic Research Newsletter* 54: 18–24.

VERDCOURT, B.

1970. "Studies in the Leguminosae-Papilionoideae for the Flora of Tropical East Africa. IV", *Kew Bulletin* 24: 507–569.

ANIMAL DOMESTICATION AND EARLY ANIMAL HUSBANDRY IN CENTRAL, EAST AND SOUTH EUROPE

SANDOR BÖKÖNYI
Hungarian Academy of Sciences

From the viewpoint of the development of animal husbandry of the whole of Europe, Southeast Europe had a key position. First of all, the first domestic animals coming from Southwest Asia landed there first, then the first domestication of the economically most important domestic species (cattle and pig) of the European temperate belt first happened there, and finally the typical domestic fauna consisting of five animal species (cattle, sheep, goat, pig, and dog) from which the whole continent got its domestic stock later took shape there.

A well-known principle of archæozoology is that an animal species could only be domesticated there where its domesticable wild form existed. Starting out from this principle, one can divide the five Neolithic domestic species into two groups: the wild forms of sheep and goat did not live in the European mainland in post-Pleistocene times, thus in the time of domestication, but those of cattle, pig and dog did. At the same time, the wild forms of all five Neolithic domestic species occured in Southwest Asia in the early Holocene.

All this means that the small ruminants (the caprovines) could not be domesticated in Southeast Europe; for their place of origin one has to look in Southwest Asia from where they got to Europe already in domesticated form. The three other species could be and undoubtedly were domesticated in Europe too, another question is whether their domestication happened in Europe earlier or later than in Southwest Asia.

With dog and pig the question is not simple because attempts of their domestication took place both in Europe and Southwest Asia well before the advent of the Neolithic, the classical period of domestication; they occurred in fact in the Mesolithic or even in the late Paleolithic (Turnbull and Reed 1974: 84; Bökönyi 1976: 20; Nobis 1984: 74; Altuna and Mariezkurrena 1985: 89, 110). The common characteristics of both species was that their food requirements were very similar to those of man and consequently they could easily survive on kitchen remnants. Nevertheless, the fact that they needed the same food than man did became an obstacle for their large-scale domestication.

Ideal subjects of domestication were the caprovines (sheep and goat) because they could survive on a fodder rich in cellulose (straw and hay as byproducts of agriculture); thus they produced meat from such a food which could not be used by man or dog and pig respectively on the one hand, and which was one of the most common organic materials on the terrestrial world.

Their domestication went hand-in-hand with agriculture in Southwest Asia (where their wild forms lived in abundance) in the early Neolithic. According to our most recent knowledge the earliest possibly domesticated sheep were found in Zawi Chemi Shanidar, northeastern Iraq from the early 9th millennium BC (Perkins 1964: 1565) but the earliest undoubtedly domesticated sheep remains came from Ali Kosh, western Iran, *c.* 7500 to 6750 BC (Hole and Flannery 1967: 171). Two sites of western Iran, Asiab (Bökönyi 1973: 71; 1977: 19) and Ganj Dareh (Perkins 1973: 279) produced the remains of the earliest domestic goats from the 8th millennium BC. Nevertheless, domestic goats were surely found in Jericho (Clutton–Brock 1971: 48) and probably in Ali Kosh (Hole and Flannery 1967: 171) at about the same period too.

In comparison to these three species, the earliest domestic pigs were found a little later, in Qalat Jarmo, northeastern Iraq, *c.* 6750 BC (Flannery 1961: 65; Stampfli 1983: 454) and the first domesticated cattle appeared in Çatal Hüyük, Anatolia, at about 6400 BC (Perkins 1969: 177).

This complete Neolithic domestic fauna consisting of the above five species reached southeastern Europe *c.* 6500 BC coming from the Near East. It is true that two of its species, pig and cattle, could also be domesticated in the European mainland but one actually does not find any traces of such a domestication before the Near Eastern one.

Anyway, the earliest domestic fauna, thus the first animal husbandry of Europe that was first found in the southeastern corner of the continent was based on the caprovines, sheep and goat, that had their origins in Southwest Asia. In fact, it was so in every early neolithic site of Greece independently of its geographical setting, in Argissa-Magula (Boessneck 1962: 27) and Achilleion (Bökönyi, in print); in Thessaly, Nea Nikomedeia in Greek Macedonia (Higgs 1962: 271), Lerna (Gejvall 1969) and the Franchthi Cave (Payne 1973: 59); in the Peloponnese and Knossos (Jarman and Jarman 1968: 241) in Crete. Pigs were far less common, and cattle and dogs were quite rare.

This domestic fauna obviously lived well in Greece since there the geographic and climatic conditions were very similar to those of Southeast Asia, the homeland of its most important species. How successful this animal husbandry was there, the facts show that the inhabitants of these sites were hardly forced to hunt and fish in order to complete their animal protein needs on the one hand and to domesticate to increase their domestic stock in the other.

This caprovine-based animal husbandry started its advance to the north, northeast and west already in the second half of the 7th millennium BC. This advance was probably helped by the "climatic optimum" of the early Neolithic that meant an average temperature 3 to 4 degrees C higher than now and thus making possible to southern types to reach the southern territories of Central and Eastern Europe. Still before the end of the 7th millennium BC this animal husbandry reached Yugoslav Macedonia (Bökönyi 1976b) and in the middle of the 6th millennium also Central Yugoslavia (Bökönyi 1976c) and Bulgaria (Dennel 1972). Right before the end of the 6th millennium BC this type of animal husbandry arrived in the Carpathian Basin (Bökönyi 1969, 1974, 1984) and reached Moldavia and southwestern Ukraine in the 5th millennium BC (Tringham 1969).

It is quite interesting that while in the whole Balkans the complete range of Neolithic animal husbandry was flourishing, in inaccessible places like the Iron Gate gorge of the Danube River the surviving Mesolithic population was still living on hunting-fishing-gathering and only made isolated attempts of dog or pig domestication (Bökönyi 1970, 1975; Bolomey 1973). Westwards it occurred in southern Italy (Bökönyi 1979–82, 1983) in the 5th millennium BC and southern France around 4000 BC (Poulain-Josien 1975; Bökönyi 1983) and a little later it even reached western Spain too.

These were the northernmost regions where this type of animal husbandry of basically Near Eastern character extended. Such parts of Europe which are located north or northwest of these regions could only get domestic animals through import from the above-mentioned areas; nevertheless, it happened very quickly because all five Neolithic domestic species could be found in the whole temperate belt of Europe in the course of the 4th millennium BC.

At the end of the first third of the Neolithic, took place interesting changes in the animal husbandry of the southern part of Europe. It is well known that animal husbandry strongly depends on the environmental conditions even now, one can just imagine how strong this dependence was in the early phases of its development. In Greece and the southern edge of France this animal husbandry did not change much because there the environment was very favorable. But in the temperate belt of Europe, the caprovine-based animal husbandry was alien and unable for further development. It was alien because its main species, sheep and goat, did not find favorable living conditions and did consequently not breed well there. It was unable for further development because the wild forms of its above-mentioned leading species did not live in the European mainland, and the stock of caprovines could therefore not be increased with local domestication. And this was an essential disadvantage because under the primitive conditions of the Neolithic animal husbandry the natural offspring of the domestic animals alone was not able to provide enough food for the growing human population and to increase the number of the domestic stock at the same time.

Thus man had to look for such domestic animals as leading species which both fit well into the environmental conditions of temperate Europe and had locally domesticable wild forms at the same time. He found them in cattle and pig, and at the end of the early Neolithic he started the switchover to these latter species. From this time on, cattle and pig domestication grew enormously in importance in most parts of Europe. In certain regions, e. g. in the Carpathian Basin, real cattle domestication centers developed. As a result cattle population markedly increased and cattle became the leading species allover the temperate belt of the continent. Pig domestication never reached that of cattle in importance but even in this way it was considerable and made pig as frequent as caprovines or even more numerous. This picture can very well be observed in the sites of the Linear Pottery Culture and its related groups where cattle generally took the leading part and pig and caprovines alternatively stood in the second place.

Up to the end of the Early Neolithic, domestic animals had only one use (not speaking about the sporadic use of dogs as sacrifical animals, hunting companions, herd or watch dogs etc.): their meat. Then man discovered their secondary products and uses such as milk, wool, draught power etc. This was almost another "revolutionary" step and started a "domestication fever" in the second half of the Neolithic when man tried to increase the number of his domesticates by all means (Bökönyi 1971: 643, 1974: 217, 1984: 32; Sherratt 1983: 90). The best examples of the large-scale domestication of the Late Neolithic can be found in the Tisza, Herpály and Lengyel Cultures of the Carpathian Basin in the 4th millennium BC.

Thus man ate the meat of all his domestic animals including that of the dog too, as it is proved by the occurrence of skulls with opened brain-cases (brain was a delicacy even in those times) and of extremity bones broken up for the marrow. In fact, man stopped eating dog flesh only at the end of the Bronze Age, and this was probably the first food avoidance in Europe.

In the whole, Neolithic cattle was the main meat supplier animal species even in such areas where sheep and goat outnumbered them by far. One must not forget that the meat quantity of a cow was equivalent to that of seven caprovines or four to five pigs. In this way, about 40% of all of the domestic animal meat came from cattle even in caprovine-dominated animal husbandries. In husbandries where cattle was most frequent this ratio could go up to 75–90%. In the Early Neolithic caprovine meat was also important but from the Middle Neolithic onward pork became the second most important meat after beef (Bökönyi 1973: 14).

With the Late Neolithic, an important period of development of animal husbandry came to an end. Hunting and domestication lost most of their importance because animal husbandry became strong enough, and with the increase of human population agricultural land also increased destroying the habitats of wild animals. Thus domesticable wild cattle tended to disappear.

Among the domestic animals new species —horse, and ass— appeared which were intentionally domesticated for work and the first elements of conscious animal breeding occurred in Mesopotamia. This certainly influenced European animal husbandry too, though not with the takeover of the new methods but with the improvement of certain domestic species through new waves of improved breeds (sheep).

REFERENCES

ALTUNA, J., and MARIEZKURRENA, K.

1985. "Bases de subsistencia de los pobladores de Erralla: Macromamíferos", Altuna J., Baldeón, J. and Mariezkurrena, K. (eds), *Cazadores magdalenienses en la cueva de Erralla* (Cestona, País Vasco), *Munibe* 37, San Sebastián: 87–117.

BÖKÖNYI, S.

1969. "Archæological Problems and Methods of Recognizing Animal Domestication", Ucko, P.J. and Dimbleby, G.W. (eds), *The Domestication and Exploitation of Plants and Animals*, London, Duckworth: 219–229.

1970. "Animal Remains of Lepenski Vir", *Science* 167, no. 3926: 1702–1704.

1971. "The Development and History of Domestic Animals in Hungary", *American Anthropologist* 73, no. 3: 640–674.

1973. "Some Problems of Animal Domestication in the Middle East", Matolcsi, J. (ed), *Domestikationsforschung und Geschichte der Haustiere*, Internationales Symposion in Budapest 1971, Budapest, Akadémiai Kiadó: 69–75.

1974. *History of Domestic Mammals in Central and Eastern Europe*, Budapest, Akadémiai Kiadó.

1975. "Vlasac: an Early Site of Dog Domestication", Clason, A. T. (ed), *Archæozoological Studies*, New York, American Elsevier: 167–178.

1976a. "Development of Early Stock Rearing in the Near East", *Nature* 264, 5581: 19–23.

1976b. "The Vertebrate Fauna from Anza", Gimbutas, M. (ed), *Neolithic Macedonia* I, Los Angeles, University of California: 313–363.

1976c. "The Vertebrate Fauna of Obre", *Wissenschaftlichen Mitteilungen d. Born. Herzeg*, Landesmus IV A: 55–154.

1977. *Animal Remains from the Kermanshah Valley, Iran*, BAR Supplement Series 34, Oxford.

1979-82. "The Early Neolithic Fauna of Rendina", *Origini* XI: 345–354.

1983. "Animal Bones from Test Excavations of Early Neolithic Ditched Villages on the Tavoliere, South Italy", Vassano, S. M. and Manfredini, *Studi sul Neolitico del Tavoliere della Puglia, Indagine territoriale in un' area-campione*, BAR International Series 160, Oxford, 237–249.

1984. "Die frühneolithische Wirbeltierfauna von Nosa", *Acta Archæologica Hungarica* 36: 29–41.

in print *The Early Neolithic Vertebrate Fauna of Archilleion*, Thessaly.

BOLOMEY, A.

1973. "The Present State of Knowledge of Mammal Exploitation during the Epipaleolithic on the Territory of Romania", Matolcsi, J. (ed), *Domestikationsforschung und Geschichte der Haustiere*, Internationales Symposion in Budapest 1971, Budapest, Akadémiai Kiadó: 197–203.

BOESSNECK, J.

1962. "Die Tierreste aus der Argissa–Magula von prækeramischen Neolithikum bis zur mittleren Bronzezeit", Milojcic, V., Boessneck, J. and Hopf, M.: *Die deutschen Ausgrabungen auf der Argissa–Magula in Thessalien* I, Bonn, Habelt: 27–99.

CLUTTON-BROCK, J.

1971. "The Primary Food Animals of the Jericho Tell from the Proto-Neolithic to the Byzantine Period", *Levant* III: 41–55.

DENNEL, R. W.

1972. "Stone Age Farming in Bulgaria", *The Illustrated London News*, September, 1972: 61–62.

FLANNERY, K.V.

1961. Skeletal and Radiocarbon Evidence for the Origins of Pig Domestication, MA Paper, University of Chicago, Chicago.

GEJVALL, N.-G.

1969. *Lerna I. The Fauna*, Princeton, American School of Classical Studies at Athens.

HIGGS, E. S.

1962. "The Fauna of the Early Neolithic Site at Nea Nikomedeia (Greek Macedonia)", *Proceedings of the Prehistoric Society* XXVIII: 271–274.

HOLE, F., and FLANNERY K. V.

1967. "The Prehistory of Southwestern Iran: a Preliminary Report", *Proceedings of the Prehistoric Society* XXXIII: 147–206.

JARMAN, M. R., and JARMAN H. N.

1968. "The Fauna and Economy of Early Neolithic Knossos", *Annals of the British School of Archæology at Athens* 63: 241–264.

NOBIS, G.

1984. "Die Haustiere im Neolithikum Zentral-Europas", Nobis, G. (ed), *Der Beginn der Haustier-haltung in der "Alten Welt", Die Anfaenge des Neolithikums, vom Orient bis Nordeuropa*, 9, Köln and Wen, Boehlau: 73–105.

PAYNE, D.

1973. "Animal Bones (from the Franchthi Cave)", *Herperia* XIII: 59–66.

PERKINS, D.

1964. "Prehistoric Fauna from Shanidar, Iraq", *Science* 144: 1565–1566.

1969. "Fauna of Çatal Hüyük: Evidence for Early Cattle Domestication in Anatolia", *Science* 164: 177–179.

1973. "The Beginnings of Animal Domestication in the Near East", *American Journal of Archæology* 77: 279–282.

POULAIN-JOSIEN, TH.

1975. "Les animaux domestiques en France a l'époche néolithique", in *L'homme et l'animal*, 1er Colloque d'Ethnozoologie, Juin 1975, Paris: 409–415.

SHERRAT, A.

1983. "The Secondary Exploitation of Animals in the Old World", *World Archæology* 15, 1: 90–104.

STAMPFLI, H. R.

1983. "The Fauna of Jarmo with Notes on Animal Bones from Matarrah, the Amuq, and Karim Shahir", in Braidwood, L. *et al.* (eds), *Prehistoric Archæology along the Zagros Flanks*, Oriental Institute Publications 105, Chicago Oriental Institute, The University of Chicago: 431–483.

TRINGHAM, R.

1969. "Animal Domestication in the Neolithic Cultures of the Southwest Part of European USSR", in Ucko, P. J. and Dimbleby, G. W. (eds), *The Domestication and Exploitation of Plants and Animals*, London, Duckworth: 381–392.

TURNBULL, P., and REED, C. A.

1974. "The Fauna from the Terminal Pleistocene of Palegawra Cave, a. Zarzian Occupation Site in Northeastern Iraq", *Field Anthropology* 63, no. 3: 81–146.

THE MOST ANCIENT SOUTH: A REVIEW OF THE DOMESTICATION OF THE ANDEAN CAMELIDS

JONATHAN D. KENT
Metropolitan State College

Introduction

The relationship between humans and the domesticated animals is one that is between one population of intelligent beings and another. Each has its own strategies for producing the food it needs from what is available and for reproducing successfully. Each seeks to avoid its enemies. The interesting aspect of this relationship is that there is usually a strong interdependence of each upon the other as these productive and reproductive strategies are followed.

In the Andean regions of the Americas, this strong interdependence is characteristic of Native American pastoralists and the domestic camelids, llamas and alpacas, with which they interact. For example, among Aymara-speaking herders of the southern central Andes of Peru, there is the following expression: "... *hiwasaha uywa uywataña, uka uywaha hiwasaru uyusitu*" which translates to "we nurture the animals, and the animals nurture us" (Palacios R. 1977). But, this is a special kind of nurturing. The Aymara term *uywaña* means care or nurturing in the way a mother would nurture her child. Another term, *awatiña*, means pasturing and feeding of animals too, but from afar, i. e. not with family-like intimacy. It is thus a term used when herding animals such as sheep and cattle —animals of European origin— and is never used to describe the herder/camelid interrelationship (*Ibid.*: 5–6). The origins of this remarkable close interaction comprise the main subject of this paper.

There is a suite of animal characteristics that both herders of llamas and alpacas and hunters of wild vicuñas and guanacos have found to be of great value. Such characteristics include:

1. the possession of physiological adaptations to both high altitude and cold;
2. a gregarious social organization that includes an established dominance hierarchy within groups and territorial defense among different groups;
3. a variety of usable products from the bodies of the animals including:

a. wool	e. fat
b. meat	f. bones
c. hides	g. dung
d. sinew	h. internal organs;

4. for llamas, and only rarely alpacas, physiology and behaviours permitting their use as burden-bearing animals; and,
5. use as sacrificial offerings.

The animals have derived a number of benefits from their association with humans as well:

1. food provisioning, especially during times of cold stress, environmental degradation, overgrazing, or when transported to areas in which adequate pasturage is not readily available;
2. nursing during illness and care of weak or stressed neonatal animals; and,

3. protection from natural predators, especially during the first few months of life.

Interest in animal domestication within the Western intellectual tradition may be traced back at least to Biblical times. More recent interest has followed Darwin's attempt to explain animal variability and its relationship to human selection (Darwin 1900, first edition 1874, vol. I: 3). Humans were not seen by Darwin as the cause of this variability, but rather as agents that could select, preserve, and accumulate the variations already at hand. Lacking from Darwin's investigation of this problem were osteological and contextual data from archæological sites that could show changes over time in the animals that would become domesticates, and data on climatic change that would permit sorting out or controlling for environmental change as opposed to cultural change as forces of phenotypic variability in future domesticates.

V. Gordon Childe was among the pioneers in examining such data. His food production revolution, defined as including deliberate cultivation of food plants and the "...taming, breeding, and selection of animals...", was linked to the melting of glaciers in northern Europe and Eurasia (Childe 1953: 23). The result was the drying up of grasslands in southwestern Asia and northeastern Africa and what Childe called (Childe 1929: 42):

> enforced concentration in cases or by the banks of ever more precarious springs and streams...Animals and men would be herded together round pools and wadis that were growing increasingly isolated by desert tracts, and such inforced juxtaposition might almost of itself promote that sort of symbiosis between man and beast that is expressed in the word "domestication".

By 1953, Childe had dropped the catastrophic aspects of the climatic changes in this region. He points to a lack of catastrophic climatic change being involved in the creation of the Oklahoma dust bowl. He still believed, however, that the drought conditions observable in the region under consideration in modern times was a result of a "...gradual process to which man and his stock may well have contributed" (Childe 1953: 17).

Noteworthy in Childe's way of thinking is the assumption that proximity in space would eventually lead to the development of a symbiotic relationship between humans and their former prey. The notion that familiarity breeds symbiosis is still pervasive in the archæological literature in general, and as will be discussed below, in the majority of work dealing with Andean camelid domestication.

Background

The earliest domesticated organism in South America was a plant. Beans (*Phaseolus* sp.) appear to be the earliest, being found in Guitarrero Cave, in a highland valley of central Peru, at about 9500 BP (Kaplan *et al.* 1973; Lynch *et al.* 1985). In the New World, there is a much greater predominance of domesticated plants relative to domesticated animals than is true for the Old World. As in the Old World, domestication of New World plants and animals was a post-Pleistocene process that followed the shrinkage of mountain glaciers between 12,000 and 10,000 BP in the Andes (Wright 1977, 1983; Mercer 1982). There is considerable debate regarding whether this shrinkage reflects world-wide warming, world-wide reduction in precipitation, changes in global weather patterns, and/or local changes in these parameters (Mercer and Palacios M. 1977; Wright 1977; Servant *et al.* 1981; Hastenrath and Kutzbach 1985).

It was not until several thousand years after the last major glaciation that any evidence for animal domestication appears. Five types of animals are now believed to have been domesticates in pre-Columbian America:

1. dog (*c.* 10,400 BP in North America; Lawrence 1967);

2. guinea pig (*c.* 6000–3000 BP in South America; Wing 1977);
3. Muscovy ducks (possibly as early as 2500 BP in Ecuador; Hesse 1980);
4. turkey (*c.* 1770 BP in Tehuacán, Mexico; Flannery 1967); and,
5. camelids, the subject of this paper.

In this paper, I briefly review the evidence pertaining to the origins of camelid domestication. In so doing, I will pay particular attention both to the kinds of evidence we have for this process and to the nature of our inferences from this evidence. The overall theme is the nature of interactions among humans, the camelids, and the environment.

The South American Camelids

In the Andean high elevation grasslands, or puna, the domestication of camelids began sometime prior to five millennia ago. Involved in the process was a shift from exclusive dependence upon hunting–gathering centered around procurement of a variety of large mammals to an economy centered around the hunting of wild camelids and deer and the herding of domesticated camelids (Matos M. 1975; Pires-Ferreira *et al.* 1976; Wing 1972, 1975a, 1983; Kent 1982; Rick 1980, 1983; Wheeler 1984).

Our best available paleontological evidence indicates a North American origin for the camelids, who probably migrated to South America about three million years BP (Webb 1978).

The migration from North to South America over the Panamanian land bridge (Marshall *et al.* 1979: 275) corresponds to great climatic and tectonic changes during the late Pliocene/early Pleistocene —most notably the major uplift of the major mountain chains of both continents and the creation of vast semi-arid grasslands in areas formerly savanna/woodlands (Berry 1926; Patterson and Pascual 1972: 251). Most of the South American pre-Pleistocene camelid fossils are found in non-Andean localities, primarily eastern Argentina (Cardozo 1975: 74, map 3). Although some sampling and/or preservation bias might partly account for this distribution, it would appear that the earliest South American camelids were not initially adapted to the high elevations, and possibly that their presence at high elevations is a post-Pleistocene phenomenon. Thus, the earliest high-altitude camelids may have entered their present range contemporaneous with and perhaps because of human hunters.

There is a complete lack of agreement among taxonomists regarding camelid systematics. This of course affects our interpretation of their phylogeny and early distribution. The major schools of thought group the extant camelids either within one or two genera, with either separate species or grouped within a single species.

The result of this confusion has been a proliferation of theoretical phylogenies which can be divided into four major groups (fig. 1):

a. a wild guanaco is ancestral to the modern guanaco, the modern llama, and the modern alpaca (Cook 1925; Herre 1952; Fallet 1961; Zeuner 1963; Herre and Röhrs 1977; Otte and Venero 1979): fig. 1 (a);
b. a wild guanaco is ancestral to the modern guanaco and llama, while a wild vicuña is ancestral to the modern vicuña and alpaca (Antonius 1922; Latcham 1922; Krumbiegel 1952; Steinbacher 1953; Capurro S. and Silva 1960; Bustinza M. 1970): fig. 1 (b);
c. a wild guanaco is ancestral to the modern guanaco and llama, then a llama × vicuña cross results in the modern alpaca (Hemmer 1975): fig. 1 (c), and
d. the llama of today descended from an extinct wild llama-type animal while the alpaca of today descended from an extinct wild alpaca-type animal (López Aranguren 1930; Cabrera 1932; Cabrera and Yepes 1940): fig. 1 (d).

Thus, domesticated llamas of today are either:

1. descended from a guanaco, or
2. descended from a wild Pleistocene, llama-like form.

Similarly, domesticated alpacas of today are either:

1. descended from a guanaco;
2. descended from a wild Pleistocene alpaca-like form;
3. descended from the llama which was, in turn, descended from a guanaco, and the alpaca resulted from a vicuña × llama cross; or,
4. descended from the vicuña.

The adoption of any one of these scenarios has occasionally unjustifiably influenced the interpretations of archæological faunal remains as will be discussed later in this paper.

An obvious line of investigation, namely biochemical studies that would yield cladistic information on the distances among the extant forms is only just beginning. Miller *et al.* (1985) have been studying blood polymorphisms, but their work has not yet produced results that can resolve the phylogenetic problem. Studies such as DNA–hybridization, DNA–sequencing and amino acid sequencing have not been carried out.

All of the camelids interbreed occasionally, and most crosses occur without any human intervention. Most of the hybrids are fertile (Fernández Baca 1971, 1978). This success may be relatively recent and the result of deliberate and intense human selection during the early twentieth century (see Kent 1982: 23–29).

Biogeographical and ethnohistoric studies have been hindered by a tremendous amount of post-Hispanic disturbance of both herders and camelid environmental zones. Documents are of little help because the Spaniards, who had never seen the New World camelids before, gave them names such as *"ovejas indias"* or *"carneros de la tierra"*, indiscriminately assigning either of these names to whatever camelid happened to be around (Latcham 1922: 93-95).

Recent advances in the study of camelid tooth morphology (Wheeler 1982: 12–13) and osteometry (Wing 1972; Miller 1979; Kent 1982: 142–179) have allowed zooarchæologists to differentiate among the four extant camelid forms with greater certainty. The extent to which these discrimination techniques are valid for the time of the origins of domesticated camelid forms will only be resolved as they are applied to progressively older archæological populations.

Based on the present situation, it seems that the four varieties of South American camelids must be placed within one genus and probably within one species. Whether the present zoological taxonomy is to reflect the historical, pre-twentieth century situation is a matter of considerable debate. Nevertheless, it is certainly the case that in the languages and conceptual systems of native South American groups such as Aymara, Tehuelche, and Quechua, the distinctions among (at least) these four main camelid varieties was, and is, clear and unambiguous (see Kent 1982: 23–29).

The Archæological Record of Camelid Domestication

Writing more than a decade ago, Wing summarized the zooarchæological evidence pertaining to camelid domestication. She concluded that as early as 5000 BP, domesticated camelids were to be found both in the highland valley of Ayacucho and in the puna site of Lauricocha (Wing 1977: 848; see fig. 2, nos. 1 and 2). In the Ayacucho Valley, the earliest evidence from Pikimachay is dated to the Chihua period, 6550–5100 BP.

At Lauricocha, Cardich (1960) highlighted two aspects relevant to domestication of camelids. First, he was one of the first Andean archæologists to save and attempt identification of faunal remains from an early preceramic site (in this region, prior to about 3800 BP). Second, he was the first to draw attention to a major shift in the fauna that had occured between his first phase, Lauricocha I, now dated to between 9500–8000 BP, and Lauricocha II, between 8000–5000 BP. This was a change from a predominance of deer bone (*Hippocamelus antisensis*) in Lauricocha I to a predominance of camelid bone in Lauricocha II (Cardich 1960:108–109).

He did not connect these finds to the domestication process, believing that the camelids of Lauricocha II were hunted. As will be discussed below, this shift from a predominance of deer (or cervids) to a predominance of camelids is suggestive to some investigators of an intermediate stage in the domestication process. If they are right, we would have to reconsider the interpretation of the remains from Lauricocha II.

Following Cardich's pioneering work, faunal analysis became part of most archæological work in the Andes. Wing has been the most prolific, and some of her most important comments on camelids follow.

In the puna, camelids and cervids normally make up 97–100% of the identified fauna from archæological sites (Wing 1982: 7). A linear trend in puna sites is observable such that between 12,000–7500 BP, the camelid:cervid ratio is about 50:50. This is followed by a dramatic increase in camelids between 7500–4500 BP. Finally, from that time to the time of the Inca State, camelids comprise 80% of the identified fauna (*Ibid.*: 13).

A very *dissimilar* picture emerges from highland valley sites. There, camelids are quite rare or absent from 12,000–7000 BP, remaining scarce in the following period 7000–4500 BP. An exception is Pikimachay in the Ayacucho Valley (fig. 2, no. 2) where a four-fold increase in the relative frequency of camelids is observed for this period. Finally, between 4500–3750 BP, camelids do increase in the highland valley sites and from then on are more abundant than deer everywhere except in the eastern inter-Andean highland valley in which the site of Kotosh is situated (fig. 2, no. 4). At that site, deer remain important from 3750–2540 BP (*Ibid.*: 16). On the coast, camelids only become common after 2450 BP (*Ibid.*).

It is clear from Wing's summaries that the greatest concentration of prehistoric camelid utilization and its early intensification occured in the puna, not in the highland valleys or coast. The precise center or centers of camelid domestication certainly have not all been identified. Latcham had suggested (1922) that the evidence for camelid domestication could be found in the Peru–Bolivia *altiplano*, and Wing (1975a) also urged more work in the Lake Titicaca area. To date, the necessary systematic surveys of the Titicaca Basin have not been carried out.

In the late 1960's and early 1970's Ramiro Matos M. and Kent Flannery began a joint examination into the origins of agriculture in the puna around Lake Junín (fig. 2, no. 3). They were joined by Herb Wright who was studying the relationship between agricultural origins and environmental change in both the Old and New World (Wright 1977). It soon became apparent that, like at Lauricocha, there were very ancient deposits which documented the transition from deer hunting to intensification in the use of camelids and the eventual adoption of camelid pastoralism (Matos M. 1975; Wing 1975b).

It should be noted that at that time, many Old World and some New World prehistorians, social anthropologists, and ethnohistorians were denying the existence of indigenous camelid pastoralism in the Andean region (see Flores 1983). Others, like Wright and Flannery, were seeking data with which to make comparisons between the development of plant and animal domestication in the Old World and New World. The

Junín puna was identified, largely through the efforts of Matos M., as an area that would provide the appropriate data.

Excavations at the Junín puna site of Pachamachay were carried out in 1969/1970, and the fauna analyzed by Wing (1975b). This early analysis was expanded upon by analyses of fauna from a number of Junín puna sites dating to between 12,000–4000 BP, work carried out by Jane Wheeler and colleagues (Pires-Ferreira *et al.* 1976; fig. 2, no. 3).

A model of puna camelid utilization emerged from this work, one that saw increasing intensification in camelid utilization over time as had earlier investigators. But intensification does not necessarily entail a transition to domestication, and it was Wheeler and her colleagues who explicity rendered this interpretation (Pires-Ferreira *op. cit., Ibid*). A number of sites were included in their analysis, lending additional weight to their model. Included were remains from a second season at Pachamachay (6200–3750 BP); Uchcumachay Cave (13,500–between 6200 and 4500 BP); Panaulauca Cave (9000–7500 BP, in this early work, now extended back to 9700 BP and forward to *c.* 1500 BP [K. Moore, personal communication]); Acomachay A (4500–3750 BP); the Lauricocha data mentioned above; and Telarmachay (a rock shelter dated to 4500–3750 BP in this early work, now extended by Lavallée's work to between 8810–3630 BP [Wheeler 1984]; see below).

The earliest levels, from Uchcumachay, show stone tools associated with extinct forms of deer and horse. The next period, 9000–7500 BP, shows the 50:50 cervid:camelid ratio mentioned by Wing. In subsequent periods, the camelid relative frequency becomes progressively higher so that by 4500 BP, the camelid:cervid ratio is 90:10. Domestication was thought to have occurred between 4500–3750 BP.

Based on these interpretations of the evidence, a model of camelid domestication was constructed (fig. 3). Initially, there were "generalized hunters" who were followed by those practicing "specialized hunting of Camelidæ". This was based on an "increasing knowledge of camelid territorial and social behavior". At some point, "increasing control over camelid territories" led to "control of semidomesticated Camelidæ". Between 6200–4500 BP, the change from semidomesticated to domesticated camelids occurred, culminating in the herding of domesticated camelids based on "increasing human control over breeding in semidomesticated Camelidæ". The final stage of the process is the "appearance of distinct breeds of domestic Camelidæ" (Pires-Ferreira *op. cit.*).

This model is currently being tested by a number of investigators, including more detailed excavation and faunal analyses by Lavallée, Julien, and Wheeler at Telarmachay (Lavallée *et al.* 1982; Wheeler 1984: 78). Although there is no change in the basic model, Wheeler now believes that the transition from hunting of camelids to herding them took place between 6800 and 4500 BP in the area of Telarmachay, and that it was an *in situ* development (Wheeler 1984).

An important feature of the Telarmachay work is that Wheeler applies the differences in tooth morphology observed among modern camelids to the assemblage (Wheeler 1982: 12–13). In doing so, however, she found it necessary to make certain assumptions about the nature of camelid phylogeny (see fig. 1). In the earliest levels at Telarmachay which, according to her model predate the appearance of domesticated camelids, she notes that spatulate incisors were found. These are characteristic of both modern domesticated llamas and modern wild guanacos (Wheeler 1982). Because her model demands only wild camelids at this time, Wheeler identifies these spatulate incisors as having come from guanacos (Wheeler 1984). In addition to the seeming circularity of this approach, it also demands eliminating the possibility that pre-domesticated llama-like animals existed [fig. 1 (d)]. It is not that Wheeler's argument is unreasonable, it is simply that its basis has not been demonstrated. It should be noted that Tonni and

Lanza (1976: 137) have rejected the arguments on which model (d) is based, but they too have not been able to demonstrate the validity of any other model. To put it briefly, we still do not have a solid basis for claiming that the 8000–year old spatula-shaped camelid incisors are from guanacos rather than from either an unknown extinct form or a pre-domesticated llama.

A second aspect of this model is that it is gradualistic in that it invokes a slow transition from hunting to herding. The gradualist paradigm is one which is implicit in most zooarchæological approaches to domestication in the Old World models as well. For example, Reed has stated (1960: 119) that most animal species are probably present among and important to the human populations exploiting them for some 2000–3000 years before we have any "unimpeachable evidence" for their domestication. However, when we look closely at Telarmachay, an interesting paradox arises if one adheres to this gradualist, or slow-transition paradigm.

In the period following intensive hunting of wild camelids (presumed to be vicuñas and guanacos), incisors similar to those of modern alpacas are found, leading to the inference that domestication has been achieved (Wheeler 1984: 79). One question that arises is: How does this new morphological trait appear if it is the result of an *in situ* process? In the gradualist model, we expect evidence for changes in animal management practices prior to the appearance of the distinctive dental feature. However at Telarmachay, Wheeler notes that the important changes in the camelid:cervid ratio coincide with the appearance of this distinctive feature rather than preceding it. Thus, it seems that new management practices and domesticated alpacas appear simultaneously.

Another reflection on management practices is the nature of the age structure of the archæological camelids. At the same time as these other changes, there is a dramatic increase in the relative frequency of fetus/new born animals. Wheeler argues that such an increase is due primarily to the changes in herd management practices at this time. The data support this claim, as we see this age grade increased from levels of 30–40% to levels of 40–70% during this same period (6000 to 5500 BP; Wheeler 1984: 77, fig. 2).

If all these changes are indeed contemporaneous or even approximately so, then gradualism does not characterize the appearance of domesticated alpacas in the Junín puna. In one period they are not present, and in the next they are. Even rough contemporaneity of changes in herd management and the appearance of domesticated animal distinctive features seem to favor instead a punctuated equilibrium model of camelid domestication (à la Gould and Eldredge 1977).

To sum up so far, it seems that the most likely locus for the earliest camelid domestication was in the central Andean puna. This occurred between 6000 and 5000 BP, as alpacas or alpaca-like animals are identified along with contextual and biological evidence for new herd management practices. These changes occurred rapidly, and I suggest a punctuated equilibrium model for camelid domestication in this area.

Why Domestication?

Why were alpacas the first domesticated camelids? It should be noted that the Telarmachay evidence is in full agreement with osteometrical discrimination of camelid bones from Pachamachay, showing alpacas as the earliest identifiable domesticated camelid between 5000 and 4200 BP (Kent 1982). Today, alpacas are not only meat providers, but are the major camelid wool source. This raises additional questions: Why did hunter–gatherers who had successfully adapted to the puna for several thousand years need high quality (presumably) wool? Why at this time and not earlier?

One possibility is that alpacas were domesticated to produce an item that would serve as a trade commodity. What were people on the Junín puna obtaining in return?

In level 4 at Uchcumachay, Richard Burger identified several pieces of obsidian from the Quispisisa source located far to the south in Huancavelica (fig. 2, no. 5; Pires-Ferreira *et al.* 1976:484). The level dates to between 6200 and 4500 BP (*Ibid.*). Thus, I suggest that an alpaca wool–obsidian exhange was operating, and that it originated either as a result of or was a cause of alpaca domestication.

Camelids and Climate

After 12,000 BP in the Junín puna there is little evidence for major climatic change (Wright 1983). It should be noted that there may be two major oscillations correlated with El Niño-like climatic conditions lasting for extended periods in the southern central Andes, one between 7000 and 5000 BP and the other between 2700 and 1600 BP (Mercer and Palacios 1977; Servant *et al.* 1981; Thompson *et al.* 1984). Such oscillations have not yet been documented for the Junín area.

These oscillations may have influenced the adoption of camelid pastoralism in first millennium BC sites near the Salar de Atacama of Chile (Druss 1980 and personal communication; fig. 2, no. 6). In that area, the sort of transhumance that Lynch (1967) has proposed as a predominant adaptive strategy in the central Andes may have been operating, but the applicability of a transhumance pattern for the majority of the central Andes was criticized by Cardich (1976). Indeed, Lynch himself no longer believes that the pattern has more than regional applicability (Lynch 1980: 310–311).

In the absence of significant climate change in the Junín puna after 12,000 BP, we are left with some interesting problems. No camelids are recorded prior to 9000 BP at Uchcumachay, so it is likely that camelids were not in the central Andean puna much before 10,000 BP. If they were not there, how did Pires-Ferreira *et al.*'s generalized hunters obtain the "increasing knowledge of camelid territorial and social behavior" that accompained the transition to specialized hunting (fig. 3)?

The gregariousness, dominance hierarchies, and territoriality of wild camelids were characteristics that made them amenable to domestication by humans. However, these are characteristics that would have made them easier prey for hunters than deer. Yet, this must have been true prior to the time at which we see a shift away from deer predominance in the faunal assemblages. Were deer disappearing? If so, they were not doing so because of any established climatic change. There was no desiccation or enforced concentration of man and beast in oases involved here. Were puna hunters gradually increasing their knowledge of camelid territorial and social behavior as Pires-Ferreira *et al.* (1976) suggest? If it took them several thousand years to obtain this knowledge it is a wonder these hunters survived for as long as they did.

Finally, we can ask, as Reed has (1977: 560), why better knowledge of one's prey species would lead to semidomestication or domestication at all. On the contrary, as Reed points out, it should lead to better hunting of the prey. Even the best hunters, he notes, maintain close social bonds with their prey populations, frequently taming young animals. The knowledge of behaviors thus obtained helps them to learn the best ways to track and hunt the animal. From this perspective, it is difficult to disagree with Reed's observation that it is not the mere juxtaposition of hunter that somehow results in domestication. Rather it is the attitude of the hunter that must change. The reasons for this change should direct future model building related to Andean camelid domestication.

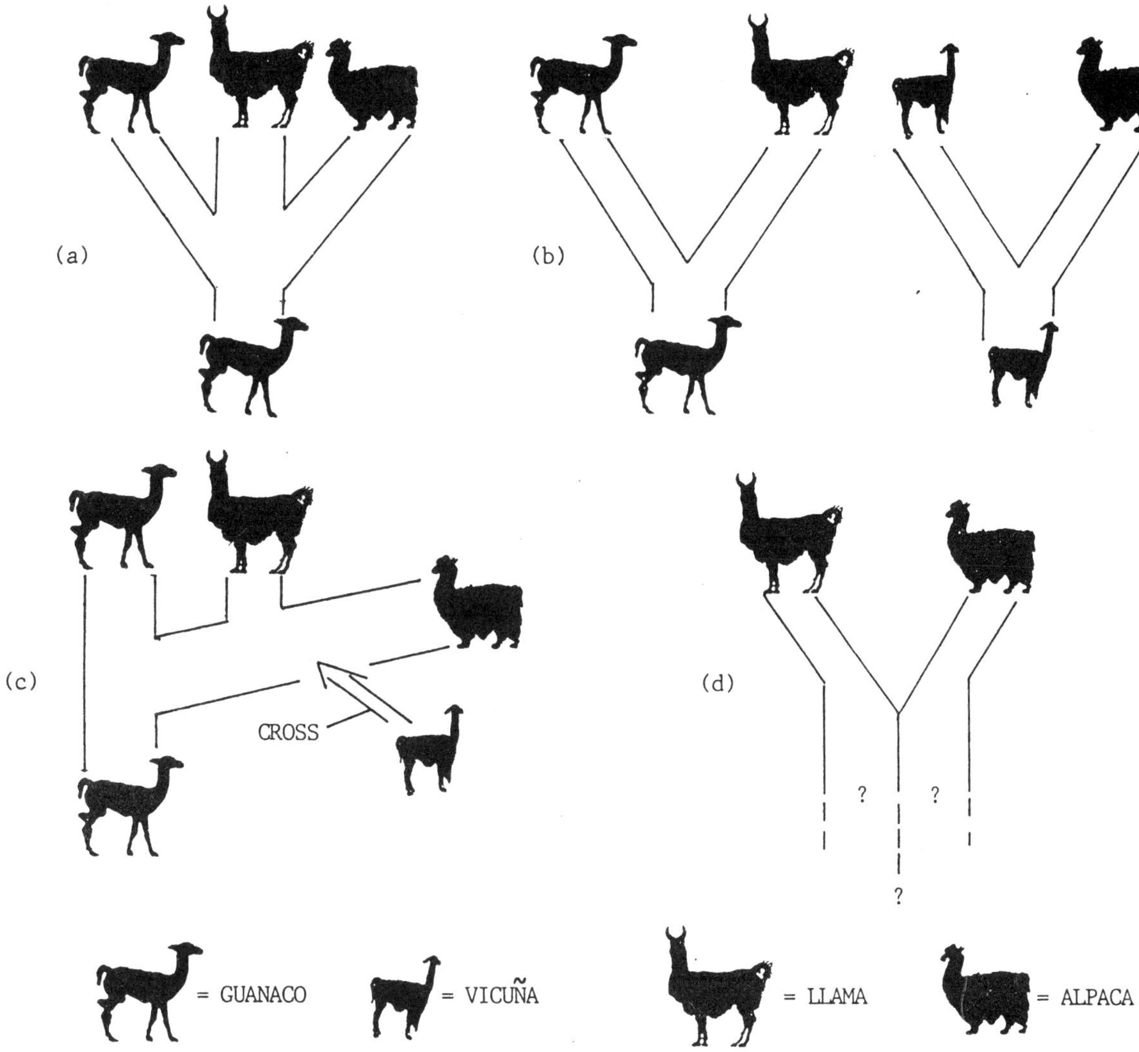

FIGURE 1

Alternative Models of Domesticated Camelid Phylogeny (See text for explanation).

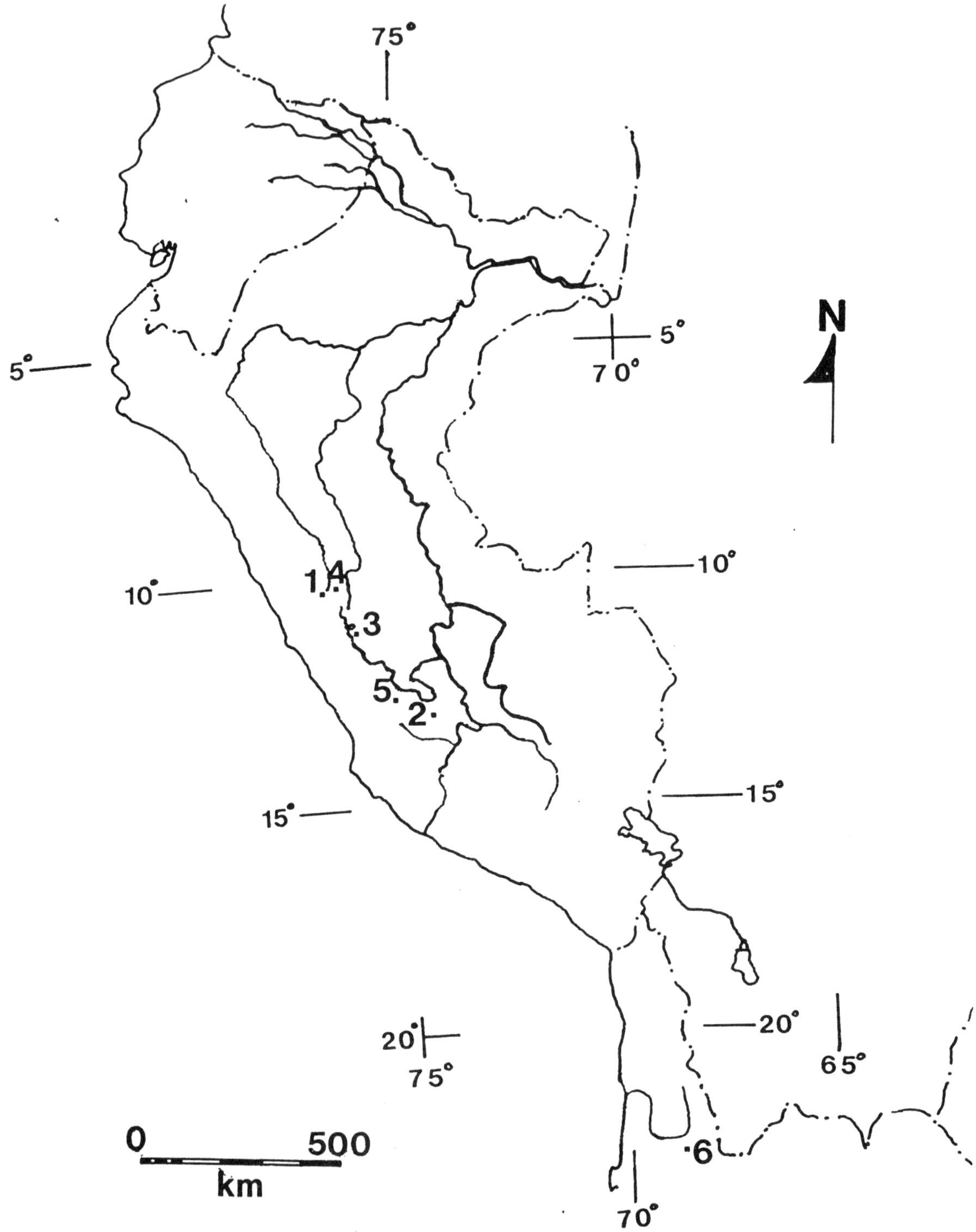

FIGURE 2

Location of Sites Discussed in Text: 1. Lauricocha; 2. Ayacucho Valley (Pikimachay); 3. Junín puna (Pachamachay, Panaulauca, Uchcumachay, Acomachay A, Telarmachay; 4. Kotosh; 5. Quispisisi Obsidian source, Huancavelica; 6. Salar de Atacama site.

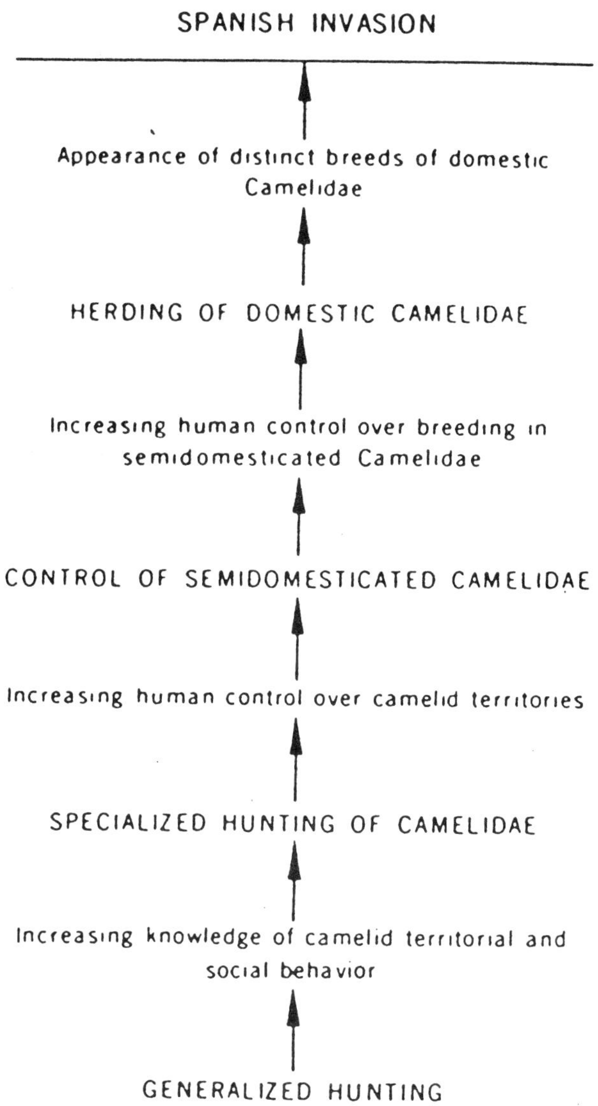

FIGURE 3

Developmental Model for Camelid Domestication in the Central Andean Puna (from Pires-Ferreira *et al.* 1976; 489).

REFERENCES

ANTONIOUS, O.

1922. *Stammesgeschichte der Hausterie*, Jena.

BERRY, W.

1926. "The Age of Uplift of the Andes", *Hrvatsko pirodoslovno drustvo u Zagrebu Glasnik* vol. 38/39: 3–29.

BUSTINZA M., J. A.

1970. "Contribución a la diferenciación específica de los camélidos sudamericanos", *Anales de la primera convención sobre camélidos sudamericanos (Auquénidos)*, Universidad Nacional Técnica del Altiplano, Puno, Perú: 26–28.

CAPURRO S., L.F. and F. C. SILVA

1960. "Estudios cromatográficos y electroforéticos en camélidos sudamericanos", *Investigaciones zoológicas chilenas* 6: 49–64.

CARDICH, A.

1960. "Investigaciones prehistóricas en los Andes Peruanos", *Antiguo Perú: espacio y tiempo*, Libreria-Editorial Juan Mejía Baca, Lima: 89–118.

1976. "Vegetales y recolecta en Lauricocha: algunas inferencias sobre asentamientos y subsistencias preagriculturas en los Andes centrales", *Relaciones de la sociedad argentina de Antropología* X, n. s., Buenos Aires: 27–41.

1978. "Recent Excavations at Lauricocha (Central Andes) and Los Toldos (Patagonia)", *Early Man in America from a Circum-Pacific Perspective*, edited by Alan L. Bryan, Occasional Papers no. 1, Department of Anthropology, University of Alberta, Edmonton: 296–300.

CARDOZO GONZALEZ, A.

1954. *Origen y filogenia de los camélidos sudamericanos*, Academia Nacional de Ciencias de Bolivia, La Paz.

CHILDE, V.G.

1929. *The Most Ancient East: The Oriental Prelude to European Prehistory*, Alfred A. Knopf, New York.

1953 *New Light on the Most Ancient East.* Frederick A. Praeger, New York.

COOK, O.F.

1925. "Peru as a Center of Domestication", *The Journal of Heredity* 16: 32–46, 94–110.

DARWIN. C. R.

1900. *The Variation of Animals and Plants under Domestication*, 2nd edition, 2 vols., D. Appleton and Company, New York, [Original first edition published 1874].

DRUSS, M.

1980. "The Oasis Theory Revisited: Variation in the Subsistence-Settlement System of the Final Preceramic Chiuchiu Complex, Northern Chile, and its Implications for the Process of Andean Camelid Domestication", Paper presented at the 33rd Northwest Anthropological Conference, 27–29 March, Bellingham, Washington.

FALLET, M.

1961. "Vergleichende Untersuchungen zur Wollbildung südamerikanischer Tylopoden", *Zeitschrift für Tierzuchtung und Zuchtungsbiologie* 75: 34–56.

FERNANDEZ-BACA, S.

1971. *La Alpaca: reproducción y crianza*, I.V.I.T.A. Boletín de divulgación no. 7, Lima.

1978. "Llamoids or New World Camelidæ", in *An Introduction to Animal Husbandry in the Tropics*, 3rd edition, edited by G. Williamson and W.J.A. Payne, Longman, London: 499–518.

FLANNERY, K.V.

1967. "Vertebrate Fauna and Hunting Patterns", in *The Prehistory of the Tehuacan Valley*, vol. I, edited by Douglas S. Byers, University of Texas Press, Austin: 132–177.

FLORES OCHOA, J.A.

1983. "Pastoreo de llamas y alpacas en los Andes —balance bibliográfico", *Revista Andina* 1 (1), Cuzco: 175–218.

FRANKLIN, W.L.

1982. "Biology, Ecology and Relationship to Man of the South American Camelids", in *Mammalian Biology in South America*, edited by M. A. Mares and H. H. Genoways, Pymatuning Laboratory of Ecology, Special Publication Series, vol. 6, University of Pittsburgh: 457–489.

GOULD, S. J. and N. ELDREDGE

1977. "Punctuated Equilibria: the Tempo and Mode of Evolution Reconsidered", *Paleobiology* 3: 115–151.

HASTENRATH, S. and J. KUTZBACH

1985. "Late Pleistocene Climate and Water Budget of the South American Altiplano", *Quaternary Research* 24: 249–256.

HEMMER, H.

1975. "Zur Herkunft des Alpakas", *Zeitschrift des Kölner Zoo* 18: 59–66.

HERRE, W.

1952. "Studien über die wilden und domestizierten Tylopoden südamerikas", *Zoologische Gärten* N.F. (19), Heft 2/4, Leipzig: 70–98.

HERRE, W. and M. ROHRS

1977. "Zoological Considerations on the Origin of Farming and Domestication", in *Origins of Agriculture*, edited by Charles A. Reed, Mouton Publishers, The Hague: 245–279.

HESSE, B.

1980. "Archæological Evidence for Muscovy Duck in Ecuador", *Current Anthropology* 21: 139–140.

KAPLAN, L., T.F. LYNCH and C.E. SMITH, Jr.

1973. "Early Cultivated Beans (*Phaseolus vulgaris*) from an Inatermontane Peruvian Valley", *Science* 179: 76–77.

KENT, J. D.

1982. *The Domestication and Exploitation of the South American Camelids: Methods of Analysis and Their Application to Circum-Lacustrine Archæological Sites in Bolivia and Peru*, Ph. D. Dissertation, Washington University, University Microfilms, Ann Arbor.

1984. "Archæology and Development: A Titicaca Basin Example", *Revista Diálogo Andino* 3, Arica, Chile: 59–72.

KRUMBIEGEL, I.

1952. *Lamas*, Neue Brehmbücherer, Leipzig.

LATCHAM, R. E.

1922. "Los animales domésticos de la América precolombiana", *Publicaciones del Museo de Etnología y Antropología de Chile* t. III, no. 1: 1–200.

LAVALLEE, D., M. JULIEN and J. WHEELER

1982. "Telarmachay: niveles precerámicos de ocupación", *Revista del Museo Nacional* XLVI, Lima: 55–133.

LAWRENCE, B.

1967. "Early Domestic Dogs", *Zeitschrift fur Säugetierkunde* 32: 44–59.

LOPEZ ARANGUREN, D.J.

1930. "Camélidos fósiles argentinos", *Anales de la Sociedad Científica Argentina* 109: 15–35, 97–126.

LYNCH, T.F.

1967. *The Nature of the Andean Preceramic*, Occasional Papers of the Idaho State University Museum no. 21, Pocatello.

1980. "Guitarrero Cave in its Andean Context", in *Guitarrero Cave: Early Man in the Andes*, edited by T. F. Lynch, Academic Press, New York: 293–320.

LYNCH, T.F., R. GILLESPIE, J.A.J. GOWLETT, and R.E.M. HEDGES

1985. "Chronology of Guitarrero Cave, Peru", *Science* 229: 864–867.

MARSHALL, L.G., R.F. BUTLER, R.E. DRAKE, G.H. CURTIS, and R.H. TEDFORD

1979. "Calibration of the Great American Interchange", *Science* 204: 272–279.

MATOS MENDIETA, R.

1975. "Prehistoria y ecología humana en las punas de Junín", *Revista del Museo Nacional* XLI, Lima: 37–74.

MERCER, J. H.

1982. "Holocene Glacier Variation in Southern South America", *Striae* 18: 35–40.

MERCER, J.H. and O. PALACIOS M.

1977. "Radiocarbon Dating of the Last Glaciation in Peru", *Geology* 5: 600–604.

MILLER, G. R.

1979. *An Introduction to the Ethnoarchæology of the Andean Camelids*, unpublished Ph.D. Dissertation, Department of Anthropology University of California, Berkeley.

MILLER, W.J., P.J. HOLLANDER, and W.L. FRANKLIN

1985. "Blood Typing South American Camelids", *The Journal of Heredity* 76: 369–371.

OTTE, K.C. and J.L. VENERO G.

1979. "Análisis de la craneometría diferencial entre la vicuña (*Vicugna vicugna*) y la alpaca (*Lama guanicoe pacos*)", *Studies on Neotropical Fauna and Environment* 14: 125–152.

PATTERSON, B., and R. PASCUAL

1972. "The Fossil Mammal Fauna of South America", in *Evolution, Mammals, and Southern Continents*, edited by A. Keast, F. C. Erk, and B. Glass, State University of New York Press, Albany: 247–309.

PALACIOS RIOS, F.

1977. "*... Hiwasaha Uywa Uywataña. Uka Uywaha Hiwasaru Uyusitu"; Los pastores Aymara de Chichillapi*, Tesis de Magister de Ciencias Sociales, Pontífica Universidad Católica del Perú, Lima.

PIRES-FERREIRA, J. W., E. PIRES-FERREIRA, and P. KAULICKE

1976. "Preceramic Animal Utilization in the Central Peruvian Andes", *Science* 194: 483–490.

PIRES-FERREIRA, E.

1981/82. "Nomenclatura y nueva clasificación de los camélidos sudamericanos", *Revista do Museu Paulista*, n.s., vol. XXVIII, Universidad de São Paulo, Brasil: 203-219.

REED, C.A.

1960. "A Review of the Archæological Evidence on Animal Domestication in the Prehistoric Near East", in *Prehistoric Investigations in Iraqi Kurdistan*, edited by R. J. Braidwood and B. Howe, Studies in Ancient Oriental Civilization no. 31, Oriental Institute, University of Chicago: 119–145.

1977. "A Model for the Origin of Agriculture in the Near East", in *Origins of Agriculture*, edited by C. A. Reed, Mouton Publishers, The Hague: 543–567.

RICK, J. W.

1980. *Prehistoric Hunters of the High Andes*, Academic Press, New York.

1983. *Cronología, clima y subsistencia en el precerámico Peruano*, Instituto Andino de Estudios Arqueológicos, Lima.

SERVANT, M., J.-C. FONTES, M. RIEU, and J.-F. SALIEGE

1981. "Phases climatique arides holocènes dans le sud-ouest de l'Amazonie (Bolivie)", *Comptes Rendus des Academie des Sciences*, Paris, t. 291 (18 mai 1981), Série II: 1295–1297.

STEINBACHER, G.

1953. "Zur Abstammung des Alpakka-Lama pacos", *Säugetierkunde Mitteilung*, Band I, Heft 2: 78–79.

THOMPSON, L. G., E. MOSLEY-THOMPSON, and B. MORALES ARNAB

1984. "El Niño-Southern Oscillation Events Recorded in the Stratigraphy of the Tropical Quelccaya Ice Cap, Peru", *Science* 226: 50–53.

TONNI, E.P. and J.H. LAZA

1976. "Paleoetnozoología del area de la Quebrada del Toro, Provincia de Salta", *Relaciones de la Sociedad Argentina de Antropología* vol. X, n. s., Buenos Aires: 131–140.

WEBB, S. D.

1978. "A History of Savanna Vertebrates in the New World: Part II: South America and the Great Interchange", *Annual Review of Ecology and Systematics* 9: 393–426.

WHEELER, J. C.

1982. "Aging Llamas and Alpacas by their Teeth", *Llama World* 1: 12–17.

1984. "La domesticación de la Alpaca (*Lama pacos* L.) y la Llama (*Lama glama* L.) y el desarrollo temprano de la ganadería autóctona en los Andes Centrales", *Boletín de Lima* 36: 74–84.

WING, E.S.

1972. "Utilization of Animal Resources in the Peruvian Andes", in *Andes 4: Excavations at Kotosh, Peru*, edited by Selichi Izumi and Kazuo Terada, University of Tokyo Press: 327–352.

1975a. "Hunting and Herding in the Peruvian Andes", in *Archæozoological Studies*, edited by A. T. Clason, North-Holland Publishing Company, Amsterdam and Oxford: 302–308.

1975b. "Informe preliminar acerca de los restos de fauna de la cueva de Pachamachay, en Junín, Perú", *Revista del Museo Nacional* XLI, Lima: 79–80.

1977. "Animal Domestication in the Andes", in *Origins of Agriculture*, edited by C. A. Reed, Mouton Publishers, The Hague: 837–859.

1982. "The Domestication of Animals in the High Andes", in *Adaptations and Evolution in Biota of High Tropical Montane Systems*, edited by M. Monasterio and F. Viulleumier, Oxford University Press, in press.

1983. "Domestication and Use of Animals in the Americas", in *Domestication, Conservation and Use of Animal Resources*, edited by L. Peel and D. E, Tribe, World Animal Science, A: Basic Information, no. 1, Elsevier, Amsterdam: 21–39.

WRIGHT, H.E., Jr.

1977. "Environmental Change and the Origin of Agriculture in the Old and New Worlds", in *Origins of Agriculture*, edited by Charles A. Reed, Mouton Publishers, The Hague: 281-318.

1983. "Late Pleistocene Glaciation and Climate Around the Junín Plain, Central Peruvian Highlands", *Geografiska Annaler* 65A: 35–43.

1984. "Late Glacial and Late Holocene Moraines in the Cerros Cuchpanga, Central Perú", *Quaternary Research* 21: 275–285.

ZEUNER, F. E.

1963. *A History of Domesticated Animals*, Harper and Row, New York.

COMMENTS ON THE TRANSITION OF HUNTING-GATHERING GROUPS INTO EARLY AGRICULTURALISTS: THE CASE OF THE SOUTH OF JALISCO, MEXICO

MA. DOLORES SOTO
Universidad Nacional Autónoma de México

Introduction

One of the transformations which caused a revolution in the history of humanity, was the shift from an appropiation economy into a production one; with the rise of agriculture, man started to recreate the world, to turn the natural media into a cultural one (Palerm 1972: 13). The effect of this discovery has been characterized as "a revolution thanks to which man stopped being a parasite... thus turning himnself into an emancipated creator of the limitations within his environment" (Childe 1928).

As we all know, this change in the means of obtaining food resources did not take place simultaneously all over the world; moreover, even today there are groups which have not incorporated it into their culture. We also know that its origin was not unique, as it took place, under different conditions, in various places and —of course— not simultaneously. It is also a fact that some human groups reached this discovery on their own, as the culmination of a process which included both environmental as well as cultural factors which brought about its achievement and that in other societies agriculture was learned through contact or relations with other groups which already practiced it. On the other hand, it is known that traditional ways of life associated with the rise of agriculture as, for example, sedentism, are not necessarily related or present a causal relationship.

However, in spite of the important role played by the agricultural revolution in the development of humanity, and of the great number of studies on this subject, we still know very little about the way in which it took place, which were the social, cultural, geographical and biological factors that allowed its rise and acceptance as an economic system. Up to now, several theories coexist that intend to explain how, when and why it took place and, and today none of them is thoroughly accepted. Nevertheless, I believe it is interesting to carry out an analysis of the main postulates which explain the rise of the agricultural revolution, and although there are different explanations for this, each one offers arguments that —together with others— give us a clear view of the mechanism which brought about the domestication process.

Analysis of the Main Theories of the Origins of Agriculture

Among the main theories on the origins of agriculture, there are those that point out how well-acquainted were the hunting-gathering tribes with seeds, with the agricultural cycles and with the climatic regime that influences plants' growth as the main causes of the rise of agriculture (Braidwood 1960; Sauer 1952).

According to Braidwood's point of view (1951), agriculture is the culmination of a continuously increasing process of cultural differentiation, specialization and knowledge of the habitat; it takes place within the nuclear zones with potentially domesticable resources. He proposes a series of cultural levels: The first is a quite extensive stage of food gathering and hunting; at the beginning, man possesses generalized tools and, later on, these become more standardized. Afterwards, there comes a food-gathering and hunting stage within a restricted area and, here, we already have tool differentiation among the various regions. Still later, we have a stage of food-gathering combined with selective hunting, as well as the appearance of regional variation as far as tool forms are concerned, and there is already a seasonal differentiation in activities. At some very specialized places, food gathering was so effective that it allowed the establishment of permanent or semi-permanent settlements. And, last of all, we have the food-production stage, which was achieved in the form of incipient farming within the area which represented domesticable material and, later on, there will appear the first agricultural villages. This model is rather a description of the regularities he was able to observe in his professional practice more than an explanation of the causes. To him, the process of change is inherent to human nature. Transition does not take place any earlier because culture was not ready to receive it (Braidwood and Willey 1962: 342) and the manifold rise of the agricultural revolution suggests that its high possibility of success is due to man's evolution, as well as to a peculiar combination of environmental circumstances. But, as Binford points out (1968), the trends observed in cultural evolution need an explanation.

There is a general agreement of opinion among anthropologists in the sense that the food-gathering and hunting groups had this knowledge, as it is of primary importance for their means of subsistence; moreover, ethnographic examples of present groups (Lee 1968) show that hunters–gatherers are acquainted not only with the fact that plants grow from seeds, but due to their contact with agricultural groups, they are also acquainted with the general principles of agriculture. As Flannery points out (1968: 68 and 1985: 240), "we do not know of any gathering/hunting group so primitive and ignorant as to be unaware of the fact that plants grow from seeds". Obviously, this knowledge is much older than the domestication process and was already a part of the group of cultural traditions of the hunting-gathering groups. It is certainly obvious, but —nevertheless— a prerequisite for carrying on the domestication process. That is to say, even though this postulate cannot be accepted as "the cause" of the origin of agriculture, it evidently constitutes a previous or preliminary condition for this change to take place. Among the basic postulates of continuity and cultural change there are different ways of interpretation and integration to the culture of "universal" principles itself, which acquire their own peculiar characteristics according to the cultural context in which they are immerse. The problem lies, then, not in the discussion of man's familiarity with plants, but in some aspects of this previously acquired knowledge which acted directly upon domestication/cultivation, how this was achieved and whether specific demonstrations of the knowledge were applied by each group to the cultivation process; thus, we may establish general standards for this economic change. As Cohen points out: "agriculture is, in a way, more the product of the context than just a new technology inserted by diffusion within a disorderly series of recipient cultures"; consequently, although it is true that the understanding of the context is not in itself an explanation of the causes of the agricultural revolution, it is utterly necessary to comprehend it in order to analyze the regional differences, thus being able to reach valid generalizations.

Some other theories are based on climatic phenomena which affected the existing fauna and flora, and that moved human beings to look for new alternatives as far as food was concerned. Archæological evidence shows, in my opinion, that these explanations are valid up to a point, but they refer mainly to local environments, not applicable, in general terms, to the whole process of the rise of agriculture. Climatic changes are clearly regional, having different effects upon each segment of our planet, according to the latter's own specific configuration. It is then very difficult to imagine that a change in climate may prove to be of the same magnitude throughout the different regions wherein agriculture arose, the reason for this being that climate is only a factor —although a most important one— in the plant/vegetable cycles; factors such as kind of soil, altitude above sea level, systems of procurement and loss of soil nutrients, topography, types of soil matrix, development, etc. cause a climatic change to reverberate in a different way in each region, according to the total number of factors which interfere with the growth and development of vegetation. A clear example of the above is, in later times, the process we are now dealing with, where, in the phase called "Manantial de Zohapilco" (Niederberger 1976: 269) a decrease in rainfall occurs and the botanic data available point towards a greater abundance in maize production. Hypotheses on climatic change are useful for the sites where climatic changes took place, as a description of the context wherein the agricultural revolution took place, but they are not an explanation thereoff.

On the other hand, there is a series of theories that mention demographic pressure as the cause for the rise of agriculture (White 1959; Binford 1968; Flannery 1969; Carneiro 1972, etc.); nevertheless, as Bender points out, it has not been demonstrated which was first, the egg or the hen, that is, whether demographic pressure forces man to start crop raising, or whether it is the adoption of this new economy that propitiates a population increase, or else, whether both positions are correct.

Binford's model (1968) establishes that, in order for a shift in the subsistence economy to take place, it is imperative that the group be submitted to pressures that motivate it to reach a new balance in its system. He suggests, using ethnographic parallels to prove his point, that at those sites wherein environment and population are stable, there may be demographic increase whilst resources remain balanced but beneath their capacity; but should there any serious alteration within the environment itself or within the group's demography take place, these would exert pressures strong enough to cause the transition in the food-obtention system. This concentration lies more on demographic pressures than on environmental ones. Demographic pressure views this not in relation to living space, but with regard to the amount of food resources available. There is a difference between internal and external demographic pressures. Internal ones may occur in optimum areas wherein there is a good supply of food resources, thus allowing an important population increase which may be solved either by means of an artificial control or through migration towards marginal areas; however, this latter choice may bring about external pressures, since the area in question may belong to other group's territory. The resulting clash may be avoided either by a drastic decrease in birth rate and life expectancy or by means of a more intensive exploiation of food resources in order to maintain demographic increase. These external demographic pressures may be recurrent and, consequently, they may provide an explanation for successive adaptations. Again, in this case, we can see that the three answers set forward by the author are possible ways of facing pressures: artificial control, increase in the obtention of living resources, and migration; they are not —however— applicable, as a whole, to every situation faced by the Neolithic revolution, and they must be understood separately in each case, i. e. why does any given group chose one of the alternative and none of the others.

Flannery (1985 and 1986) points out that the archæological data show that demographic pressures are not true in every site showing archæological evidence of the rise of agriculture; in regions such as Middle America, the archæological indicators we have up to now show, as Flannery points out (1986, 1985: 241), that population density was too low to be considered as the cause for the shift in the subsistence economy.

Flannery (1985: 242) suggests that, contrary to models based on agents alien to man, climatic changes at sea level and plant mutations are one of the factors for the lack of balance, etc. and points out the idea that agriculture springs as a way of making nature more predictable, that is as "a flexible and diversified strategy aimed at facing an unpredictable succession of rainy years, dry years and average ones", taking advantage of the rainy seasons in order to experiment with plants and, should the results prove satisfactory, gradually incorporate them to their economy during the dry years. It is nevertheless difficult to conceive that within a group of early agriculturers with an equally primitive agricultural technique, the crops raised might represent any kind of security at times of prolongued drought. Another problem would be the development of an efficient storage system. As an example, during later periods, there is the statement made by Lázaro de Arregui (1980: 82–3), Chronicler of Nueva Galicia during the early XVI Century, regarding the problem of crop storage, given the lack of an adequate storage technology.

One of the ideas on the origins of agriculture which has not received much attention is Sauer's proposal (1952: 27) that during the early domestication period food production was certainly not one of the main reasons for having cultivated plants. It has been suggested that, possibly, the first crops raised consisted of plants which were considered magical, medicinal or useful in other ways, which would enable the owner to gain knowledge and experience regarding agricultural techniques before cultivated plants became basic for sustenance. In this sense, Flannery's proposal with regard to the rise of agriculture as a system for predicting nature is very important, as there would already be previous experience regarding crops which are not directly related with the fulfillment of food necessities. In this sense, looking for botanical remains which could help us either to prove or reject the applicability of this theory is of outstanding importance.

Presently, one of the most widely accepted currents is the systemic approach, wherein culture is viewed as an adaptive system; artifacts and social organization forms are seen as necessary tools in the adaptation process regarding environmental situations, both biologically and culturally speaking. Culture is made up by a series of articulated parts or subsystems which include environment, economic activities, technology, social organization, religious beliefs, etc., all of which are intimately related, that is, they are part of a whole. This implies that any alteration in any of the subsystems, no matter how slight it may be, will require a readjustment in all the other sybsystems (Flannery 1968). This approach also presupposes that the system tends to remain in balance, a dynamic balance, wherein all the system is constantly changing (Clark 1968: 50).

Finally, this approach also implies that societies tend to obtain optimum responses with a minimum effort, it being understood, of course, that this is a relative concept, and that each group will understand in a different way the meaning of the term minimum effort (Bender 1975: 27).

Flannery applied this model to Mesoamerica and suggest the existence of a knowledge of natural cycles, both with regard to plants as well as to animals within their different ecological niches, and considers five main food-obtention systems, three concerning plants and two concerning animals, knowledge which will regulate their movility and size during the various seasons of the year. These groups may have occasionally incorporated into their activities an early crop of no importance at the beginning. However, it is feasible that one of these crops could have an extraordinary genetic potential, thus vigorously responding to cultivation. He suggests that teosinte could have played this role, ever-increasing in size and productivity. When beans and gourd joined it, the nourishing potential of these crops became enormous and, consequently, this new system entered into conflict with the seasonal cycle. Eventually, the increased productivity and population growth will force the group to acquire a new balance with this new shift in their food-obtention system.

This ecosystem model is a very particular one within the approach to systems, where our attention lies upon the inter-relations between human organism and its environment, and thus between cultures and their environment (Sahlins 1964: 132). As Bender points out (1975: 29), this presents the limitation of not taking into account the non-material aspects of the system as, for example, religious explanations, which have no place within a merely economical explanation. Flannery, when using this model (1969), sets forth three hypotheses:

1. Transition into food production only takes place when there is a domesticable resource potential present. This affects not only the place where domestication occurs, but also that which is going to be domesticated.
2. The technical ability level must be high enough for a systematic and intensive utilization of resources. Technology includes, besides tools and techniques, the knowledge, ability and method to reach the desired objectives and avoid failure under different circumstances, and,
3. Just like Sauer and Braidwood, he assumes the existence of an extensive economic basis which, together with the technical ability and environmental availability of a wide variety of resources, will allow the development of agriculture. The above is important since it presupposes an intimate knowledge of the resources available, a variety of ecological niches to which plants may be transplanted once experimentation has begun.

Harris (1969) has elaborated a predictive model concerning environmental potential. He points out that, in some instances, there are hunting groups specialized in the intensive exploitation of a limited number of species located in specialized ecosystems characterized by a low diversity of plants and animals, but there is a great amount of each species. On the other hand, other groups may be found within generalized ecosystems wherein there is a wide variety of plants and animals, each of them represented by few individuals of each species. He states that this latter type is generally associated with transition areas. Since most of the plants are adapted to climates with well defined seasons, man is consequently exploiting a wider variety of adapations which enables the survival of plants in periods during which growth is affected by adverse conditions such as cold and drought.

All these theories focus their attention upon the importance of ecology and how it is related to culture. Pumpelly and Childe suggest that agriculture has its origin at an optimum ecological zone ("oasis") during a period of climatic change, whereas accoridng to the theories of Braidwood, Sauer and Binford optimum zones are not a feasible place of origin, instead, they are inclined towards marginal ecological zones. Notwithstanding the differences, each of these theories refers solely to one zone of origin of agriculture *per se*. Flannery's theory on the other hand, considers numerous zones wherein the cycles of regional exploitation —taking into account both optimum as well as marginal zones— contributed directly to the agricultural process. It is obvious that all these theories cannot be correct, but instead of favoring any of them at this moment, we believe it is more important to notice that if, historically speaking, these different theories are true, then they should leave patterns of archæological, paleobotanic, paleontological and paleoecological remains, different and identifiable.

Due to the above, most of these theories, if not all of them, may be tested if we are able to find a regional context ideal for carrying out archæological and paleoecological studies. This region must have:

1. Wild relatives of an important cultigen.
2. Ecological variety which includes both optimum and marginal zones.
3. A stratified cultural history covering from the hunting gathering stage up to incipient agricultural settlements.

Another problem where little work has been carried out and which is most important to understand the regional factors which take part in the domestication/cultivation process and to help to wholly understand the origin of this process is to determine when plant cultivation developed in an autonomous way, and when this was due to the stimulus of other regions which already dominated agricultural technology. There is enough archæological evidence showing that not all domestic species were originated in the same place; moreover, there is evidence enough to believe that several domestic species were domesticated independently at different places and times, starting from different wild races or species.

The Jalisco Case

Jalisco clearly seems to have all these necessary prerequisites. In the first place, recent botanical studies made by Guzmán (1982) and Doebley (1983) among others, have shown the widest variety and concentration of wild relatives of maize in Jalisco. Secondly, Jalisco counts with numerous macro and micro-ecological zones, including optimum regions ("oasis"), such as the Lake of Chapala, Sayula, etc. and a series of valleys, as well as marginal zones —hills, mountain ranges, etc.— with a wide range of intermediate ecological zones, such as valleys between highlands, plateaus, etc. In the third place, there is enough archæological evidence which enables us to infer that the cultural history of this region goes as far back as the Paleoindian period. If human occupation in this region is substantially continuous, then the cultural stages necessary for this investigation will be present in the region. And if the botanical evidence showing that the wild teosinte found in the Western part of Mexico is the ancestor of maize is correct, then the cultural sequence of this region should include an early agriculture stage.

Based upon Harris' proposals (1972 and 1977) we hereby submit as a work hypothesis that spatial distribution of resources across the landscape may influence the mobility structure of the local human population, affecting both its demographic growth rate as well as its probable response to the demand increase. The diversity of botanical species present in this region (Rzedowski and McVaugh 1966; Secretaría de Patrimonio y Presupuesto 1980), enable us to assume the rise of sedentism and plant cultivation. The presence of different ecosystems (lacustrine, swamp, woods/forest, plateau, dell/ravine, etc.) provides, for the region under study, an important amount of local and seasonal variation of food resources which favored the shift in the sustenance economy.

On the other hand, we assumed that, in order to reach the agricultural stage, besides the aspects already mentioned by the above indicated authors (variety of ecosystems, genetic mutation of plants at the adequate moment, demographic pressure, human transportation of vegetables with domesticable potential, etc.), there must exist an accumulation of techniques to increase the growth rate of given resources already known by hunters/gatherers —and which they used only when environmental and cultural conditions were adequate for a change in the sustenance enconomy. As Cohen points out (1977: 30), that means: "That the adoption of agriculture seems no more than another adaptation within a long series of ecological adaptations to the population increase".

Therefore, our research will be centered upon the study of the relationship between the dimension of human populations which inhabited our region and the productivity or productive potential of the various economic strategies employed by the same groups. In order to achieve this, it is utterly necessary to explore our knowledge of the factors which rule the behavior of hunters/gatherers, emphasizing aspects such as seasonal cycles and resources, as well as data concerning environmental changes, working and productivity costs of wild plants.

Considering that the geographical data of this region point out towards a wide variety of microenvironments —characteristic which was mentioned by Flannery (1969), Sauer (1952) and Braidwood (1951) as necessary for the rise of agriculture and based upon Guzmán's biological data (1979), the project in question includes part of the municipalities of Autlán, Ciudad Guzmán, El Grullo, El Limón, Sayula, Tonaya, Tuxcacuexco, Venustiano Carranza in the State of Jalisco, Mexico.

On the other hand, recent studies carried out by biologists, emphasize the importance of the region chosen for this research. Guzmán (1979) states that teosinte or wild maize is represented, within the State of Jalisco, by at least three taxa:

a) *Zea perennis*, considered extinct during a long time in the typical locality, in Jalisco, and recently rediscovered in 3 additional localities in this same state.

b) *Zea diploperennis*, a new entity reported in five localities of the Manantlán Mountain Range, in the south-west of Jalisco, and

c) *Zea mexicana*, found in seven localities, six of which are in the southwest of Jalisco and the remaining one in the northeast of the state. The strongest evidence establishing the analogy between maize and its wild relatives, the annual and perennial teosinte, lies in morphology and cytogenetics (Guzmán 1979: 31). A remarkable fact resulting from the works of Guzmán is the wide variety of teosinte in the southwestern part of Jalisco, where there are perennial species: *Zea perennis* and *Zea diploperennis*, and two races of the annual *Zea mexicana*. It should be remarked that the gender which is more closely related to *Zea*, the *Tripsacum*, also shows a wide variety of species in the same region, where the following entities have been found: *Tripsacum lanceolatum, Tripsacum pilosum, Tripsacum maizar, Tripsacum dactyloides* and *Tripsacum laxum* (Guzmán 1979: 75–76).

This extensive presence, both of maize as well as teosinte and *Tripsacum* in the southeast of Jalisco is important, due to the fact that a unanimous agreement has not been reached yet between biologists and maize genetists as to which is the role of the two latter species in the origin of cultivated maize (Guzmán 1979; Galinat 1971; Beadle 1977; Mangelsdorf 1974). Perhaps carrying out a study in the region where both species are found in a natural way may help to solve the problem.

Guzmán's works led him to conclude that:

> the state of Jalisco is the probable geographic place of origin, distribution and basic diversification of *Zea*. The distribution carried out by various factors (wind, water, animals) and the fact that an outstanding role was played by Prehispanic groups, brought about a geographic specialization and thus contributed to the separation of subspecies and varieties. Out of a population complex of annual teosintes sprung modern maize, the progressive diffusion of which led to a wide genetic diversity, presently recognized for 30 types in Mexico. Teosinte has largely contributed to such diversity, issuing a constant flow of genes... (Guzmán 1979: 82).

On the other hand, Mangelsdorf has suggested that cultivated maize comes from a variety of wild maize now extinct. This author, together with MacNeish, suggest that the oldest cobs from the Tehuacán maize were really wild maize. However, other botanists differ from this point of view. Recently, Doebley (1983) has suggested that, up to now, no sample of wild maize has been discovered, whether alive or fosilized. This same author suggests, after studying the morphology of several maize and teosinte varieties both from Mexico and Guatemala, that the origin of maize is to be found in teosinte, which agrees —with some reserves— with Iltis' new theory in the sense that the female corn cob evolves from the tassel of tesosinte male flowers as a result of a catastrophic transformation. Doebley holds that the morphology of early cobs found at Tehuacán is similar to a transformed male tassel.

This problem regarding maize is a most controversial subject: there are several theories trying to explain it:

1. Wild teosinte is the ancestor of cultivated maize.
2. Wild maize is the ancestor of both cultivated maize as well as wild teosinte.
3. Wild maize gives origin to cultivated maize.
4. Iltis' theory on the catastrophic mutation of teosinte into maize.

The theory which holds that wild teosinte is the direct ancestor of maize is based on the analogy, both morphologically as well as genetically speaking, between both varieties. In fact, maize and teosinte easily form natural hybrids with grains easily reproduced. However, according to Doebley, the corn of maize clearly showing the female part with the edible grains is very different from this same part in the teosinte. The idea is that of an evolution of this critical element of the biological theory regarding cultivated plants.

It is well known that information regarding prehispanic Mexico is very poor, at least with regard to the transition period from nomad hunting/gathering tribes to sedentary agricultural groups; in the specific case of the Western part of Mexico there is almost no information whatever on this respect. On the other hand, if we consider that biologist still don't agree over the biological theories concerning the origin of one of the most important Mesoamerican crops, nor over the interpretation of botanical remains found by archæologists, since "the small corn ears, one inch long, found in the Tehuacán caves and which Mangelsdorf interpreted as wild corn, may just as easily be interpreted as stages in the transition of teosinte into maize through human selection" (Flannery 1985: 251). Therefore, we still have to submit more archæological evidence on this process.

REFERENCES

BEADLE, G.W.

1972. "The Mystery of Maize", *Field Museum of Natural History Bulletin* 43: 2–11.

1977. "The Origins of *Zea mays*", in Charles A. Reed: *Origins of Agriculture*, Mouton Publishers, The Hague: 615–635.

BENDER, B.

1975. *Farming in Prehistory. From Hunter-Gatherer to Food Producer*, John Baker, London.

BINFORD, L.R.

1968. "Post-Pleistocene Adaptations", in S.R. Binford and L.R. Binford (eds): *New Perspectives in Archæology*, Aldine, Chicago: 421–449.

BRAIDWOOD, R.J.

1951. "From Cave to Village in Prehistoric Iraq", *Bulletin of the American Schools of Oriental Research* 124: 12–27.

BRAIDWOOD, R.J. and G. WILLEY.

1962. *Courses Toward Urban Life*, Viking Fund Publications in Anthropology no. 32, Aldine Publishing Company, Chicago.

BYERS, D.S. (ed)

1967. *The Prehistory of the Tehuacan Valley*, vol. 1: Environment and Subsistence, University of Texas Press, Austin.

CARNEIRO, R.

1968. "The Transition from Hunting to Horticulture in the Amazon Basin", *Proceedings of the VIII Congress of Anthropological and Ethnological Science*: 234–243.

CHILDE, G.V.

1929. *The Most Ancient East: The Oriental Prelude to European Prehistory*, Trubner and Company Ltd, London

1944. *Progress and Archæology*, Watts and Company, London

1952. *New Light on the Most Ancient East*, Routledge and Paul Kegan, London.

CLARK, D.L.

1968. *Analytical Archæology*, Methuen and Company Ltd., London.

COHEN, M.N.

1977. *La crisis alimentaria de la prehistoria*, Alianza Universidad, Madrid.

DOEBLEY, J.F.

1983. "The Maize and Teosinte Male Inflorescence: a Numerical Taxonomic Study", *Annals of the Missouri Botanical Garden* 70 (1): 32–70.

FLANNERY, K.V.

1968. "Archæological Systems Theory and Early Mesoamerica", in B. Meggers (ed): *Anthropological Archæology in the Americas*: 67–76.

1969. "Origins and Ecological Effects on Early Domestication in Iran and the Near East", in P.J. Ucko and G.E. Dimbleby (eds): *The Domestication and Exploitation of Plants and Animals*, Gerald Duckorth and Company, London: 73–100.

1985. "Los orígenes de la agricultura en México: las teorías y la evidencia", in Teresa Rojas Rabiela and William T. Sanders (eds): *Historia de la agricultura. Epoca prehispánica S. XVI*, Colección Biblioteca del INAH, Instituto Nacional de Antropología e Historia, México.

FLANNERY, K.V. (ed)

1986. *Archaic Foraging and Early Agriculture in Oaxaca, Mexico*, Academic Press, New York.

GALINAT, W.C.

1971. "The Origin of Maize", in H. Roman (ed): *Annual Review of Genetics* 5, Palo Alto, California: 447–478.

GUZMAN, M.R.

1979. *El Teosinte en Jalisco: su distribución y ecología*, Tesis, Universidad de Guadalajara.

HARRIS, D.R.

1969. "Agricultural Systems, Ecosystems and the Origins of Agriculture", in P.J. Ucko and G.E. Dimbleby (eds): *The Domestication and Exploitation of Plants and Animals*, Gerald Duckworth and Company, London: 3–15.

1972. "The Origins of Agriculture in the Tropics", *American Scientist* 60: 180–193.

1977. "Alternative Pathways Toward Agriculture", in C.A. Reed (ed): *Origins of Agriculture*, Mouton, La Haya: 179–243.

ILTIS, H.H.

1983. "From Teosinte to Maize: the Catastrophic Sexual Transmutation", *Science* 222 (4626): 886–893.

LAZARO DE ARREGUI, D.

1980. *Descripción de la Nueva Galicia*, Colección Historia, Serie Crónicas de Occidente no.1, Unidad Editorial, Secretaría General, Gobierno de Jalisco, Guadalajara, México.

LEE, R.B.

1968. "What Hunters Do for a Living, or How to Make Auto on Scarce Resources", in R.B. Lee and I. Devore (eds): *Man the Hunter*, Aldine Publishing Company, Chicago.

MANGELSDORF, P.C.

1974. *Corn, its Origin, Evolution and Improvement*, Harvard University Press, Cambridge.

NIEDERBERGER, C.

1976. *Zohapilco. Cinco milenios de ocupación humana en un sitio lacustre de la Cuenca de México*, Instituto Nacional de Antropología e Historia, México.

PALERM, A.

1972. *Agricultura y sociedad en Mesoamérica*, Sepsetentas no. 55, México.

PUMPELLY, R.

1908. *Explorations in Turkestan: Prehistoric Civilizations of Anau*, Publications of the Carnegie Institute of Washington 73, two vols., Washington.

RZEDOWSKI, J. and R. McWAUGH

1966. "La vegetación de la Nueva Galicia", *Contributions from the University of Michigan Herbarium* 9 (1), University of Michigan, Ann Arbor, Michigan.

SAHLINS, M.D.

1964. "Culture and Environment", in S. Tax (ed): *Horizons of Anthropology*: 132.

SAUER, C.O.

1952. *Agricultural Origins and Dispersals*, The American Geographical Society.

SECRETARIA DE PATRIMONIO.

1981. *Síntesis geográfica de Jalisco*, Secretaría de Programación y Presupuesto, Coordinación General de Servicios Nacionales de Estadística, Geografía e Informática, México.

WHITE, L.
1959. *The Evolution of Culture*, McGraw Hill, New York.

PROSPECTS FOR COLLABORATION BETWEEN ARCHEOLOGY AND ETHNOHISTORY IN RESEARCH ON THE EVOLUTION OF MESOAMERICAN AGRICULTURE

TERESA ROJAS RABIELA
Centro de Investigaciones y Estudios Superiores en Antropología Social

Recently I felt it necessary to examine the bibliography produced in the last four decades on the evolution and charactersitics of pre-Hispanic agriculture in Mesoamerica. The object of this was to summarize what is known and make it available to a wider public than specialists by means of a collective work to be published by the Ministry of Education.

After reading and analyzing the publications in which archeologists, botanists, historians and ethnohistorians set out and analyze the results of their investigations, I am able now to propose some fields of study and suggest some problematical areas concerning Mesoamerican agriculture and its history that it seems important to research in a co-ordinated manner. Before discussing these I would like to mention some of the difficulties I had with the existing bibliography, in particular archeological.

Firstly I must say that I found it extremely difficult to examine what archeologists, both Mexican and foreign have produced, especially in what concerns the subject in hand. This is not simply because of the technical language or jargon used in the works (which in fact was a real obstacle at times) but rather because most of them are very narrow in their outlook. Only a few try to make general or comprehensive interpretations that take into account, at least from time to time, the "state of the question", in other words the advance in knowledge about, in this case, the general evolution of agricultural societies in Mesoamerica or about some area or particular question related to this. It is also rare to find general theories at least in an explicit form, that might guide the reader as to the general orientation of the work.

In short, there is a notable lack of works that consider the advances made in our knowledge that we owe to direct empirical investigations, and this problem seems particularly acute in the case of works in Spanish. The result is that the gap between the production of knowledge and the integration and diffusion of it to society which, after all is its audience, is becoming ever wider. I believe that in the case of studies on Mesoamerica, Mexico and the various Central American countries should be the first to hear the results of research.

Another problem I found in the production of scientific material on Mesoamerica is the low level of collaboration among specialists and the poor feedback among the several disciplines. I feel this problem is more acute in the case of Northamerican archeologists working in Mesoamerica than among their Mexican counterparts, but it exists even among these, despite the fact that the academic tradition in Mexico has always been one of greater coordination and collaboration. I think this is a very sad state of affairs since archeology has the opportunity of inquiring into the long period before the Conquest, but it is difficult for it to establish certain aspects of a socio-cultural nature that leave very few direct clues in physical remains. Ethnohistory has a much more restricted vision of time but one that is very rich in the socio-cultural field.

One possible explanation of the little collaboration and feedback between these two disciplines could be that the specialists in the fields do not ask themselves the same questions or approach the same problems and it is this that produces results and conclusions that are dissimilar or even incompatible. There is no doubt that the problem must be analyzed more thoroughly, but it seems that we have been unable to reach any agreement about planning or at least co-ordinated research on the basis of certain suggestions or certain hypotheses and conceptual frameworks to orientate investigation.

To sum up, my proposal is that in the case of a study of the evolution of agriculture and of the economic basis of pre-Hispanic societies, both the possibilities and need for collaboration are great, urgent and promising. This should not be postponed any longer.

From the wide range of possible subjects for study linked to this question I shall limit my comments to some related to the agriculture of the area and its history, viewed through the changes in technology, the domestication and diversification of the plants grown.

As I noted earlier, the regional and general evolution of agriculture (and of the economy) is a phenomenon that can only be understood through archeological data. The period covered by the documents written in the sixteenth century which we ethnohistorians use is very short and goes only as far as the Toltec period. In regard to this late period known through written documents it has been proposed that pre-Hispanic agriculture was based not on the richness of complexity of its tools but on the perfection of both its labor–intensive agricultural methods and of the wide and varied range of domesticated plants (Rojas 1985: 7). Also that "Highly developed forms of coordination and organization of work replaced or at least made up for the lower technical development of tools" (*Ibid.*).

These theories which as remarked earlier are based on late information about the pre-Hispanic period, need to be put to the test and if necessary improved on if we wish to demonstrate their validity for earlier periods or stages. To these must be added others that consider basic questions such as the natural conditions of the territory where the agricultural cultures or societies of Mesoamerica developed. These conditions were particularly favorable, existing in a territory that was very varied both in climate, soil, vegetation and hydrography. This generally resulted in the simultaneous exploitation of microclimates at relatively short distances from each other both horizontally and vertically. This diversity, important in any of its aspects (soil, climate, vegetation or hydrography) was especially important in the case of vegetation. It was rich and varied, characteristic of Mexico's position on the northern limits of the tropics, with climates that varied from those typical of medium latitudes to subtropical and tropical ones (West 1964: 364).

I just wish to mention very briefly, not dwell on, how unwise it is to use technological criteria exclusively to evaluate or classify the level of development of Mesoamerican societies or of their agriculture. In 1964 V. Gordon Childe himself wrote with very solid arguments that technological criteria are "really misleading" when archeological data of societies different from those for which it was originally designed and in which it had a chronological meaning is evaluated sociologically or anthropologically. Referring to the "Three Ages" system he said "...it was originally designed as a chronological framework for the classification of the pre-historic artifacts of a single homogenous area: Denmark" (useful also in England, Switzerland and other comparable geographical units) (Childe 1959: 307).

For Childe "To label an artifact or a society 'Neolithic' for example says absolutely nothing about when the artifact was made or when this society lived" (1959: 307).

In other words, it is useless as a chronological principle, nor does it say anything about the "cultures, that is, the societies whose purposes and needs determined the characteristics of the artifacts" (Childe 1959: 307). He goes on to say that "the tools and weapons whose material serves as a criterion for classification in no way exhaust the labor-force in the society".

The transport available and even natural resources can also exercise an influence on social structure. Secondly it is not the labor–force that is the immediate determinant but the method of production, the economy within which these forces can operate. If archeological data is to be really useful in social science it must be classified nowadays on a new, less superficial basis (Childe 1959: 309).

In the case of Mesoamerica, of course it is wrong to consider the features of artifacts as a chronological criterion or economic classifier and in fact this becomes an obstacle in understanding this unique, independent endogenous process. However, the fact must not be hidden that the use of this criterion and this type of interpretation is very widespread and also deeply rooted. On another occasion I commented on how common it is among some Mexican writers and historians to describe ancient indigenous societies as primitive, technologically backward, Neolithic or in other similar terms, mechanically adopting models of evolution (produced by nineteenth-century unilinear evolutionism) that have been superseded (Rojas 1985). This is why it is so regrettable to find this sort of classification in such recent, widely read works as for example Enrique Semo's *Historia del capitalismo en México* (1973: 29) (*The History of Capitalism in Mexico*). It does nor help us to understand either the development or the characteristics taken by the productive potential of the pre-Hispanic era or of the production methods of these societies.

So, this author writes that "At the beginning of the sixteenth-century the material culture of the Aztecs was in a state of transition from the Neolithic to the Bronze Age. However in productive potential there were important differences between it and the conditions in which most of the peoples of Europe and Asia went through the same stage. Wheat, barley, rye and rice were unknown; cattle and the associated occupations did not exist. The wheel was not used either in transportation or in irrigation. The plow, that made its appearance towards the tenth century BC was not known among the Aztecs..." (Semo 1973: 29).

We must admit, as I pointed out at the beginning of the paper, that much of the responsibility for these types of concepts and deformations of known facts lies with the scientists dedicated directly to the relevant investigation and its dissemination to us archeologists and ethnohistorians who have not done what is necessary or done enough for these viewpoints and interpretations to be either rejected or corrected.

Two recent studies by R. A. Donkin (1979) and the present writer (Rojas 1982) organize and classify the wealth of historical data contained in sixteenth century sources referring to the inventory of tools and implements used in Mesoamerican agriculture. These works, although preliminary, bring out several points that are worth looking at carefully. First, the usefulness of making broad studies of subjects that are apparently very specific, but which are important in the wider context of the interpretation of the characteristic features of the type of agriculture and the level of agricultural development that had been reached at a given moment, in this case the period of contact between the Old and the New World.

Second, the need for something similar to be attempted for previous stages based on archeological data gathered from studies carried out in different regions of Mesoamerica. An ethnohistorian working on these questions finds, on analyzing the inventory of pre-Hispanic agricultural implements recorded in sixteenth century written sources, that only a few of them are mentioned by archeologists in their publication as existing in earlier periods, so that it is very nearly impossible to trace their history or their possible regional distribution earlier than the sixteenth century.

It will eventually be possible to clarify many points (not only purely technological) on the basis of archeological research such as the sequence in which the different types and classes of tools appeared, their distribution and regional variations, their modification or persistence as time went by, the type of use, specialized or multiple, to which they were put, their association with different types of agricultural systems, among the main ones.

Archeological literature rarely deals with these problems specifically and when it does it generally presupposes that there was little change in the types of implements ever since the moment they appeared, and consequently very little innovation in the type, although there was in the techniques by which they were made. Attention is drawn to the persistence of "types of equipment and techniques" in the succession of Mesoamerican cultures from very early times (Armillas 1985: 28; Sanders 1957: 338).

As regards the types of *implements* used in agriculture, historical documents give us a picture showing more variety than we generally accepted, but I wonder whether contemporary archeologists know when these originated and what the distribution of the main ones was. This information would be a basis for testing the validity of statements about little technological change or of theories as to a certain insufficiency in the economic and technological base that enables a certain point to be reached in development but not beyond without a great crisis (Palerm, in Kirchhoff 1983: 51).

Turning to the inventory and characteristics of the *plants grown* in the area, there is much to be discovered and many possible areas of collaborations (some already explored) not only between archeologists and ethnohistorians but also between the latter and botanists. In the question of the process of domestication of the plants grown there are many points that archeology could clarify and ethnohistory enlarge upon, for example the origins and dispersion of the basic plants characterized by their wide distribution and low degree of specialization (I am referring principally to corn, beans, squash, amaranth and chili). In the sixteenth century, according to existing ethnohistorical and botanical studies these plants, as a result of constant selection and hybridization existed in many varieties adapted to a wide range of microclimates located practically all over the agricultural lands of Mesoamerica at that time. It would be important to define something similar about these species with a narrower distribution that were grown in more restricted areas, for example cotton, agave, cacao, tobacco, sweet potato, yucca or *huacamote*, sage (*chía*) and avocado among the main ones (Rojas 1985: 225–250).

It is of fundamental importance for an understanding of the evolution of agriculture, to discover more about the gradual increase in the yield of cultivated plants through the process of domestication, which ethnohistory can only document partially for the moment of contact and later centuries. Quite a number of very useful studies have been published on this subject, but they are still only partial from the point of view of region and period (referring above all to Tehuacan and the Valley of Oaxaca) (see especially Byers [ed] 1967; Flannery and Marcus [eds] 1983; García Cook 1985, and Winter 1985). This is an important subject, since one of the theories to explain the phenomenon of the rise of civilization and urbanism (with the different specializations it implied) is that of the gradual increase in agricultural productivity, based not so much on the transformation or improvement of tools and equipment (and the consequent increase in labor productivity) but on the advances made in agricultural methods and techniques and in the greater yield of cultivated species.

The change and characteristics of the ***pre-Hispanic diet***, its regional variations, the ways of converting and preparing agricultural products are, among others, fields where archeology and ethnohistory can collaborate to advantage.

The evolution of ***agricultural methods and techniques***, including ***irrigation*** and the ***systems of cultivation*** is inseparable from that of the processes mentioned above (tools and equipment, plants cultivated). All are components of the history of the formation of the variety of complex agricultural systems of Mesoamerica.

The study of how these systems developed, of their origin, diffusion and diversification, as well as of the factors that played a part in their transformation has been linked to the ideas of Childe about the agricultural revolution and the urban revolution in the Old World, to the multilinear evolutionism suggested by Julian H. Steward, as well as the theory of societies of an Asiatic or hydraulic type formulated by Karl A. Wittfogel. These ideas established the bases for a reinterpretation of the information existing in the 1940's and 50's on the evolution of civilization and also for a search for new information. Multilinear evolutionism and the model of Wittfogel, the two most influential theories, attributed a fundamental and causal role to irrigated agriculture in the rise and development of urban civilization, as it led to the intensification of agriculture, the increase of potential and the production of the necessary surplus.

Other, later lines of interpretation, some derived from those mentioned above and other completely new ones, such as that of the economist Ester Boserup (1965) had positive effects on research, especially that into the economic bases of Mesoamerica conducted by archeologists and ethnohistorians (see the account given by Palerm in 1970, 1972: 166). The most interesting problem in this field was that "...of the capacity of agriculture to produce sufficient social surplus to maintain, on the one hand, an elementary division between town and country and, on the other, a genuine system of social stratification" (Palerm 1972: 170). At that time there was a battle (and still is today to a large extent) against prejudices and preconceived ideas based on the extrapolation of data about current native agricultural systems on the past (as in the case of the badly names "cornfield" [*milpa*], in fact long fallow cultivation system or swidden) and on out-of-date patterns of unilinear evolution.

Research into documentary sources done by Angel Palerm, Pedro Armillas, William T. Sanders and other writers during the 50's and continued ever since then into technological and economic aspects, especially of pre-Hispanic agriculture, was the basis for destroying, as Palerm so rightly noted, "The myth of purely subsistence farming...capable of generating only very limited surplus" (Palerm 1972: 173). There was still much to do of course. The question of the age of agriculture rested in the hands of archeology. From those years on it was helped by these ideas and suggestions, and the indications and remains of intensive systems of cultivation and of irrigation works "...began to appear in many places..., numerous agricultural terraces were located, as were ponds, reservoirs and dikes for diverting rivers" (Palerm 1972: 174; for a summary see Rojas 1985).

One of the problems that archeology had to solve was whether the complex systems of agriculture "with water management" (not irrigation, to use a wider concept that includes both environments where water was extremely scarce, in deserts and arid zones, and others with a plentiful supply, in the humid tropics and marshy areas) had been the foundation from which the societies of Mesoamerica progressed from a natural economy to a political one (from mere subsistence to the generation of productive and social surplus). Also, it had to determine exactly when and where this occurred.

Today, many of the questions that arose regarding the type of irrigation works, their age, the "technical and geographical scale" of water management and its socio-political consequences have been documented archeologically and the matter partially solved in the case of some regions. We have reasonably satisfactory reconstructions of the agricultural and socio-political evolution of the valleys of Tehuacan, Oaxaca, Nochixtlan, Mexico, Puebla and Tlaxcala, the Cuicatlan *cañada* and also of some zones in the Maya area such as Peten, Belize and Tikal[1].

In almost all of these regional investigations the most common and favored formula has been a combination of archeology, botany and ethnography, while ethnohistorical research by region is less frequently found. For some reason that is not completely clear to me, many archeologists have preferred to use ethnographical analogy rather than data from sixteenth-century sources.

With ethnohistorical studies as a basis, those who have dealt with the question of pre-Hispanic native agriculture have reached some significant conclusions. Among these I include those of Palerm and Armillas in particular as well as other writers such as Donkin (1970, 1977, 1979), Meyer L'Epée (1943), and Eva Hunt (1972). In a recent work (Rojas 1985) I have been able to define and go deeper into a series of obscure points related to the pre-Hispanic agricultural methods and techniques used in the various types of cultivation systems.

From this analysis the existence becomes clear both of intensive agricultural systems in which the land was worked in various degrees of intensity and of an inventory of tools and equipment that is wider than was previously believed; of a set of practices of combinations and rotation of crops that increased agricultural productivity and methods of handling crops that were individual or particular (from sowing to harvesting and selection of seed, passing through the stages of weeding, watering, fertilizing and the conservation of the humidity) that led to a higher yield per plant and per surface area, as well as being the basis for the creation of numerous varieties and sub-varieties with which it was possible to achieve a better management and use of an environment that differed horizontally (in soil, vegetation, humidity) and vertically (altitude and the consequent differences in climate, soil and vegetation).

It can today be firmely stated that if natural diversity was the dominant characteristic of the regional countryside of Mesoamerica, variety and multiplicity of agricultural methods was the dominant characteristics of agricultural technology.

In spite of all the progress that has been made, there still remains a lot to do in these areas. I would like to conclude by mentioning some of the fields that it has not been possible to study properly, maybe because of their difficulty and the lack of pertinent data.

1. The regional distribution of the types of agricultural systems in the sixteenth century (and also in historical perspective), both from the point of view of technology and the intensity of frequency of land use and the crops grown in each type.
2. The amount of land necessary for subsistence and for the production of socially necessary surplus in each of the regions (always suppossing the combined and simultaneous use of various types of agricultural systems and a diversification of crops).
3. The yield of the different types of crop per seed or per seedling planted and per area of land sown, in the different types of systems (according to intensity of agriculture and the technology employed).

---

[1] Without claiming to be exhaustive, the following is a list of the main works in which these regional syntheses have appeared. On Tehuacan: MacNeish 1967: 290–309; the Valley of Oaxaca, Nochixtlan, Tehuacan and Cuicatlan: Flannery 1983 and Marcus 1983; on Nochixtlan: Spores 1969; Tehuacan, Puebla, Tlaxcala and Mexico: García Cook 1985; on Oaxaca: Winter 1985; for the Maya area see Sanders 1973; Harrison and Turner II (eds) 1978, Flannery 1982, Dahlin 1985.

4. Ways of organizing agricultural work, the exclusivity of these to agricultural activity or the extension of them to other activities (public works, war, ritual services).
5. Productivity, number of days invested, lenght of the working day, products obtained in the various types of agricultural systems.
6. The regional distribution of the types of plants cultivated, the absolute and relative importance, specific methods, cultivation, rotation with other species, seasons and cycles of cultivation.
7. Ways of converting and preparing agricultural produce.

REFERENCES

ARMILLAS, P.

1949. "Notas sobre sistemas de cultivo en Mesoamérica. Cultivos de riego y humedad en la cuenca del río Balsas", *Anales del INAH* t. III: 85–113.

1961. "Land Use in Pre-Columbian America", in L. Dudley Stamp (ed): *A History of Land Use in Arid Regions*, UNESCO, Arid Zone Research XVII: 255–276.

1971. "Gardens on Swamps", *Science* vol. 174, no. 4010: 653–661.

1985. "Tecnología, formaciones socio-económicas y religión en Mesoamérica", in: J. Monjarás-Ruiz, R. Brambila and E. Pérez-Rocha (eds): *Mesoamérica y el centro de México. Una Antología*, México, INAH: 25–40.

BOSERUP, E.

1965. *The Conditions of Agricultural Growth*, Chicago, Aldine Publishing Company.

BYERS, D.S. (ed)

1967. *Environment and Subsistence, The Prehistory of the Tehuacan Valley*, vol. 1, Austin and London, University of Texas Press.

CHILDE, V.G.

1950. *Qué sucedió en la Historia*, Buenos Aires, Lautaro.

1959. "La arqueología como ciencia social", in: *Homenaje a V. Gordon Childe*, UNAM, Suplemento del Seminario de Problemas Científicos y Filosóficos, Segunda Serie, no. 17.

1964. *Evolución social*, México, UNAM, Dirección General de Publicaciones.

DAHLIN, B.H.

1985. "La geografía histórica de la antigua agricultura maya", in: T. Rojas and W. T. Sanders (eds): *Historia de la agricultura. Epoca prehispánica-Siglo XVI*, 2 tomos, México, INAH, Biblioteca del INAH 2: 125–196.

DONKIN, R.A.

1970. "Pre-Columbian Field Implements and their Distribution in the Highlands of Middle and South America", *Anthropos* vol. 65: 505–529.

1977. *Spanish Red. An Ethnogeographical Study of Cochineal and the Opuntia Cactus*, Philadelphia, The American Philosophical Society, vol. 67, part 5.

1979. *Agricultural Terracing in the Aboriginal New World*, Tucson, The University of Arizona Press, Viking Fund Publications in Anthropology no. 56.

FLANNERY, K.V.

1983. "Precolumbian Farming in the Valleys of Oaxaca, Nochixtlan, Tehuacan and Cuicatlan: A Comparative Study", in: K.V. Flannery and J. Marcus (eds): *The Cloud People. Divergent Evolution of the Zapotec and Mixtec Civilizations*, New York, Academic Press: 323–339.

FLANNERY, K.V. (ed)

1982. *Maya Subsistence*, New York, Academic Press.

FLANNERY, K.V., and J. MARCUS (eds)

1983. *The Cloud People. Divergent Evolution of the Zapotec and Mixtec Civilizations*, New York, Academic Press.

GARCIA COOK, A.

1985. "Historia de la tecnología agrícola en el altiplano central desde el principio de la agricultura hasta el siglo XIII", in: T. Rojas and W. T. Sanders (eds): *Historia de la agricultura. Epoca prehispánica–Siglo XVI*, 2 tomos, México, INAH, Colección Biblioteca del INAH 2: 7–75.

HARRISON, P.D., and B.L. TUNER II (eds)

1978. *Pre-Hispanic Maya Agriculture*, Albuquerque, University of New Mexico Press.

HUNT, E.

1972. "Irrigation and the Socio-political Organization of Cuicatec Cacicazgos", in MacNeish and Byers (eds): *The Prehistory of the Tehuacan Valley*, vol. 4, Chronology and Irrigation: 162–253.

KIRCHHOFF, P.

1983. *Principios estructurales en el México antiguo*, T. Rojas (ed), México, Cuadernos de la Casa Chata 91.

MEYER L'EPPE, C.

1943. "La agricultura de México en la época precortesiana", *Investigación económica* vol. 2: 375–385.

PALERM, A.

1972. *Agricultura y sociedad en Mesoamérica*, México, Sepsetentas 55.

PALERM, A., and E.R. WOLF

1972. *Agricultura y civilización en Mesoamérica*, México, Sepsetentas 32.

ROJAS RABIELA, T.

1982. "Los instrumentos de trabajo agrícola en el siglo XVI", *Biótica* vol. 7, no. 2: 205–222.

1985. *La agricultura indígena en el siglo XVI*, Tesis de doctorado Universidad Iberoamericana.

ROJAS RABIELA, T. and W.T. SANDERS (eds)

1985. *Historia de la agricultura. Epoca prehispánica–Siglo XVI*, 2 tomos, México, INAH, Biblioteca del INAH.

SANDERS, W.T.

1957. "Tierra y agua", *A Study of the Ecological Factors in the Development of Mesoamerican Civilization*, Ph.D. Dissertation, Harvard University.

1962. "Cultural Ecology of Nuclear Mesoamerica", *American Anthropologist* vol. 64, no. 1, part 1: 34–44.

1968. "Hydraulic Agriculture, Economic Symbiosis and the Evolution of States in Central Mexico", *Anthropological Archæology in the Americas*, Washington, D.C.

1972. "Population, Agricultural History, and Societal Evolution in Mesoamerica", in B. Spooner (ed): *Population Growth: Anthropological Implications*, Cambridge Londres, The MIT Press: 101–153.

1973. "The Cultural Ecology of the Lowland Maya: a Reevaluation", in P. T. Culbert (ed): *The Classic Maya Collapse*: 325–366.

1976. "The Agricultural History of the Basin of Mexico", in E. R. Wolf (ed): *The Valley of Mexico*, Albuquerque, University of New Mexico Press: 101–159.

SANDERS, W.T., and B. PRICE

1968. *Mesoamerica. The Evolution of a Civilization*, New York, Random House.

SEMO, E.

1973. *Historia del capitalismo en México*, México, Ediciones Era.

SPORES, R.

1969. "Settlement, Farming Technology, and Environment in the Nochixtlan Valley", *Science* vol. 166, no. 3905: 557–569.

STEWARD, J.H.

1955a. "Some Implications of the Symposium", *Irrigation Civilizations: a Comparative Study*, Washington, D. C., Pan American Union, Social Science Monographs no. 1: 58–78.

1955b. *Theory of Culture Change. The Methodology of Multilinear Evolution*, Urbana, University of Illinois Press.

WEST, R.C.

1964. "The Natural Regions of Middle America", in: *Handbook of Middle American Indians*, vol. 1: 363–383.

WINTER, M.C.

1985. "Los Altos de Oaxaca", in T. Rojas and W. T. Sanders (eds): *Historia de la agricultura. Epoca prehispánica–Siglo XVI*, 2 tomos, México, INAH, Colección Biblioteca del INAH, 2: 77–124.

WITTFOGEL, K.A.

1955. "Aspectos del desarrollo de las sociedades hidráulicas", *Las civilizaciones antiguas del Viejo Mundo y de América*, Washington, D. C., Unión Panamericana, Estudios Monográficos I: 45–54.

1966. *Despotismo oriental*, Madrid, Ediciones Guadarrama.

1972. "The Hydraulic Approach to Pre-Hispanic Mesoamerica", *The Prehistory of the Tehuacan Valley*, vol. 4, Chronology and Irrigation: 59–80.

THE URBAN REVOLUTION

OBSERVATIONS ABOUT THE THEORETICAL TERM "ARCHAIC STATE"

MANUEL GÁNDARA
Instituto Nacional de Antropología e Historia

1. *Scientific Change and Rationality*

The slogan "Let theories proliferate!" is sensible only if the new theories improve the existing ones, or at least offer the promise of improving them in a reasonable time. Scientific rationality requires that the new proposals be justified either by showing theoretical shortcomings of older proposals or evidence against the expectations of the existing theories. Unfortunately, this simple and apparently obvious rule is not always followed by archæologists. In recent years there has been a proliferation of theories on the origins of the state. But criticism of extant theories is in many cases nothing but a mockery of the real process of theoretical criticism and substitution.

Why should this concern us? Well, first, because proliferation under those conditions represents no real progress for archæology, since the basic principle of scientific rationality is violated; second, because theories and authors are unfairly treated as "refuted"; third, because when a theory is rejected as superceded, work on the tradition to which the theory belongs is discouraged, perhaps diverting the attention of academic communities from research lines which may eventualy turn to be more fruitful. Finally, because many of these "refutations" reveal logical and methodological flaws severe enough to embarass anybody claiming to be a social scientist, and act against the credibility of the discipline as a whole.

These considerations are not foreign to Childe's work which brings us here today (cf Pérez 1981). His theoretical contributions have been hailed as "verified" or dismissed as "superceded". Childe is normally included today among the theoreticians on the origins of the state. And he is included for worse. In particular, his ideas on the urban revolution and its role on the origins of civilization have been the object of supposedly disqualifing critiques.

For instance, Service —after acknowledging the global importance of Childe's work— goes on to say that the "urban" conception, with its class struggle implications, is today inadequate, despite how much more sophisticated it may be in relation to the classical marxists. Let me quote: "Childe's 'urban' conception has been mostly superceded; we know that some archaic civilizations lacked cities, while others became states before their cities developed" (Service 1978: 26–27).

There is something immediately peculiar in this quote: Service talks about civilizations without cities, when for Childe their presence is indispensable for something to be a civilization. In any event, for Service, Childe is mostly superceded. You may think that the shallowness of the criticism is a consequence of limited space (the article in question is a short one), but Childe fares no better in Services's book on state origins (Service 1975). And clearly many authors would side with Service, otherwise their own new theories would be of childean flavor and they are not; we are forced to assume that they advance them because they overcome theoretical or factual shortcomings of Childe's theory.

But please do not think that Service is alone in this: what could be called the "dogmatic falsificationist approach" seems to dominate the discipline these days, claiming more than one theoretical causality on its way. In fact, Carneiro (1970) must hold one of the first places in terms of the number of times he has been "refuted", maybe second only to Sanders, Parsons and Santley (1979), whose book seems to have become the paradigm of a refuted theory according to its critics. I have raised my doubts elsewhere (Gándara 1981) about this kind of "refutations", since they normally result from serious errors in the logic of refutation and in general from the misapplication of some relatively simple and well known epistemological and methodological principles.

However, one should not be surprised that "refutations" and "verifications" are handed out so freely when it would seem that the attending theoretical analysis reduces itself to the classification of theories (internal/external, integration/conflict, etc.), the sort Service (1978) himself holds; or to flowchart representations (disastrously circular: in some the origin of the state causes itself!) like Redman's (Redman 1975: 230). Both procedures, while maybe justified as means of exposition, are far from being sufficient. For instance, Childe's ideas on the origins of civilizations are variously classified as being "exchange", "demographic", "internal conflict" or "technological development" theories, according to who builds the classification (compare Service 1978 to Claessen and Skalnik 1978).

These deficiencies in theoretical analysis lead to what I have called, in a metaphor that upsets some because of its colourfulness, "premature empirization": giving prematurely the last word to the data, both for "refutation" and for "verification". The results are as untenable in one case as in the other. Could not we do better?

2. *Theoretical Analysis*

Theoretical analysis could go beyond classifications, which of necessity reduces the richness of a theory under the classification's criteria, and beyond flowcharts, which destroy or distort the theory's causal relations and obscure the understanding of its explanatory principles. We claim that the analysis should include at least six areas: a) a study of the features of the theory's nomological principles; b) a consideration both of its problem–situation and of its explanatory capacity and fertility; c) the identification of the kinds of entities it postulates, as well as the kind of causal relations it establishes; d) evaluation of some formal properties like simplicity and parsimony; e) the identification of the models of man, society and history it implies, as well as the social determinants of the theory; and finally, f) the empirical area, where the theory now suitably analysed and made explicit can be tested against data.

As you may have probably noted, I hold the view that a theory is a set of systematically related statements, which include at least one law-like principle and which is susceptible of corroboration; that theories are advanced to solve explanatory problems, and that we can evaluate comparatively how well they do this.

In fact, my whole proposal depends on at least three very general and simple principles: 1) that one should treat theories just as what they are, to wit, theories; 2) that a theory is what a theory says; and 3) that before you can refute or corroborate a theory you must have an analysis of what the theory says. We arrive this way to my first conclusion, one which I offer you as subject of debate: no theory may be "superceded" or "refuted" by data in the absence of a rigourous theoretical analysis.

Relax. I do not intend to carry out a full analysis of Childe's theory here; not only because of space limitations, but also because it implies many difficulties: social scientists do not normally make their theories explicit, thus making indispensable to first embrace on a careful exegesis to obtain at least a "rationally reconstructed" version of the theory. Of course, in that moment the theory becomes not only the author's but rather a shared responsibility. And in the case of Childe's proposition there is seemingly no agreement on the *locus classicus* of the theory, even when his article on the *Urban*

Revolution (1981, orig. 1950), *What Happened in History* (1973, orig. 1942) and the second edition of *Man Makes Himself* (1954, orig. 1951) are frequently pointed out.

Rather, I want to insist that, strictly speaking, data can not do anything, for better or for worse, to a theory for which we do not even know exactly what it says —and here I am not talking about the "story" the author tells, but to the nomological principles that could receive corroboration or refutation. In the absence of a detailed analysis we risk being unfair, the way Service is: he seems to think that the urban process in causally central; as he finds "civilizations" and states without cities, he claims the theory is superceded (by the data, it is understood) (Service 1978: 27).

Scholars loyal to Childe may be tempted to defend him producing data to correct Service. But doing so implies that, in the absence of a theoretical analysis, we are taking Childe not in his own terms, but in Service's. Any positive evidence would support only Service's version of Childe, and Childe would have been defended only to the extent Service's version is true to the original. Evidently, this line of reasoning hits not only Childe's detractors, but also those claiming to have "verified" his or any other theory without the benefit of a serious theoretical analysis.

3. *A Comedy of Equivocations?*

Now, I have pointed out elsewhere that in many cases these "Mickey Mouse" refutations are guilty of, among others, an elementary logical fallacy known as "equivocation". It consists, briefly, in suprepticiously substituting terms in the course of an argument, frequently the original ones with our own. We have seen that this is precisely what Service does with Childe's term "civilization".

Another example of this kind of error (which I think is be an honest error), is Wright and Johnson's "refutation" of Carneiro (1975). In this case, data supposedly shows the state emerging in a moment of depopulation and not of demographic pressure as, according to them, Carneiro (1970) would require. One could question the reliability, representativity or truth of the data, as Weiss (1977) has done, and try to "save the theory" this way. But that would imply the test had been fair; and, if we look closely, it turns out that they used their own definition, from which certain archæological data (we call them "*indicadores*" —indicators— in spanish) follow, and not Carneiro's, from which a different set of indicators would be obtained. As a result, the state emerges "earlier" than the theory predicts, and the theory is "rejected". Sanders *et al.* (1979: 366) complain of a similar error committed by Adams, when this author rejects Wittfogel using his own notion of "urban society" and not Wittfogel's notion of "oriental society"; he uses a different set of indicators which purportedly show the state to be the cause of large scale irrigation and not the other way around.

The preceding examples have a common pattern: they substitute, in the process of testing a theory T, the key theoretical terms of T with those of a different theory, T^1, which is the one the tester normally holds.

To see why this substitution is fallacious, let us review the process of archæological testing. To test a substantive theory using archæological materials, one needs to establish through "arguments of relevance", the type of data ("*indicadores*" or "indicators") which will allow us to relate terms in the substantive theory with terms in the archæological theory; these, in turn, are related by new "arguments of relevance" to terms in the antecedently understood language shared by an academic community. If this account of archæological testing is correct, then when two theories define a term differently, the arguments of relevance which derive from it will tend to be different as well, and so will be the indicators which derive from these. This is why all indicators are always indicators in relation to theoretical terms, and hence there is no such a thing as a "neutral" indicator.

Thus, if "civilization" implies cities for Childe, and the theoretical term "city" is defined archæologically through ten indicators, any test which claims to be a test of Childe would have to use that definition of "city". But if someone characterizes "civilization" differently, focusing in an institutionalized form of power, then perhaps the key theoretical term will be "government", which would have a different set of indicators that may not necessarily coincide with Childe's. Accordingly, when testing the theory, we may look for sites and periods which may or may not match the ones required by the original theory, since its original reference will have been shifted. Incidentally, this devaluation acts normally in favor of the criticism of the initial theory.

We may draw a second conclusion from all of this: given the way substantive theory articulates with archæological theory, there are no "neutral" indicators, since all depend on the arguments of relevance which derive from the key terms in the substantive theory. This is why theories of state origins which come from non-archæologists are normally at a disadvantage when they are not fully explicited and developed: in these cases it will be the archæological testing of the theory the one who contributes the required arguments of relevance, and as we have seen, this is not always done fairly.

Hence, a refutation is valid only if there are no equivocations about the key terms of the theory tested. As you can see, in this case we can make a defense of Childe which does not require a search of data to contradict Service: one of the advantages of performing a theoretical analysis, however partial, is precisely to help us solve theoretical problems treating them as such, thus avoiding falling into "premature empirization".

4. *Commensurability*

To summarize: in order to legitimately test T against the data, the theoretical terms involved should be those of T, which will then select, via the arguments of relevance of T, the relevant indicators and hence the required data.

But perhaps Service did not mean to be fallacious, but to express a basic desagreement with Childe regarding civilization: he is advancing an alternative theory. What happens when what you want is not to test T against the data, but rather to compare it with a new theory, T^1, in order to be able to select the better one of the two? Probably the theoretical terms and arguments of relevance will not coincide. How to compare, then, and thus fulfill the desideratum that the rational growth of science requires that new theories improve on old ones? Is it legitimate to compare different theories about different things? Can it be done at all?

It would be tempting to answer that it is not legitimate, and this way to rescue Childe. Service would have no right to prefer his own theory over Childe's, since they are about different things (one about the origins of civilization, the other about the origins of the state level of integration). It seems intuitive that in order for a theory to refute another theory both should be about the same things. However, this intuition can lead to absurd consequences. The first one is relativism: each theory would be equally respectable, making it impossible to choose among options.

Another absurd consequence would rise if "about the same things" was to be interpreted in the strict sense of "about the same theoretical entities"; that is, if in order for a theory to refute another theory one would demand that they dealt with the same theoretical principles. This would be the equivalent of asking that no new theories be developed, since if any new theory would have to deal with pre-existing theoretical entities, then scientific change would be practically impossible. "Phlogiston" is not the same entity as "oxigen"; not even "mass" is the same for Newton and for Kepler, according to the experts. And yet we say that in these cases the new theory refuted the previous one: That is, under certain conditions, theories with different theoretical apparatuses may enter in a legitimate relation of contrast and refutation. Besides, if we were to follow the absurd criterium mentioned above, then we would not be able to compare not only theories about civilization with theories about state formation, but we could not even begin to compare two different theories on state origins: since each would have different defintions of "archaic state", they would not be about the same things;

and even an author who changed through time his notion of "state" could not compare his previous version with his new one, a conclusion which is patently absurd.

But going back to Childe, it seems then that we are faced with the horns of a dilemma: if there is a sense in which two theories about different things can compete, and Service was to be right, then Childe's theory about the origins of civilization would be refuted by Service's theory on state origins: on the other hand, if different theories can not be about the same, then Childe is irrelevant to the current discussion about state origins, to the extent to which his theories are not about that problem. It would seem that we have to choose between a refuted Childe and an irrelevant Childe. But it is quite clear that, intuitively, we would all agree that both theories are about the same. The question is again: can two different theories be about the same thing? What is that sense that seems intuitively obvious?

5. *The Reference of Theoretical Terms: Two Solutions*

It turns out that defining this "intuitively obvious" sense has been incredibly difficult. Specialist have debated heatedly during the last two decades about which solutions may be considered acceptable. The position we propose here solves the problem but requires the explicit adoption of a materialist ontology and epistemology. This does not worry me, since I am already convinced, with Childe, that materialism is the position most congruent with day-to-day scientific practice.

a) *Causal Chains and Introductory Events*

What we need is a way of having concepts in different theories refer to the same reality, a condition necessary in order to be able to hold that succesive theories are about the same things, and hence fulfil the rule of scientific rationality; concepts which, while being defined differently for each theory, can have a "trans-theoretical" reference, to the extent they have the same reference. Putnam (1975, orig. 1973) has suggested a solution for "physical magnitude terms" that we shall adopt and adapt here: what fixates the reference of theoretical terms is their belonging to a chain that originates when, faced in a direct causal relationship with the entity responsible for it, someone introduces the term that associates to it. Let me quote:

> Benjamin Franklin knew that "electricity" was manifested in the form of sparks and lightning bolts; someone else might know about currents and electromagnets; someone else might know about atoms consisting of positively and negatively charged particles. They could all use the term "electricity" without there being a discernible "intension" [in the logician's sense] that they all share. I want to suggest that what they do have in common is this: each of them is connected by a certain kind of causal chain to a situation in which a description of electricity is given, and generally a causal description —that is, one which singles out electricity as the physical magnitude responsible for certain effects in a certain way... Thus, suppose I were standing next to Ben Franklin as he performed his famous experiment. Suppose he told me that "electricity" is a physical quantity that behaves in certain [ways, etc...] He would have given me an approximately correct definite description of a physical magnitude. I could now use the term "electricity" myself. Let us call this event —my acquiring the ability to use the term "electricity" in this way— an introducing event. (Putnam 1975:199–200)

Putnam points out that now his use of the term will be causally connected to this introductory event, and was he to teach someone else to use it, this new person would be in a causal relation to him, and via him, to the initial introducing event, and so on. Probably, in the course of this development, we could find that parts of the original description were false; in that case we would replace them with new ones, but to the extent that reference remains fixed by the initial causal connection, the term will continue to refer to the same. In the case of the term "electricity" these different descriptions would correspond to different theories, but to the extent to which these talk about the entity responsible for the original causal relation which fixed the reference in the introducing event, this term becomes "trans-theoretical" in precisely the sense we were looking for (*Ibid.*: 200–202).

Something quite similar happens with the so-called "natural kinds" terms, the use of which Putnam describes under three conditions (the last two of them will be crucial for us. A competent user of a natural kind term, such as "lemon" must know: a) how to classify the term under certain syntactical and semantical markers [that "lemon" is a noun, the name of a fruit, etc.]; b) he should be able to associate the word to a stereotype (color, taste, form, etc.); and c) he will use the word to refer to a natural kind "—-say, a natural kind of fruit whose most essential feature, from a biologist point of view, might be a certain kind of DNA". Additionally, Putnam makes clear that the stereotype gets associated to the name, but that does not imply that all members, not even the most "normal" of them, will fulfil the stereotype as if it was a specification of necessary and sufficient conditions for the membership in the corresponding class.

I am sure you have probably guessed my strategy here: the sense in which two theories are about the same, or more precisely, that their key terms can be trans-theoretical even when they get defined differently in each theory is the sense in which they both refer to the same real entity, since they are connected by a causal chain that eventually leads to the initial introductory event. In the case of "archaic state" or of other more recently introduced terms, there would be a chain that leads —obviously with changes in the respective descriptions— to the term "civilization"; this, in turn, leads to an initial introductory indexical event, in which the stereotype (Putnam's second condition) of an ancient civilization was specified. Historically, the development of archæology displaced in time and space this stereotype, from classical antiquity to the older civilizations of the Near East (Mesopotamia and Egypt). As research progressed the term was then extended to cover the other four cases (India, China, Mesoamerica, and Peru), conforming thus the six cases which are, no more nor less than the paradigmatic, exemplary or stereotypical cases of the term "archaic state". To the extent that different theories refer to terms linked in this chain all the way to the initial introductory event, or point to the six exemplary cases, they are talking about the same, independently of the coincidence (or lack thereoff) of their respective definitions.

b) *Continuity in the Explanatory Problem–Situation*

We can now propose something similar to Putnam's causal chains and introductory events for the case of the continuity of explanatory problem–situations, that is for the sets of "why–questions" the theory is attempting to solve. Two succesive theories hold necessarily a causal relation in the sense mentioned before as soon as the new theory either assumes and solves the problem–situation of the one before, or has to show that such problematic is no more than a subset of a larger and more interesting problematic, or that the new one represents an unavoidable reduction of the previous one. The extreme case would be one in which the older problematic is totally rejected; but even then the causal connection is preserved the moment one has to justify the apparent relevance it seemed to have, and why it should be abandoned in favor of the new one. In all these cases it is likely that there will be a modification of the stereotype associated to the key terms.

In the case in which it is shown that the older problematic is a subset of the new one, what I think would automatically happen is that new law–like principles would have been found, which were capable of explaining the previous law–like principles. In his third condition, Putnam talked about "essential" features of a natural kind; I prefer to talk about "degrees of causal efficacy". In the case of the type of succession mentioned, it would be shown that elements previously considered crucial in the definition of key terms in the older formulation are actually effects of deeper causal processes. Stated in a different way, while the new terms are capable of accounting for the older ones, the reverse does not hold. This would allow us to understand what Service and Adams have in mind when they insist in that while cities require of states to emerge, the reverse does not necessarily holds: there can be states without cities.

6. *Final Considerations*

If what I have said so far is at least plausible, then several consequences or results follow from it, which I think could be of interest for some of the current polemics on state origins:

First: The distinction between "primary state" (that which emerges in the context of non-state societies) and "secondary state" (that which emerges in the context of existent states) is no longer simply optional, but turns out to be indispensable for preserving both the reference and the explanatory problematic of state origins. Criticisms of the distinction have stated it is arbitrary, unnecessary and artificial, since the state supposedly emerges anytime where certain structural conditions are present, regardless of their being primary or secondary, and hence it is the same to study Sumer or Bonapartist France; but clearly the term "state" does not have the same causal chain in archæology and in political science (even when they are related). I do not think that it is legitimate that "state" be applied to ethnographic situations; Service's "primitive states" are not coterminous with the original "state", the one which perhaps we should call "archaic state" in order to maintain this distinction.

Second: Nor could the stereotypical or exemplary cases be increased in an arbitrary fashion: it would be necessary first to show that the new candidates are ontologically similar to the known ones. It is illegitimate in that sense to use an ethnographic community of present–day Oaxaca, as Susan Lees does, to test a theory about the archaic state (in Sanders *et al., op. cit.*: 367). On the other hand, I would not be surprised if the dates in the exemplary cases would go back in time or some new case could show up. However, it is crucial that all cases be established with precision, since they constitute the original reference for the theory, and hence, the battlefield for competing theories. It is hence indispensable to discuss the arguments of relevance for the archæological identification of the state; it is definitively not a waste of time to question whether a determined period or site are evidence of the archaic state.

Third: If the idea of degrees of causal efficacy proposed here is correct, then the anarchic proliferation of definitions for the state could be controlled, by simply showing that the traits some new definitions priviledge turn to be just effects of the lawlike principles of other theories. Thus, if a theory T1 posits as a defining trait an administrative apparatus with four levels of hierarchy, while a theory T2 posits class conflict as central, and T2 is capable of generating from it the administrative apparatus, then T2 is preferrable to T1.

Fourth: A direct consequence of the previous result is that polemics about state definitions cease to be simple and boring terminological discussions, since each definition would imply a hypothesis, and competing hypotheses can then be evaluated. Each new proposal would have to be at least promissing in relation to the advancement of the solution of the explanatory problematic. Otherwise it is pointless to proliferate definitions and classifications, the way Claessen and Skalnik do (1978).

All of these results are intimately related to our starting points: first, that proliferation of theories is justificable only if it implies progress, otherwise we betray the principle of scientific rationality; second, that since a theory is what a theory says, you cannot refute nor corroborate it without determining first what the theory says. The problem of the rise of class society, which is in my opinion what lays behind the problem of the stately form, is a question of great importance not only archæologically but also politically speaking. I believe Childe was perfectly aware of this. Accordingly, I think it is a problem that has to be treated with the seriousness and rigour it deserves. Theoretical analysis may have something to contribute to this task, as I hope to have shown.

REFERENCES

CARNEIRO, R.

1970. "A Theory of the Origins of the State", in *Science* 169: 733–738.

CHILDE, V.G.

1970. *Los orígenes de la civilización*, Fondo de Cultura Económica, México.

1973. *Qué sucedió en la historia*, La Pléyade, Buenos Aires.

1981. "La revolución urbana", in Pérez 1981: 265–278.

CLAESSEN, H. and P. SKALNIK

1978. "The Early State: Theories and Hypotheses", in Claessen and Skalnik (eds): *The Early State*, Mouton, The Hague: 3–30.

COHEN, R. and E. SERVICE

1978. *Origins of the State*, ISHI Press. Philadelphia.

GÁNDARA, M.

1981. "El falsificacionismo dogmático en Arqueología: Las 'refutaciones hawaiianas'", en *XV Mesa Redonda de la S.M.A.*, San Cristóbal Las Casas, Chis.

PÉREZ, J.A.

1981. *Presencia de Vere Gordon Childe*, INAH, México.

PUTNAM, H.

1975. "Explanation and Reference", en Putnam, H. (ed): *Mind, Language and Reality*, Philosophical Papers. vol. 2, Cambridge University Press, London: 196–214.

REDMAN, CH.

1978. *The Rise of Civilization*, Freeman and Company, San Francisco.

SANDERS, W.; J. PARSONS, and R. SANTLEY

1979. *The Basin of Mexico*, Academic Press, New York.

SERVICE, E.

1975. *The Origins of the State and Civilization*, Norton, New York.

1978. "Classical and Modern Theories of the Origins of Government", in Cohen and Service (eds): 21–34.

WEISS, H.

1977. "Periodization, Population and Early State Formation in Khuzistan", in Levine, L. and T. Young (eds): *Mountains and Lowlands*, Bib. Mesopot. 7, Undena.

WRIGHT, H. and G. JOHNSON

1975. "Population, Exchange and Early State Formation in Southwestern Iran", in *American Anthropologist* 177 (2): 276–298.

REVOLUTIONS REVISITED

S.E. VAN DER LEEUW
Cambridge University

"La réalité est généralement plus complexe que ne laisse entrevoir la pénurie de notre documentation. Dès que nos renseignements deviennent plus nombreux, les solutions simples son dépassées" (R. Dussaud, *Syria* 25, 1946–8:36).

Preface

This paper does not look at Childe's two "revolutions" from the point of view of archæology, trying to decide when they occurred, how quickly they changed human societies, what triggered them, and/or what their consequences might have been in specific areas. I suspect that most of this symposium will be dedicated to these issues, where they will be put forth by people who have invested much more in researching the details of the periods and phenomena concerned.

Rather, I would wish to present a somewhat more removed view which deliberately focusses on theoretical questions involved, and on some of the consequences of the inevitable theoretical choices for archæology; I will thus use the "Revolutions" as examples, rather than as the focus of my argument.

Inevitably, therefore, some of you will feel that my ideas are irrelevant, or worse, that they attack fundamental tenets in archæology. So be it. I am not trying to convince anyone. One can only convince oneself.

Introduction

Scientists, whether in the hard or in the social sciences are usually not concerned with observing how they influence the present, or the world would have been a more user-friendly place. Even when they see the immediate consequences of their discoveries, they do but rarely take distance enough to fathom the effects on the longer term, and in a wider area.

Anthropologists are involved in two places, two cultures, but usually only in the present or very recent past. They have to deal with different perspectives, but not with differences in the way in which different perspectives change or have changes. Archæologists and other historians have to deal with just that...

In view of that unique aspect of the historical sciences, their unity, and the possibility to apply uniformitarian principles within the realm of these sciences, depends on assuming that the change of change is a universal phenomenon.

If we are to compare phenomena with a different position in time and space, we need to find a dimension in which they are comparable. Clearly, the phenomena which we observe in different places and at different times are very different, and comparing them would mean comparing apples and oranges. Moving one level of abstraction up, one could propose to compare processes, to compare the way phenomena change. Many have done so; in many ways the transition to a "processual" archæology was born from this need. But twenty years of processual archæology have taught us that processes differ between times and places, and that they are not to be assumed comparable, unless discussed at so general a level as to be useless.

If we are nevertheless to proceed with our work, there seems to be only one way open, *i. e.* to assume the comparability of the way processes change. At first sight that is as risky a business to invest another twenty years of research in as the investigation of processes themselves. But on further consideration, the assumption proposed is none other than to assume that all humans learn in the same way, and that seems fairly easy to accept.

In the first part of this paper, I will try to illustrate how people learn by taking ourselves, archæologists, as an example and looking at the way we have learned from certain of our theories in order to develop others. And I will notably take as an example the two areas of research which were so close to Gordon Childe's heart, the neolithic and the urban revolution.

Archæologists do not usually use the history of their own discipline in developing theories on the past. Until recently, the history of archæology was anecdotal or purely descriptive. Hardly was it up to standards set by the discipline which is known as the "History of Ideas". Standards which seem appropriate for a mature discipline such as the History of History, the History of Science, and more recently, the History of Anthropology. Because there is no such History of Archæology, there is no literature for me to lean on in discussing the development of ideas concerning the transition to an agricultural and/or herding society, or that towards an urban society. That I nevertheless attempt to base a major part of the argument on the history of research has a practical reason. All of you are more or less conversant with the main aspects of the theories involved and, more important, you are also participants in that history.

As archæologists, our present concern with change has no doubt often made us ask ourselves: "Why did these folks look for another solution, why did they (suddenly or not) change the way in which they went about 'it'?" Or in other words: "Why and how did they learn?" As participants in our own history, we have at one point or other had to ask ourselves, "How does my theory, my idea, fit in with the others?" In which way is it different? Probably we have also changed our minds more than once on our ideas, on what we really thought. Thus, we have also been confronted with the question: "What makes me change the way in which I make up my mind?" "How do I learn?" and "How do I learn how to learn?" (*cf.* Bateson 1972).

After such a discussion of the history of theories on the neolithic and urban revolution, I will then try to develop a model of human learning in a societal context which is very simplified, but which I think might serve the purpose of the final part of this paper: thinking about modelling the ways in which organisational changes in societal development have occurred, and have been overcome.

Theories on the "Neolithic Revolution"

Childe originally argued that the origins of agriculture in the Near East were due to palæoclimatic changes occurring after the last Ice Age, notably incipient desiccation (1925). This desiccation would have led to population concentration around water (oases), and thus to local population pressure and a search for alternative subsistence means.

Braidwood then went ahead and tried to test the hypothesis, with essentially negative results. Not only had there been earlier fluctuations in climate, which did not result in husbandry or cultivation, but there was no concrete evidence for the kinds of climatological fluctuation posited by Childe. On the other hand, similar tendencies towards readaptation had been occurring before the end of the Pleistocene (Braidwood and Willey 1962). Braidwood proposes as an alternative the thesis that cultivation is the expected outcome of a long natural development (a directional evolutionary trend) constrained in its occurrence by the presence of the right kinds of plants and animals (*i. e.* in the "hilly flanks" of the Near Eastern core area). From there, it spread to other areas in Europe and Asia.

The environmental determinist theme had other protagonists, such as Clark (1932) and Gimbutas (1956) eventhough there are differences in the degree to which the spread of cultivation and husbandry can be tied in with cultural diffusionism (*cf.* more recently Ammerman and Cavalli Sforza 1978). The main advances seem to be in the collection of new data, a more accurate dating of the phenomenon *per se*, and a refinement of what is meant by the "Neolithic Revolution" in practical terms.

With Boserup (1965), population pressure is introduced as an external "prime mover" towards agriculture. The introduction leads to a violent argument, which still rages in many ways today, between theories which invoke "push" (such as population pressure, *cf.* Cohen 1975, 1977) and those which invoke "pull", such as technological advances which trigger changes in the subsistence mode, through least cost innovations, and therefore eventually allow an increase in the number of people which can depend on a certain amount of land for their subsistence needs (Green 1980). As Green (1980) argues, this debate is not to be resolved, and its major advantage has been the introduction and further elaboration of various detailed arguments, notably the relation between sedentism and cultivation (*e. g.* Harris 1977 as a result of population-pressure "push" due to increased dependency on specialised means of subsistence), or between a much wider range of variables, and along several possible trajectories (Redman 1977 as a result of least-cost "pull").

Contributions to this debate have also felt the impact of the relatively early and less simplistic approaches of Binford (1968) and Flannery (1973), essentially based on systemic juxtaposition of a number of variables, moving away from determinism towards a position in which more importance is attached to human choice and its effects. The former places the transition to cultivation and husbandry in a much longer time perspective, and stresses possibilism as opposed to determinism, and "...internal structural differences in integration between communities made up of non-analogous components integrated in different ways" (1968: 324). Thus, stresses between the amount of food available and the number of people dependent upon it are resolved through cultural means (including decreasing population) at a level below the level of depletion, so that there is no need for technological innovation. The main variable which Binford newly introduces is the existence of regional differences in demographic structure within the area which he considers, the Near East. Differences which are due to a combination of ecological and cultural factors, and which generate three kinds of situations; closed regional systems (where the pressure builds up, and cultural regulators are developed), open donor systems (where emigration regulates pressure alongside the cultural regulators, and which are often highly mobile) and open recipient systems (which adaptational modes from adjacent open donor systems may expand into as far as is suitable and where, then, sudden pressures for regulation and/or increasing efficiency of production might occur. In the latter, changes could thus be triggered.

Flannery attempts a detailed and in part comparative study of the origins of cultivation and domestication in four areas: the Near East, Southeast Asia, Mesoamerica and Peru (1973). His models are directly related to the areas concerned for the specifics of the process of each case. In general terms, what is involved is essentially a deviation amplification model which causes a change in the means of production, the division of labour and social stratification as a result of the interaction of a number of ecological and cultural variables.

With Reed's (1977) contribution summarising a hefty volume of specialist reports concerning all areas of cultivation and husbandry, we are most acutely aware that the problem has, as it were, diversified into a large number of issues, each of a different nature, which I cannot summarise here due to constraints of time and space.

I will end this summary with Rindos' recent and encompassing contribution (1984) which is, as far as I understand it, a plea for evolutionism in the original Darwinian sense, *i. e.* an accumulation of large numbers of decisions, mainly short-term and usually unconscious, interacting with non-human processes. None of these did in themselves "steer" towards "Neolithisation". Cultivation and husbandry were, rather, the unpremeditated and unpredictable outcome. As a result, they can only be understood *a posteriori* on the basis of statistical processes of many kinds, both stochastic and sequential, among them ecological, evolutionary, and cultural processes.

Theories on the "Urban Revolution"

In Childe's original argument (1925), the rise of the City and that of the State in the Near East coincide. A major prerequisite for the "Urban Revolution" was the fact that the prior introduction of agriculture allowed for the creation of a surplus in food, which in turn permitted societies to support people not actively busy in food production, such as the hierarchy necessary to keep the city/state together. This argument has been severly damaged by Boserup (1965), Carneiro (1970), and Sahlins (1972).

Another major correlate in Childe's theories was a growth in population density. This side of the argument has been expanded on by numerous others who each propose different reasons for the increase. Thus, Carneiro proposes *external warfare* which requires ever growing masses of warriors, and hence ever growing cooperation and organisation. Material traces of these developments are found in art and in the defences around the cities. Flannery, however, points out that these traces postdate the earliest state formation (1972). Similarly, others assume *internecine warfare* as a prime mover, which begged for the establishment of authority (Haas 1982). Against both variations on the population growth/warfare theme, one may adduce that there is a considerable amount of variation in population densities in both chiefdoms and early states. Wright and Johnson (1975) even propose a population decline immediately prior to the formation of the state.

Steward, Wittfogel (1949; 1957) and others propounded that irrigation was the task which provided the impetus for the cooperation of large bodies of people which in turn led to the formation of the State. Sanders and Price (1968) later modified this thesis and applied it to parts of Mesoamerica. A number of cases is known, however, where central control over water did not lead to state formation or urbanisation, such as on Bali.

Long-distance trade has been proposed by many, among others collected by Sabloff and Lamberg-Karlovski (1975). The control over prestige goods, as one of the signs of power, could have been an aid to the rise of a ruling class, while the concomitant administrative duties might have elaborated the administrative system as such. Wright, however, points to at least one case where trade came after "state formation" (1969).

All the above attempts have in essence referred to one "cause" for the urban revolution/state formation. In essence, these theories were usually developed based on one example, such as the Near East, or Mesoamerica (and more recently Southeast Asia, *cf.* Hagesteijn 1985). Comparison of datasets in different parts of the world has led to theories which involve several "causes" or variables. These may take two forms, *i. e.* "multilinear evolution" where different variables are held responsible for the same result, but in different places/periods, and "combined interaction of variables".

A good example of the former approach are both Cohen's and Claessen's approaches to "Early State Formation". They essentially sketch various stages of social development which societies seem to have to pass through in order to reach the level of the "State", but the mechanism which cause the changes involved to happen are left open, and vary from case to case (Cohen 1978, 1981; Claessen and Skalnik 1978, etc.). The approach is a direct application of the more general multilinear evolutionist approaches of Steward (1955), Service (1971, 1975) and Fried (1967).

Adams' (1966) view of Early State Formation is an example of the theory that many variables combine to interact. He compares Mesoamerica with the Near East and finds in both an overall similarity in development, which may be represented as a succession of three phases, theocratic, military and political. The large number of differences seems to be satisfactorily brought under these umbrellas by assuming the following process. Various forms of subsistence (herding, collecting, irrigated cultivation) led to both increased redistribution and increased differences in wealth, mainly because the latter form of subsistence resulted in other products and high concentrations of produce on the fertile soils. Warfare led to the genesis of warlords, and provided slave labour. Part of this labour accelerated the trend towards craft specialisation, which required even more redistribution, and hence management. The growing differentiation in wealth and that between warlords and "ordinary" people led to social stratification. All these processes converged on "State Formation".

Renfrew's contributions have stressed the fact that many subsystems of a cultural system may be in independent operation in a more or less stable manner without generating any growth in complexity of the scale usually implied when we speak of "Early States". He argues that "takeoff" is not generated by the existence of such subsystems itself, but by a positive feedback interaction between two or more of the subsystems concerned (1972). His contribution is thus one of lifting the discussion to a higher level (the nature of the subsystems involved is no longer material). His area of study, the Cyclades islands, may be important in this respect, both because there is extensive documentation on the interaction in island biogeographical systems and because these islands are naturally confined systems in the sense that communication on them goes on at a different rate than communication between them. Thus, we see "takeoff" in an area with limited permeability of boundaries.

Wright and Johnson develop the argument that individual people can only handle a certain amount of information. If, therefore, the amount of information which flows through the system grows, that society must create the most efficient pathways to cope with that growing amount of information. This is done through horizontal differentiation and vertical hierarchisation. Specialisation reduces the number of people involved in a decision. Hierarchisation increases the length of communication channels, but reduces the amount of information flow. The difference between states and other kinds of social organisation is in this approach the number of organisational levels in the group. The nature of the information processed is of less relevance (Wright 1969; Johnson 1973, 1978; Wright and Johnson 1975). With this approach, we go one step further up the ladder of generalisation, because no longer are specific subsystems concerned, or specific combinations of subsystems. Increasing complexity seems to follow from increase in information, whatever the reason for that increase is.

Comparison

In comparing these theories, what seems most essential has been summarised in the quote on the title page of this paper; there is a relationship between the nature of the theories concerned and the amount of information that was available when they were conceived. I think we could distinguish the following phases, referring between brackets to the discussion in the last section:

1. Search for origins: location of the beginnings in space and time;
2. External explanation of the new phenomenon as it occurs in one location (for example Childe 1925; Braidwood 1968; Carneiro 1970);
3. Generalisation of an explanation over several occurrences (for example Wittfogel 1957);
4. Design of alternative external explanations, each relating to their own aspect of the data, and originally applicable to one area, but extended to others (for example trade vs. warfare vs. population increase, etc.);
5. Introduction of multilinear evolutions (design of external explanations which are considered equivalent possibilities under one umbrella (for example Carneiro 1973; Claessen 1982);
6. Introduction of closed systemic explanations for the phase "before" and the one "after" the change, leaving the actual mechanism of change to a considerable extent implicit (for example Binford 1968);
7. Introduction of open systemic explanations which incorporate externally caused changes (for example Flannery 1973; Rindos 1984);
8. Introduction of morphogenetic systems modelling to account for change, which include both external and internal variables which, in conjunction, are held responsible for change (for example Flannery 1972; Renfrew 1972).

There are a number of aspects of this process which bear further scrutiny. Such a scrutiny is best organised at three levels, discussing (a) the whole trajectory as described in steps 1 to 8, (b) what happens, as it were "during the rise and fall" of a single theory, and (c) what makes up individual decisions.

1. *Long-Term Change*

Here, the following points seem of importance. Looking for the origins of a phenomenon such as "cultivation" or "states" is in a sense also looking for clear-cut definitions, *i. e.* of the categories "pre-cultivation" and "cultivation", "pre-state" and "state", etc. When reviewing the research on these and similar topics, it also becomes clear that with an increase in the amount of information involved, and thus in the number of dimensions in which it is attempted to define these categories, the definitions become less and less clear, and one finds a large number of phenomena and connections between them in many dimensions in its place. Recently, Feinman and Neitzel (1984) found this when they attempted to make a typology of sedentary pre-state societies in the Americas. They find the diversity to be continuous and many-dimensional, and without clear sub-types. O'Shea attempted to make a typology of tribes, and concluded that it was possible to distinguish and define clearly in any one of a number of dimensions, but impossible to keep clarity when more than one dimension was involved at a time (personal communication).

Second, the reverse is also true: the concepts with which the phenomena are confronted are increasingly sophisticated, are better circumscribed and refined. It is my contention that the impact of a number of the researches done by those present was directly related to the "fuzziness" of the concepts involved. After all, we do not know whether the people whose remains we study defined their concepts with a clarity and incisiveness akin to our own. And even if they did, the chances are inherently small that our concepts tally with theirs. The more narrowly defined ours are, the smaller the chances that we catch one of theirs with one of ours.

Third, the larger number of phenomena, and of dimensions of the problem involved reflects the increase in information gained as a result of the search, and when one (intuitively) correlates the various phases of the process of theory formation with the amount of information gained, there seems to be a positive correlation between the amount of information and the structure of the theory used (compare the effect of the use of new equipment in "hard science"). This relationship is clearest when one looks at the various theories concerned from the point of view of their effectiveness in storing information, *i. e.* in the amount of redundancy with which they store the data. The first, intuitive and vague, explanations ("diffusion" etc.) did not really account for any data in a "hard" way. The next set tried to "bind" a number of observed characteristics of the transition towards cultivation (c.q. statehood) into a framework. The main mechanism which seems to be involved is that it costs much less memory space to store and/or manipulate one or more transformations (such as a unilinear or a multilinear causal chain, or the design of a system), the initial state(s) (such as the situation before the occurrence of cultivation) and the manner in which to calculate the outcome(s) (such as the result of the process of "neolithisation") than it does to store all the possible relevant initial states and outcomes (*i. e.* all separate relevant aspects of the society before neolithisation and after, such as their modes of subsistence and storage, their kinds of living quarters, their kinship structures, their exchange mechanism, and last but not least the exceptions, etc.). (The shock of suddenly saving memory storage space is what we experience as the satisfaction of an explanation.)

Fourth, this relationship too works two ways. The amount of redundancy (the effectiveness of the model) is pushed up as more information is gained but equally, the more effective models allow one to process more information, so that a change to a more effective model eventually also increases the amount of information involved.

2. *The History of a Single Theory*

In a recent book, Keene and Moore (1983) have argued that there are phases to be distinguished in the life-span of archæological models. Notably, I think one could see the following: (1) the discovery of a new model, (2) the quick spread of it, which proceeds by testing its validity on a number of phenomena and areas of research, (3) the realisation that it fits fewer cases than anticipated, and is less powerfull than was thought, (4) refining the model and focussing it on those phenomena with which it seems to deal best, (5) acceptance and use of the model, almost without any further doubt, as a tenet for the development of further models, (6) growing doubt as to whether the model is indeed as effective as required and (7) replacement of the model by a newer, more encompassing one.

In the process, the nature of the research changes. Generally, the model will be arrived at inductively. Once it exists, most research will proceed deductively in attempting to test its applicability (the "hammer phase" of Keene and Moore).. As more data accumulate that cannot fit the model, research will —again— revert more and more to inductive attempts, and the collection of more data, in order to find the model's successor.

3. *The History of Individual Decisions*

At any moment, there is always the choice how we will deal with new information. In principle, that choice is always free in the academic community (if we bar for the moment the social constraints, which I will reintroduce in a moment in a different guise). But it is a corollary of this situation, that the choice whether newly discovered phenomena are indeed used to support the model ("fitted into it") is in part dependent on the point of the cycle where the confrontation between model and phenomena occurs. There are stretches of time when everything seems to fit into the model almost without any problem, and there are moments when nothing seems to fit, when we say that the field is "in crisis", that we don't know what we are all about. (Socially speaking, it is in the former situation that the "bosses" grow up who dictate research and funding, because they have a stretch of "clear sailing" and can accumulate power, and it is the latter situation that the "rebels" undertake to change the system after a certain delay.)

"In crisis" need therefore not be defined as "overcome by what happens around us", but rather as a "temporary lack of sufficient information processing capacity (*i. e.* sufficiently effective models) to deal with the information which needs to be dealt with in order to make things go smoothly". The importance of this reverse viewpoint can hardly be overstressed in this context, as will become clear in the next pages. Moreover, it is theoretically predictable how the decision will go. But what is not predictable, is what the new model will look like once it emerges from the "crisis".

To summarize, I can probably present the following simple statement: "The occurrence of change is dependent upon the state of the balance between the amount of information which needs to be processed and the organisation (*i. e.* information processing capacity) which is available to do the processing".

Information Processing and Storage

In this chapter and the next, I would like to give the above argument a somewhat more solid basis by examining how it relates to a current model of perception and concept formation.

Perception may be defined as the awareness of uniformity (similarity, continuity, homogeneity), and of contrast (dissimilarity, discontinuity, heterogeneity). Evidently, uniformity and contrast are relative to one another, and can only be perceived in conjunction. Because the two are inseparable, human information-processing always operates on two levels: the level of context which stresses the underlying similarities, and the level of specifics, which stresses the dissimilarities (and in this I differ fundamentally from those who argue that universals should not be a part of our interpretation of specific cultural phenomena). To give an archæological example, categorisation can work at the level of the artifact category, where similarities between, for example, pots or houses are stressed, and also at the level of the type, where dissimilarities between different kinds of pots or houses are stressed. Thus, "no class (category) can be a member of itself or of the class of nonmembers" (Russell 1903; Bateson 1972).

Following a range of experiments, the psychologist Tverski and his associates (Tverski 1977; Tversky and Gati 1978; Kahnemann and Tversky 1982) have concluded that:

1. Similarity and dissimilarity should not be taken as absolutes.
2. Categorisation (judging in which class a phenomenon belongs) occurs by comparing the subject with a referent. Generally, the subject receives more attention than the referent.
3. Judgement is directly constrained by a context (the other subjects or other referents surrounding the one under consideration).
4. Judgement of similarity or of dissimilarity are also constrained by the aims of the comparison. For example, similar odds may be judged favourably or unfavourably depending on whether one is told that one may gain or lose in making the bet.

On the basis of these conclusions, I have argued for the following model of perception (*cf.* fig. 1):

a. A first comparison always takes place outside of any applicable context, so that there is no referent, nor any specific aim. Thus, there is no specific bias towards similarity or dissimilarity. If there is any bias at all, it is due to intuition or to what people have learned on past occasions, which have no real application to the case at hand.

b. Once an initial comparison has led to the establishment of a referent (a relevant context or patterning of similarity and dissimilarity), this context is tested against other phenomena to establish the limits of its applicability. In such testing, the established pattern is the subject, and the phenomena are the referents. There is therefore (following Tverski's second statement) a distinct bias in favour of similarity.

c. Once the context is firmly established and no longer scrutinised, new phenomena are subject in further comparisons, and the context is the referent. Thus, the comparisons are biased towards the individuality of the phenomena, and towards dissimilarity.

d. Once a large number of phenomena have been judged in this way, the initial bias is neutralised, the context is no longer considered relevant at all, and the cycle starts again, so that further comparisons lead to establishing another context.

e. Ultimately, this process leads to the grouping of a large set of phenomena in a number of categories at the same level, which are generally mutually exclusive (establishing dimensions and categories along them). At a certain point, the number of categories is so large that the same comparative process begins again, at a higher level, which treats the groups as phenomena, and results in higher level generalisations).

The Cognitive Feedback Cycle, Open and Closed Categories

The continuous two-way process between information and information-processing capacity therefore entails a feeback cycle, which may be thought of as the following.

Suppose one has a problem. By processing information, a model gets devised for solving it. The model is retained after the solution has worked, and is to be used on other problems. As no two problems are ever the same, the solution will only partially solve the new problems. This leads to further development of problem-solving (*i. e.* information-processing) capacity. As the problem solving capacity grows, so does the size of the problems tackled. The increase in information processing capacity is therefore non-linear, bringing more and more "brains" together in an ever finer tuned manner.

The feedback cycle bears some closer scrutiny. The self-reproduction involved is never exactly identical, because when the feedback cycle re-enters the point of conceptualisation, the subject-referent balance has slightly changed from its original position at the time of initial conceptualisation. (Even if history repeats itself, our perception of it never does.) The first re-entry after initial conceptualisation, for example, will only register a minimal shift, because (as we have seen in the model on Tverski's experiments) the subject–referent balance causes the information–processing apparatus to stress similarities rather heavily. With each subsequent re-entry, the bias will shift a little more towards a balance between similarity and dissimilarity, though as long as similarity is dominant, the information will outwardly be identified with the original categorisation. Once dissimilarity is the domimant mode of information processing, each re-entry will cause increasing dissatisfaction with the original conceptualisation, so that the basis for this conceptualisation will slowly be whittled away. Eventually, there comes a point when re-entry results in a different conceptualisation. Then suddenly (or so it seems) there is a qualitative change. The difference between the old and the new is externalised after the actual conceptual transition has taken place, and when the focus is —again— on similarities.

At the lowest level such a change manifests itself, for example, as the introduction of a new category along a pre-existing dimension. But at higher levels, it may entail the introduction of new dimensions, new models or even new paradigms, alongside the existing ones or instead of them (*cf.* fig. 2). Bateson has argued how stability (non-learning), change (learning) and the change of change (learning how to learn) are nested more or less along the lines presented here (1972).

Once a new cognitive category has developed, there will be two such categories in existence, and the information-processing framework will expand. In other words, the "power of resolution" has been increased, and another variable (or another group of variables) has "suddenly" been cognised, usually in a different dimension of variability. More information may therefore be processed simultaneously.

New categories differ fundamentally in status from old ones, as the original ones have undergone a complete cycle of subject-referent shift, while the new ones have not. We might say that the original ones are truly "known" because we know the phenomena which will belong to each one as well as those which eventually will not. In the case of new categories, however, we only know what might belong to them (and not what eventually will not) because the subject-referent shift is still stressing similarities. Thus, the pre-existing categories are sharply defined, but the definition of the new ones is rather "fuzzy", and we are still learning to deal with them. One might say that newly-initiated categories or other perceptual generalisations are "open-ended" and partly-known, whereas long-used categories are no longer "open-ended" but "closed". Experience has taught us what belongs in them and what does not.

Figure 3 presents the trajectory which organisation follows with a continuously increasing amount of information flow. The graph is, for reasons which be beyond the scope of this paper, essentially the same as the graphs used by Prigogine in his monitoring the history of "dissipative structures" (Prigogine 1978, 1981; van der Leeuw 1982). It summarizes the conjunction of two kinds of "learning", "problem solving" or "behaviour", *i. e.* (1) a predictable progression based on further development of acknowledged principles, and (2) sudden, apparently "irrational" decisions taken outside the framework of those principles (*e. g.* "random" behaviour). It follows from the graph that it is possible to predict (actually by non-linear dynamic modelling) in theory *when* sudden jumps occur, but never *what* they will be. Note also that the processes thus modelled are essentially irreversible in a direction which is the reverse of the second law of thermodynamics, and the ensuing organisation is self-structured, *i. e.* its form is self-imposed and not (entirely) dependent on external constraints. In this it differs fundamentally from any descriptions of the flow of energy or matter through an individual (or a society). Because the latter commodities are subject to the laws of conservation, they are always organised externally.

Information Flow and Organisation at the Level of Societies

Intuitively, one would assume that learning at the level of a society would have to be modelled in the same manner as in the individual. After all, the society is made up of the individuals. As we will see, this is true at a certain level, but there are consequences to the fact that at this level we are talking about aggregates of people rather than individuals.

It has long been acknowledged that a society contains fusionary and fissionary tendencies (and individuals), *i. e.* (groups of) individuals who conform to the organisation of the society (*i. e.* what the society has collectively learned) and others who do not, and go their own way to a greater or lesser degree (when the degree is too great, they come into conflict with the society and have to leave one way or another). In the History of Science, the same is acknowledged when Kuhn speaks of the "essential tension" between conformism (in intellectual "schools") and innovation.

We must assume that individuals have, at least at certain phases in their lives, preferences for what will here be called "stochastic" (non-conforming) and "cartesian" (conforming to the logic of the society) behaviour. The terms come from a very stimulating study by Allen and McGlade (in press), upon which I draw heavily for parts of what follows.

A society with only "stochasts" would devolve into anarchy, whereas a society with only "cartesians" would soon become homogenous, so that exchange would dwindle, and the society would equally fall apart. The interaction (information exchange) between the two is therefore as essential as their coexistence in the society, and it is that information flow which needs to be monitored and understood, if we are ever to understand social process.

Anything around us is potential meaning, and thus potentially significant for a group of humans. "Stochasts" will hit on many unexpected things, and thus perceive a lot of meaning. They will, however, always move on, and never attempt to do anything with this information (= discovered meaning). A society which consists only of "stochasts" will build a minimum of organisation. That is left to the "cartesians". But they have the natural tendency to stick with what they already know, and make the most of that. As most sources of matter and energy are limited, they would tend to exhaust these, for example. Also, they would prefer a world which could be pictured in a minimal number of dimensions. They reduce pluriformity.

How would the two interact? The "stochasts" would make a discovery; then the "cartesians" would jump on the bandwagon and go on with applying the discovery, for a long time. All the while the resources involved, for example, would diminish, and so would the yield. Because "cartesians", however, never look at what they do not know, they would fail to notice that until the very end was in sight. Then, they would become dependent on the "stochasts" again to find them a new something to exploit. The curve is essentially the same "bifurcation curve" as shown in the last section.

If the flow of information between "stochasts" and "cartesians" were continuous, there would be place for at least two (extreme) strategies, *i. e.* collaboration and competition. To begin with the last, as soon as the "cartesians" stepped in, they would be so much more effective that the "stochasts" would have to play at their game, or run the risk of being eliminated, with possibly disastrous results for the whole society, because as soon as yields diminish, the "cartesians" would be left dry: the society would have no capacity for adaptation left.

If, on the other hand, the "stochasts" managed to avoid passing on information the "cartesians" would not stand a chance, because the stochastic strategy is essentially a better long-term strategy.

Some degree of collaboration, even if silently acknowledging each other's contribution, is the only strategy which might work in both the short and the long term. But even then, it essentially will drive the situation more and more out of balance. The rate at which this occurs depends on the degree to which the "cartesian" strategy prevails. The minimum is best in the long term.

The Feedback Cycle Between Organisation and its Environment

In the feedback cycle of an individual's learning processes, which has been discussed earlier on, it has been made clear that the second time around that a category is cognised, it is never exactly the same. Here, it seems, the recently popular ideas of Bourdieu and Giddens fit in, stating that for a society as a whole, "habitus generates behaviour, behaviour generates context, and context generates habitus" (Bourdieu 1979). One of their implications is the interaction between a society and its material manifestations (material culture, organisation of space, etc.) and another the continuity and uniqueness of each cultural tradition, possibly to the point, according to for example Hodder (1982), of excluding comparative research (we have already seen that such research *is* possible, provided one steps a level of modelling up). A third major corrollary, however is the fact that (a) organisation plays into the fluctuations of the environment, natural and cultural, and often amplifies them due to the adaptive delay inherent in human information processing (a delay which varies with the size of the organisation involved), and (b) that even after feedback, the response is *always* (theoretically speaking) less than effective because the response only takes into account what was known, not what was unknown at the time of learning. We will return to this last point in a later section; here I would like to devote some more attention to the deviations.

Figure 4 presents the interaction between organisation and its (relevant) environment if we assume continuous change of the environment [*i. e.* if we use a Newtonian model of change which is determinist and so generalised that its own inherent dynamic fluctuations are ignored —which is more or less what we have always done in studying past change— and if we ignore the feedback between the two for the moment]. In assessing what happenes from the point of view of the past population, one has to imagine oneself on the discontinuous line, whereas we, as archæologists, have always assumed that we were on the continuous one, and "events" on the other. (Hence my earlier remarks about the perspective on a crisis or "revolution".)

When the change in the environment is interactive with that of the organisation, the rapidity with which the distance between environment and organisation (or in other words perception of the environment) grows is even much greater. It seems a miracle that we stay in touch with the environment at all. That distance indicates the necessity of stochastic behaviour in relation to the state of perception and the "age" of the extant paradigm.

Some Implications for our Modelling of Past Processes

In this section, I will discuss some changes which this perspective implies for our conception of various processes in the past. Not so much the exact nature of the variables and their interactions, but the way "change of change" plays a part in them. In order to do so, I will refer to four kinds of processes, *i. e.* (a) very long term processes, essentially our conception of cultural evolution, (b) medium term processes, such as the change from one major organisational mode to another (*i. e.* the transition to cultivation, or the transition to statehood), (c) short-term processes such as fluctuations in the latter, and (d) instantaneous processes, *i. e.* human decision-making.

Prior to those discussions, however, one remark, summarised in a quote from Allen (in press): "Stable organisation can emerge and temporally dominate the underlying "chaos", but evolution takes the form of a dialogue between the macroscopic forms and flows that exist at any given time, and the microscopic (molecular) level which has temporally consented to their existence". The fluctuations which I will discuss separately in the next few pages are all present in conjunction and it is indeed that conjuction which is to be modelled if we are ever to get beyond the present state of archæological modelling. Ignoring these shorter-term and smaller amplitude fluctuations in favour of the "smoothed" larger ones has led us up the garden path where we are now (not only as archæologists).

a. *Long-Term Processes: Cultural Evolution*

What would the long-term effect be of the processes of ever increasing categorisation which I have outlined? Presumably ever increasing refinement of categories, and multiplication of the same, and (important for archæologists) of the material representations of these categories. Concomitantly, an increase in the hierarchy of conceptual levels available to human information processing: increase of abstractions and their scope.

Conversely, decrease of the "fuzziness" of definitions and concepts. Such a decrease in fuzziness is an important element in the growth of information flow in a society. Fuzzy concepts are relatively easy to transmit face-to-face, because one has many means at one's disposal, but difficult to transmit otherwise (through messenger, or pictograph or symbol or writing, etc.). Thus, the decrease in fuzziness allows the increase of the size of the interaction group, as well as its complexity and coherence.

Evidently, the increase in size of the interaction group itself pushes for more effectiveness in information processing. In this context, it is important to remember that with a linear increase in population of an interactive group, other things equal, the number of interactions goes up exponentially (*cf.* Johnson 1978). This requires ever better signals, and ever less noise in the channels. Bigger problems and better solutions.

In that respect, the develoment as envisaged here also pushes the mix of the society towards the cartesian side. Stochastic signals are essentially noise. The larger the interactive group, the more important is it that everyone conforms to cartesian rules. That in itself pushes the system into ever increasing instability in the relation to its environment. Only the admission of more noise in the channels, and thus of more conflict in the society, would counteract that tendency and reduce the society's cohesion (eventually to anarchy/entropy).

In summary, cultural evolution could be seen as follows: "Evolution is a change from a no-howish, untalkabouteble allalikeness to a somehowish and in general talkaboutable not-all-alikeness by continous sticktogetherness and somethingelseifications" (Jones, quoted in McGuire 1984: 91).

Looking very briefly at the evidence for long-term evolution of societies which we have, it is indeed tempting to see developments since the Lower Palæolithic as an increase in circumscription (in a social sense, as well as in the degree of definition of artefacts, *cf.* Gamble 1986), a decrease in stochasticity (from randomly roaming hunters in a landscape virtually devoid of other humans to spatial circumscription and eventually sedentation, increase in population and exploitation of a fixed set of resources), increasing dependency on communications (and concomitantly increases in forms of symboling behaviour, such as art) (van der Leeuw 1986).

One of the consequences of this approach is that in modelling change in the more distant past, we have to ask questions and use concepts with increasing degrees of fuzziness. Another consequence relates to the fact that we have always included an optimisation hypothesis in our modelling of (biological and) cultural evolution. In a recent paper, Allen and McGlade have pointed out that such is impossible, because optimising behaviour is always relative, in relation to other solutions. In relation to *what* actually, following our approach, is not known to the person who presumably does the optimising (*cf.* Allen and McGlade, in press). Rather, we can only use the hypothesis that people did what seemed adequate in their minds. In a way, we have here the return of the substantivist debate ["fuzziness" is very close to "embeddedness", indeed one could argue they are the same thing (van der Leeuw 1984)]. Nothing demonstrates this "adequacy hypothesis" better that Sahlins' indications on leisure time among hunter-gatherers (1972).

b. *Medium-Term Processes: Transition Between Organisational Modes*

If, in the last paragraphs, we have looked at an "ironed-out" version of evolution, at this level we are concerned with the fluctuations in the process just described: the major changes which the history of mankind has undergone, such as the two "revolutions" which this symposium is all about. I will therefore discuss what goes on at this level in somewhat more detail.

The following points seem to require making. *First*, that as soon as we are discussing processes at this level, we are no longer concerned with "change of change" at the cultural level, because we are essentially discussing the rise and fall of a single mode of organisation. We are therefore less concerned with generalities than with the specifics of the structuring to be modelled.

Second, that does not imply that the micro-scale processes which I have summarised in the last section in a general description of "evolution" are not at work at this level. Indeed, the transition between organisational modes is in the end only explicable by assuming that it *does*, but the precise effect in *qualitative* terms of the narrowing of categories is unpredictable, and can therefore not be included.

Third, there is no kind of modelling which does in any way enable us to predict any processes or phenomena. At most, modelling at this level may elucidate possibilities. As we have seen, the system may for theoretical reasons never hope to achieve close enough correspondence between what can be known and what ought to be known to enable full prediction (even if logically structured along the same lines, explanation and prediction can never be the same because the status of the cognised concepts in both differs. Explanation works with *a posteriori* (and hence "closed") concepts, and prediction with *a priori* (and hence "open" ones).

The actual modelling proposed is dynamic and non-linear. Allen and McGlade give a very enlightening example of such modelling in relation to fisheries (1986), which highlights some critical aspects. Firstly, that in order to be effective, modelling of human systems cannot be done on the basis of deterministic generalisations about what happens at lower levels. Such deterministic modelling is part of the tradition in archæology to look for similarities rather than contrasts. A tradition which, at each step, reduces the dimensions of variability considered, and thus brings us further from reality with each such step. Rather, we should move towards the study of contrasts, open up more and more dimensions of variability, and model their interaction dynamically. In Allen and McGlade's fisheries model, for example, the crucial interaction is between the fluctuations in the fish demography and the human reaction to these, in themselves minor, oscillations which have a periodicity of about four years. The deviation amplification between these two fluctuations is what keeps the fisheries system in oscillation, and brings it further and further from equilibrium rather than allowing it to dampen down (*cf.* fig 5).

The modelling as proposed by Allen and McGlade atrributes to "chance" the discovery by the system, after simulated running of 1000 (!) years, of a second "local" minimum in which it then proceeds to spend much time, (and which was untraceable in a deterministic modelling of the same variables). In the approach sketched in this paper, I do think we can do better than "chance" because of the evolutionary framework. Allen and McGlade show clearly, how the rapidity of the rate of response of effort to return is crucial in determining the range of different solutions found by the system. An increase of that rapidity drives the system to a wider range of responses (see fig. 6). Now, the rapidity of the rate of response of effort to return is tied into the organisational coherence of the framework which creates that response. As we have seen, that coherence is bound to increase with the time in which the system functions in any organisational mode (*cf.* the bifurcation curves of fig. 3). Thus, any system will of necessity change from a near-deterministic one to one very far from being deterministic (from fig. 6 a through b to c). In other words, it is inherent in any system state that it digs its own grave, oscillating heavier and heavier, over a wider polydimensional space, until it suddenly hits another local minimum. Then, very rapidly, the values involved change (such as the cultural values or, in Allen and McGlade's case, the price and the species of fish caught), only in order to set a new cycle moving. In actual fact, it must be possible to calculate the probability of drastic change in the system (its finding a new local minimum from the size and frequency of the oscillations).

Before I go on to discuss those, in the next section, I would like to emphasise that notwithstanding the limitations of the "predictive value" of models of this kind, they are extremely powerful tools in learning ***what might have gone on*** (rather that "what went on"). Their power derives in part from the fact that the options ***not*** chosen by the system may be scrutinised, so that we may learn more about that vaguest of all archæological concepts: "tradition".

Last of all, it seems that through this kind of modelling, the two major areas of present-day research, devoted to respectively the operation of the system and the context and symboling in which it functions, might be brought closer together.

c. *The Third Level: Oscillations Within an Organisational Stage*

J. Allen (1984) presents us with a beautiful study of system oscillation and diversification in his study of the pottery trade from Motu on the South Coast of Papua New Guinea. The curve he uses is presented in fig. 7. What we see is a parabolic rise, then a virtually straight crash *to a level of organisation just above that at which the system's studied trajectory started*, and the another parabolic rise to *just above the last peak*. Connecting both the peaks and the troughs would in itself give two interesting curves, *i. e.* of the increase in oscillation, respectively the increase in organisational gain of the system. His intuitive interpretation (1984) is that the crashes are due to the fact that a generation of men invests much in the control of the system, and keeps doing things a certain way. As others, not in control, adapt to the way things are run, the organisational advantage is lost, and the system has either to change or crashes. Clearly, this interpretation fits *at the level of observable anthropological and archæological data* the kinds of system oscillations we have been discussing.

But apart from the "visibility" of the processes at this level, there are a number of interesting questions. First, what is the field of tension in perception which causes these oscillations to happen? Referring back to an earlier section, I would venture that it is the increasing tension between the perceived world (or in other words, the organisation which a group has achieved in order to deal with the environment) and the environment itself (which changes as a result of the organisation with which it articulates).

Another interesting aspect of these data is the regionalisation which occurs towards the end of the trajectory. Such regionalisation has been observed to occur towards the end of a system's trajectory in one organisational mode in quite a number of cases, in various prehistoric period and areas (*e. g.* Bradley 1984), and with different interpretations. In one sense, it could be seen as a result of hypercoherence of the system, *i. e.* too heavy an information load to process by the "central" regulator, so that decentralisation is needed (*cf.* Flannery 1972). In the terms proposed here, that might be rephrased as follows: as response time (see last section) increases, we have seen that the system begins to cover a wider range of permutations very quickly (the area in multidimensional space it occupies increases rapidly). These will effect the environment differently in different dimensions. As the environment itself is inhomogeneous, the various "solutions" will "take" to different degrees in different dimensions [and one of the results is that the system, in this way, dicovers "new" dimensions in which it might takes off, to the effect that a totally new organisational mode is realised (paradigm change)]. In "real" space, the inhomogeneities of the environment are regionally determined, and "regionalisation" occurs.

Another interesting question at this level is more substantive: which features of the information processing organisation are retained during a "crash", which are rejected, and (possibly) which are transformed? Is the answer dependent on the frequency of use, maybe? Or on the "centrality" of the category to the tradition as a whole? Or on its effectiveness in communication? Or on the energy invested? Or is the answer not to be given in substantive terms? Here, contextual and symbolic questions could present us with one kind of answer, but maybe there are others, too.

d. *The Instantaneous Level: the Decision*

I have included this level mainly to point out that there are also implications of our approach at this level of the individual decision. It, too, functions in a context and it, too, is dependent on the state of the bifurcation curve. But to go into that at this point would exceed the physical limitations of this paper (but see van der Leeuw, in press).

Conclusion

Rather than end with a summing up of a number of pages which reflect an outlook, rather than data or specific ideas or otherwise summatable things, I would like to point at one further implication of all this, and so come back to where we started, the History of Archæology. Clearly, we are involved in the trajectory of the human system in two moments in time, those studied, and the one we live in.

This paper has shown yet again how difficult it is to connect these moments. Problems of different degrees of "fuzziness" of concept definition, the probabilistic nature of the approach, the many levels of oscillations, the theoretical impossibility ever to know that what it is essential to know to monitor the system's trajectory accurately... And yet there is one more problem, that is the past trajectory of the system of archæologists. We tend to, these days, treat our questions and answers, our datasets as "neogene", but they are not. We are irrevocably bound in our answers by earlier conceptualisations, by the history of research in any particular area. This requires us to adopt a more modest approach, and at the same time, a technique of slowly untangling what in a dataset is past history, what is history of research, and what is usable in both. Historians have done this since the last century...

Acknowledgements

This paper is, as any other, indebted to many. First the "*grandes*" which I have only met through their writings, such as Childe in whose honour we are here gathered. Next, to a number of colleagues of more or less my own generation, of whom I specifically want to mention Roel Brandt, Bert Voorrips, Henry Wright, Gregory Johnson, Colin Renfrew, Ian Hodder, Peter Allen, Richard Bradley, Clive Gamble, and Mike Jochim (in order of appearance), to whom I owe various interesting ideas or critical remarks. Last, there are some of my students, Ineke Abbink, Linda Therkorn and Willem Jan Hogestijn.

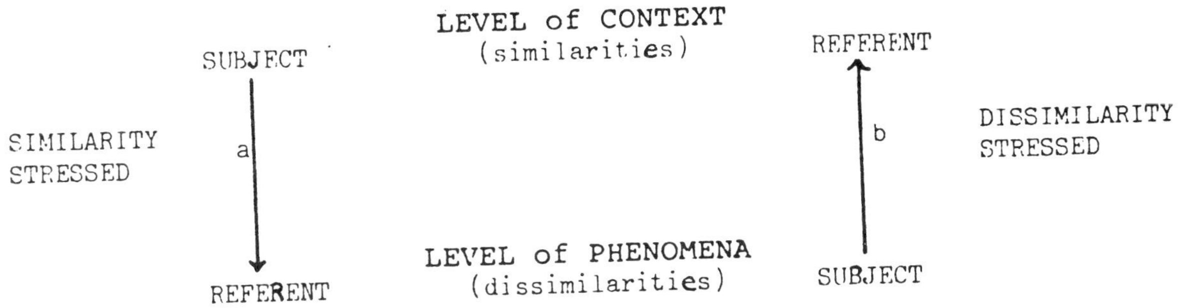

Figure 1. Schematic representation of categorisation/conceptualisation based on interaction between two levels of cognition, following the precepts of Tverski and Gati, 19 (ill, after van der Leew, 1981). In the initial phase, the level of context is the subject of testing against the phenomena, and similarities are accordingly stressed, biasing the comparison towards validation of the concept. A slow shift eventually occurs towards testing the phenomena against the (accepted) concept. This entails a shift towards dissimilarity and delimiting the boundaries of the concept (i.e. creation of a 'closed' category).

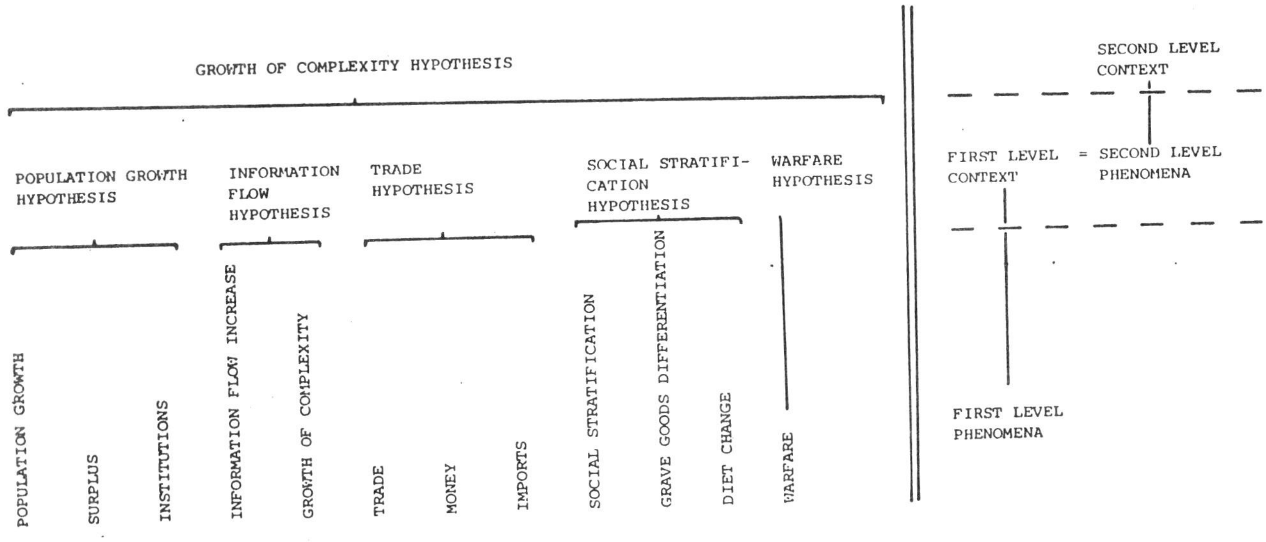

Figure 2. Relating the two levels of perception: the level of the phenomena and that of the context. As is seen from the right-hand column, the context of the lower level may serve as the phenomena at the higher level. The 'explanations' at the two upper levels may relate two entirely different dimensions of conceptualisation.

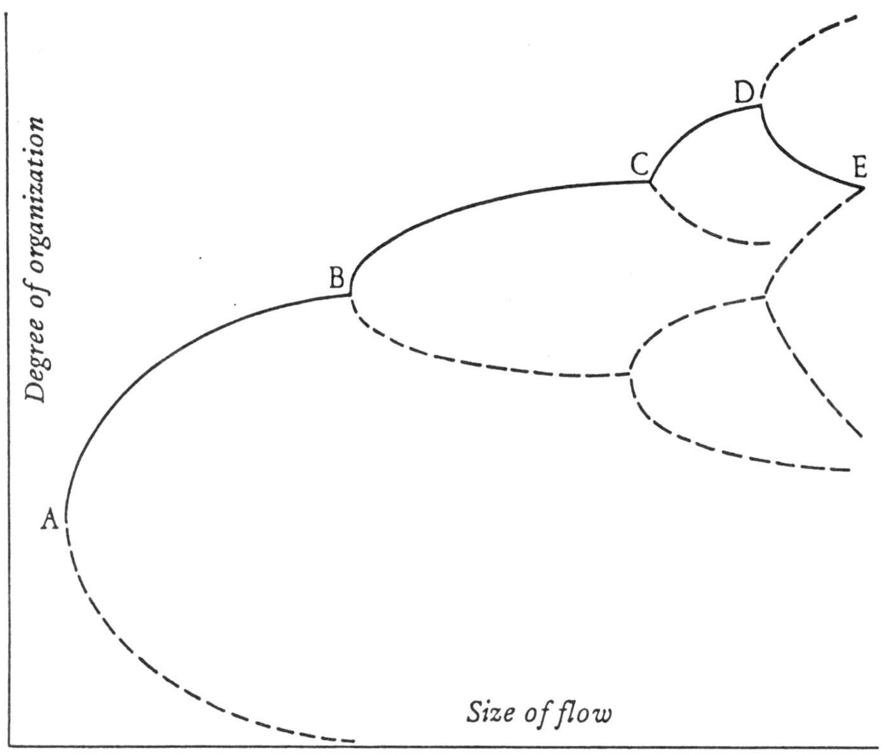

Figure 3. Relationship between the size of an information flow and the degree of organization. Note that as the former increases continuously, the latter increases discontinuously. Also, if one has to extrapolate the trajectory from observing it at point A and point E, one cannot decide with any probability what the intervening stages have been, unless one reconstructs the total trajectory: the growth of complex organizations is an irreversible process through time, and each has its own trajectory. This does not imply that there are no (intervening or final) phases of decay: D-E is such a trajectory of decreasing complexity, as are all the discontinuous lines. They represent trajectories the system could have taken at the breaking points.

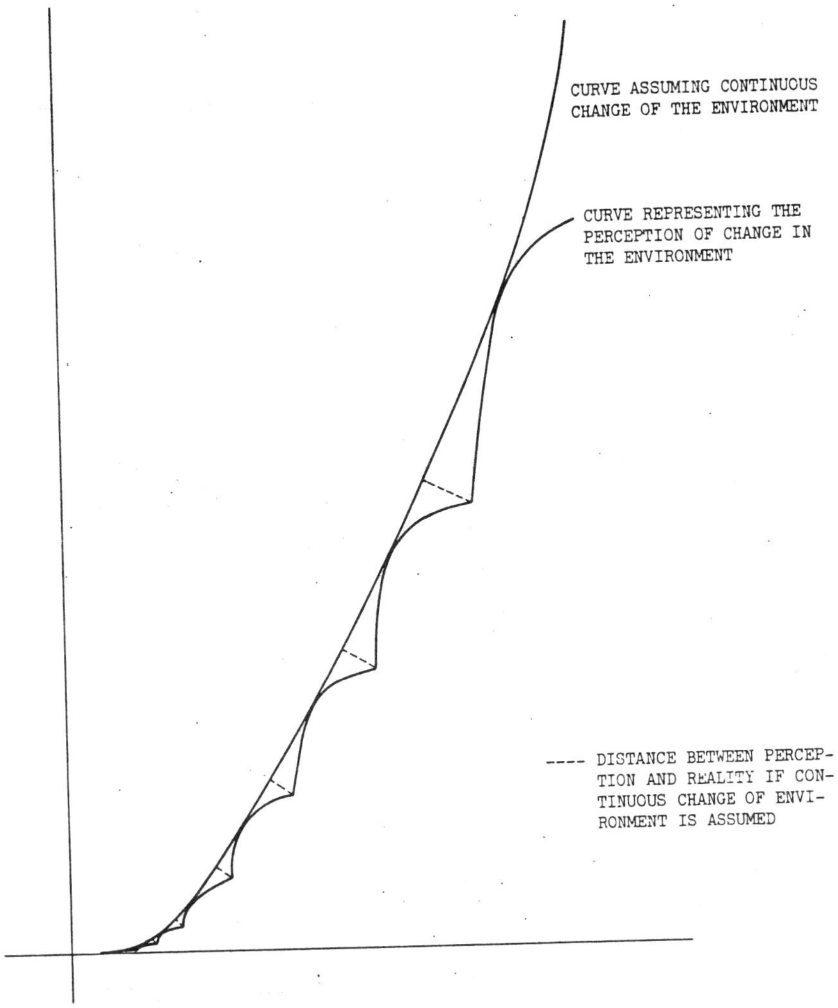

Figure 4. Imaginary trajectory of the perception of an environment (human, natural or other), if change in that environment is assumed to be continuos and nonlinear. At intervals, the distance between the reality and its perception grows bigger and bigger, until fundamental changes in the perceptual system are required (i.e. paradigmatic changes).

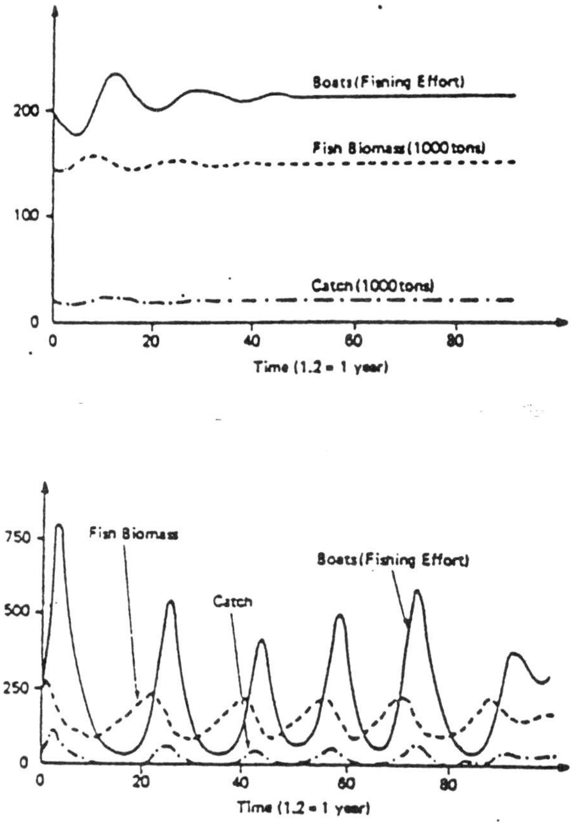

Figure 5. *a*. Newtonian approach to fish yields. The (naturally determined) curve of fish biomass is compared with the (culturally determined) curve of fishing effort. The curves are summated and the result is a flat curve of catch, with levels out the differences with occur from year to year. An apparent move towards stability results.

b. In the comparison, deviation amplification and reduction is assumed where necessary, the real year to year changes are taken into account, and there is no 'move towards stability'.

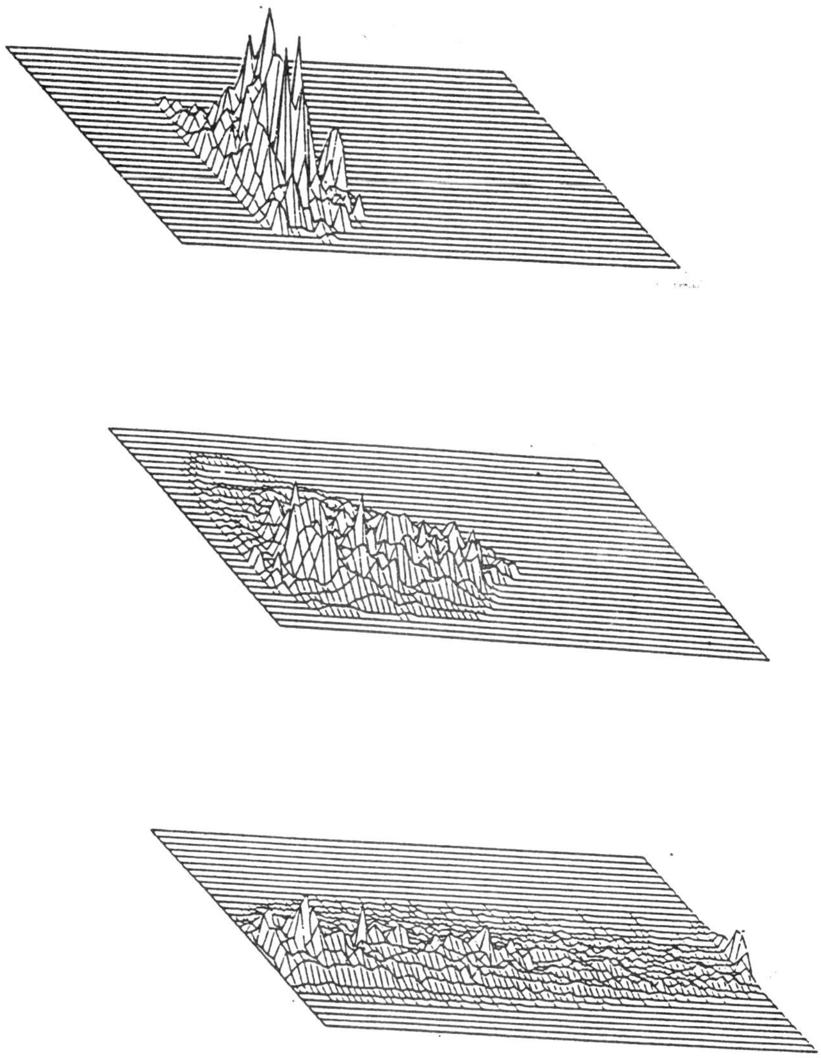

Figure 6. The probability of finding values for the variables x and y in the fishing system. (a) With slow response rates, the number of 'solutions' found is low, and the system stays in a narrow basin over all 1 000 years of the simulated run. (b) With faster response rate, the number of different 'solutions' found increases, ultil (c) with the fastest of three response rates, a large number of 'solutions' is 'considered' by the system. Thus, the chances increase that it will 'find' another local minimum (i.e. another structurally coherent set of solutions) outside the range of the present one.

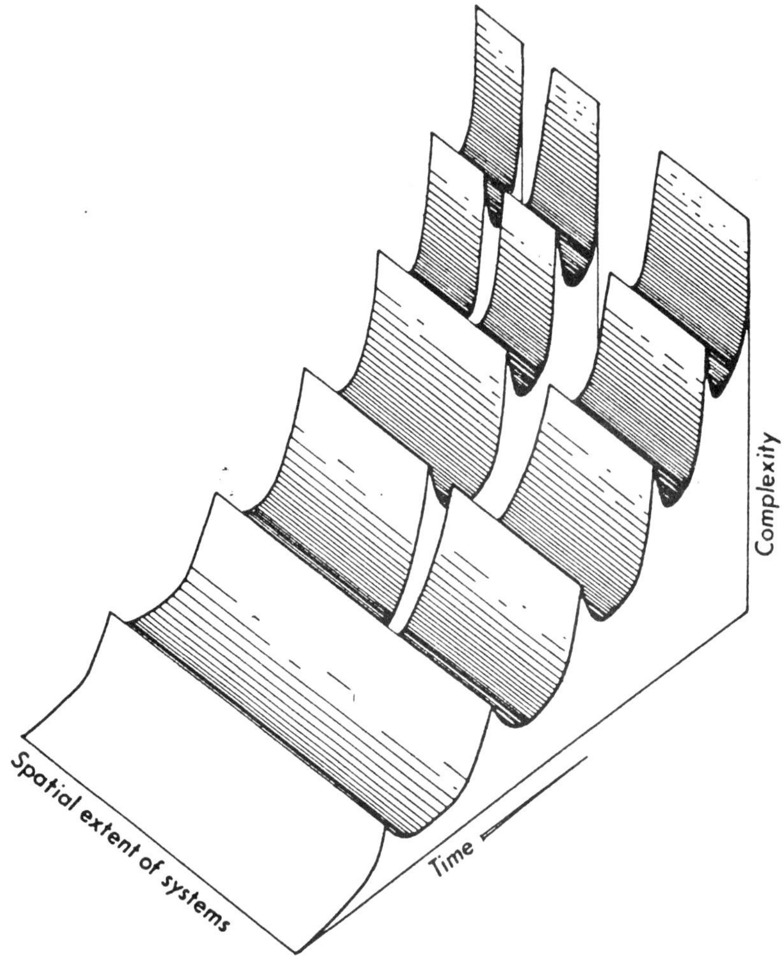

Figure 7. Degree of organizational coherence in a trading system along the South Coast of Papua. Note the regionalisation of the system (after Allen, 1984).

REFERENCES

ADAMS, R.M.

1966. *The Evolution of Urban Society: Early Mesopotamia and Prehispanic Mexico*, Chicago, Aldine.

ALLEN, J.

1984. "Pots and Poor Princes: a Multidimensional Approach to the Role of Pottery Trading in Coastal Papua", in: *The Many Dimensions of Pottery: Ceramics in Archæology and Anthropology*, edited by S.E. van der Leeuw and A.C. Pritchard, Amsterdam, Albert Egges van Giffen Instituut voor Pre- en Protohistorie: 407–464.

ALLEN, P.

in press Origami, Paper presented at a Research Seminar, Department of Archæology, Cambridge University, November 5, 1986.

ALLEN, P.M., and J.M. McGLADE

1986. "Modelling Complex Human Systems: a Fisheries Example", *European Journal of Operational Research.*

in press The Theory of Evolutionary Drive.

AMMERMAN, A., and L.L. CAVALLI-SFORZA

1973. "A Population Model for the Diffusion of Early Farming in Europe", in: *The Explanation of Culture Change*, edited by A. C. Renfrew, Duckworth, London: 341-357.

BATESON, G.

1972. *Steps to an Ecology of Mind*, New York, Ballantine.

BINFORD, L.R.

1968. "Post-Pleistocene Adaptations", in: *New Perspectives in Archæology*, edited by L.R. Binford and S.R. Binford, Chicago, Aldine: 313–341.

BOSERUP, E.

1965. *The Conditions of Agricultural Growth*, London, Allen and Unwin.

BOURDIEU, P.

1979. *Outline of a Theory of Practice*, Cambridge, Cambridge University Press.

BRADLEY, R.

1984. *The Social Foundations of Prehistoric Britain*, London, Longman.

BRAIDWOOD, R.J.

1958. "Near Eastern Prehistory: the Swing from Food-Collecting Cultures to Village-Farming Communities is still Imperfectly Understood", *Science* vol. 127: 1419–30.

BRAIDWOOD, R., and G. WILLEY

1962. *Courses towards Urban Life*, Chicago, Aldine.

CARNEIRO, R.

1970. "A Theory of the Origin of the State", *Science* vol. 769: 733–38.

1973. "The Four Faces of Evolution", in: *Handbook of Social and Cultural Anthropology*, edited by J.J. Honigmann, Chicago: 89–110.

CHILDE, V.G.

1925. *The Dawn of European Civilisation*, New York, Knopf.

CLAESSEN, H.J.M., and P. SKALNIK (eds)

1978. *The Early State*, The Hague, Mouton.

CLAESSEN, H.J.M.

1982. "Reaching for the Moon–Some Problems and Prospects of Cultural Evolutionism", in: *Archæological Approaches to the Study of Complexity*, edited by S.E. van der Leeuw, Amsterdam, University of Amsterdam.

CLARK, J.G.M.

1932. *The Mesolithic Age in Britain*, Cambridge, Cambridge University Press.

COHEN, R.L.

1978. "State Origins –a Reappraisal", in: *The Early State*, edited by H.J.M. Claessen and P. Skalnik, The Hague, Mouton: 31–75.

1981. "Evolution, Fission and the Early State", in: *The Study of the State*, edited by H.J.M. Claessen and P. Skalnik, The Hague, Mouton: 87–115.

COHEN, M.N.

1975. "Population Pressure and the Origins of Agriculture", in: *Population, Ecology and Social Evolution*, edited by S. Polgar, Chicago, Aldine: 79–121.

1977. *The Food Crisis in Prehistory: Overpopulation and the Origins of Agriculture*, New Haven, Yale University Press.

FEINMAN, G., and J. NEITZEL

1984. "Too Many Types –an Overview of Sedentary Prestate Societies in the Americas", *Advances in Archæological Method and Theory* no. 7: 39–102.

FLANNERY, K.V.

1972. "The Cultural Evolution of Civilisations", *Annual Review of Ecology and Systematics* no. 3: 399–426.

1973. "The Origins of Agriculture", *Annual Review of Anthropology* no. 2: 271–310.

FRIED, M.H.

1967. *The Evolution of Political Society: an Essay in Political Anthropology*, New York, Random House.

GAMBLE, C.

1986. *The Palæolithic Settlement of Europe*, Cambridge University Press.

GIDDENS, A.

1979. *Central Problems in Social Theory*, London, MacMillan.

GIMBUTAS, M.

1956. *The Prehistory of Eastern Europe*, Part I, American School of Prehistoric Research, Bulletin 20, edited by H. Hencken.

GREEN, S.W.

1980. "Towards a General Model of Agricultural Systems", *Advances in Archæological Method and Theory* vol. 3: 311–355.

HAAS, J.

1982. *The Evolution of the Prehistoric State*, New York, Columbia University Press.

HAGESTEIJN, R.R.

1985. "Best among Equals: Problems of Social Stratification in the Angkor State", in: *Raakvlakken*, edited by S.E. van der Leeuw, Amsterdam, University of Amsterdam; 50–71.

HARRIS, M.R.

1977. *Cannibals and Kings*, New York, Random House.

HODDER, I.

1982. "Theoretical Archæology: a Reactionary View", in: *Symbolic and Structural Archæology*, edited by I. Hodder, Cambridge University Press: 1–16.

JOCHIM, M.

1976. *Hunter-Gatherer Subsistence and Settlement*, New York, Academic Press.

JOHNSON, G.A.

1973. *Local Exchange and Early State Development in Southwestern Iran*, Ann Arbor, University of Michigan, Museum of Anthropology, Anthropological Papers no. 51.

1978. "Information Sources and the Development of Decision-Making Organisations", in: *Social Archæology: Beyond Subsistence and Dating*, edited by C.L. Redman *et al.*, New York, Academic Press: 87–112.

KAHNEMANN, D., and A. TVERSKI

1982. "The Psychology of Preferences", *Scientific American* vol. 246, no. 1: 136–42.

KEENE, A.S. and J.A. MOORE (eds)

1983. *Archæological Hammers and Theories*, New York, Academic Press.

KUHN, T.S.

1962. *The Structure of Scientific Revolutions*, Chicago, University of Chicago Press.

McGUIRE, R.

1984. "Breaking down Cultural Complexity: Inequality and Heterogeneity", *Advances in Archæological Method and Theory* no. 8: 91–142.

PRIGOGINE, I.

1978. "Time, Structure and Fluctuations", *Science* vol. 201: 777–86.

1981. *From Being to Becoming*, San Francisco, Freeman.

REDMAN, C.L.

1977. *The Emergence of Civilisation*, San Francisco, Freeman.

REED, C.L.

1977. *Origins of Agriculture*, The Hague, Mouton.

RENFREW, A.C.

1972. *The Emergence of Civilisation: the Cyclades and the Aegean in the Third Millennium BC*, London, Methuen.

RINDOS, D.

1984. *The Origins of Agriculture*, New York, Academic Press.

RUSSELL, B.

1903. *Principles of Mathematics*, second edition, photostatic reprint, New York.

SABLOFF, J., and C.C. LAMBERG-KARLOVSKY

1975. *Ancient Civilisation and Trade*, Albuquerque, University of New Mexico Press, School of American Research Advanced Seminar Series.

SAHLINS, M.

1972. *Stone Age Economics*, London, Tavistock.

SANDERS, W.T., and B.J. PRICE

1968. *Mesoamerica: the Evolution of a Civilisation*, New York, Random House.

SERVICE, E.R.

1971. *Primitive Social Organisation: an Evolutionary Perspective*, New York, Random House.

1975. *The Origins of the State and Civilisation*, New York, Norton.

STEWARD, J.H.

1949. "Cultural Causality and Law: a Trial Formulation of Early Civilisations", *American Anthropologist* vol. 51, no. 1: 1–27.

1955. *The Theory of Culture Change: the Methodology of Multilinear Evolution*, Urbana, University of Illinois Press.

TVERSKI, A.

1977. "Features of Similarity", *Psychological Review* vol. 84: 327–52.

TVERSKI, A., and I. GATI

1978. "Structures of Similarity", in: *Cognition and Categorisation*, edited by E. Rosch and B.B. Lloyd, Hillsdale, New Jersey: 79–98.

VAN DER LEEUW, S.E.

1982. "Information Flows, Flow Structures and the Explanation of Change in Human Institutions", in: *Archæological Approaches to the Study of Complexity*, edited by S.E. van der Leeuw, Amsterdam, University of Amsterdam: 230–329.

1984. "Dust to Dust: a Transformational View of the Ceramic Style", in: *The Many Dimensions of Pottery: Ceramics in Archæology and Anthropology*, edited by S.E. van der Leeuw and A.C. Pritchard, Amsterdam, University of Amsterdam: 707–774.

1986. "On Settling Down and Becoming a 'Big Man'", in: *Private Politics*, edited by M.A. van Bakel, R.R. Hagesteijn and P. van de Velde, Leiden, Brill.

in press "Risk, Perception, Innovation", in: *Innovation*, edited by S.E. van der Leeuw and R. Torrence, London, George Allen and Unwin.

WITTFOGEL, K.

1957. *Oriental Despotism*, New Haven, Yale University Press.

WRIGHT, H.T.

1969. *The Local Administration of Rural Production in an Early Uruk Town*, Ann Arbor, University of Michigan Museum of Anthropology, Anthropological Papers no. 39.

WRIGHT, H.T., and G.A. JOHNSON

1975. "Population, Exchange and Early State Formation in Southwestern Iran", *American Anthropologist* vol. 77: 267–89.

FURTHER REFLECTIONS ON RESOURCE CONCENTRATION AND ITS ROLE IN THE RISE OF THE STATE

ROBERT L. CARNEIRO
American Museum of Natural History

The origin of the state can be looked at in two ways. Each instance of it can be thought of as a unique historical event, the result of a singular set of conditions. To account for the rise of any particular state, then, would require a detailed knowledge of all the special circumstances involved. As long as the origin of states is viewed in this light, no general theory of state formation will ever emerge. If a hundred cases of state formation are known, then a hundred different explanations will be needed to account for them.

This is historical particularism. It is the point of view associated with traditional historians. But while this view is best exemplified by the discipline of history, it is by no means restricted to that discipline. It can be found in anthropology as well. Read, for example, Richard Blanton's discussion of state formation in Oaxaca and you will be struck by his predisposition to see the rise of the state in there as the result of a unique constellation of factors (Blanton 1981: 230–231). He seems resolutely disinclined to look for the common elements that might underlie *all* cases of state formation, in Oaxaca, Mesopotamia, North China, or anywhere else[1].

There is, of course, another way of looking at the matter. One can take the position that states probably arose through the operation of a small number of recurring factors. The role of each factor need not have been exactly the same in each case, but their joint operation may well have assured the rise of the state wherever these factors were present. The task, then, would be first, to ferret out these few factors, and then to assess their respective effects in each instance of state formation.

This is the point of view of science. It begins with particulars but endeavors to transcend them. It aims to reach propositions and explanations of the widest generality. To be sure, one cannot *assume* that recurring regularities exist. But on the other hand, regularities cannot be found unless they are sought for. Thus, we must be ready to look beneath superficial differences for underlying commonalities. The parallels in the origin and development of every state may indeed be disguised and obscured, but they may exist nonetheless.

At any rate, this was the position I took in 1970 in an article entitled "A Theory of the Origin of the State". Today, 16 years later, I would like to go over the essential features of the theory as I originally proposed it, and then explain how I have recently modified and elaborated it.

At the outset, though, we need to have a common understanding of what we mean by a state. The definition I favor is this:

> A state is an autonomous political unit, encompassing many communities within its territory, and having a centralized government with the power to draft men for war or work, levy and collect taxes, and decree and enforce laws. (See Carneiro 1970: 733, 1981: 69).

---

[1] Reviewing Henry Claessen and Peter Skalník's two volumes, *The Early State* (1978) and *The Study of the State* (1981), Malcolm Webb complains of the "unfortunate tendency" of contributors to these volumes "to view the cases [of state formation] *ad hoc*, with little attention to the larger ongoing worldwide history of the state as a social type..." (Webb 1984: 277).

If a society has these attributes, it is a state. If it doesn't, it isn't.

Years ago, after examining many societies in both the Old World and the New, societies with states and societies without, I came to the conclusion that all theories of state origins then current were flawed. It seemed to me that a new one was required, and I was emboldened to propose one. My theory sought to explain why autonomous villages, which during the early Neolithic period were all but universal, later gave way in certain parts of the world to successively larger and more complex political units, culminating, in certain favored areas, in the rise of the state.

Let me summarize my theory here in a somewhat different way from the one in which I originally presented it. Broadly speaking, the rise of the state is the climax of a certain *process*. This process is brought on by certain *conditions*. And these conditions are based in turn on certain *factors*.

The *process* which overcame the political autonomy of villages and welded them, first into chiefdoms, and then into states, was conquest warfare. Assigning this positive role to warfare may be unpalatable to some people, but it is nonetheless inescapable in view of the evidence. In my paper of 1970 I distinguished between *voluntaristic* and *coercive* theories of state origins, and dismissed the former as inadequate, largely because I then knew of no case of autonomous political units voluntarily relinquishing their sovereignty and joining with other similar units to form a larger whole. I still don't. Every village or chiefdom I know of, which gave up its autonomy and became part of a larger political unit, did so either through the exercise of force or the threat of it. Accordingly, I argued in 1970 that only a *coercive* theory could account for the rise of the state. I still do.

But while I identified warfare as the mechanism by which autonomies were overridden and larger political units established, I recognized that not all wars had had this effect. During early Neolithic times, for instance, warfare rarely if ever resulted in conquest. It was waged over murder, witchcraft, wife stealing and the like, but not over land. Special conditions were needed for warfare to become redirected from these objectives to the taking of land and the subjugation of people. Of these special conditions, the major one was population pressure. And this pressure, I argued, resulted from an increase in human numbers relative to the amount of land available to support them.

Obviously, population pressure was much more likely to arise where land was in short supply than where it was abundant. And this condition would have been especially liable to occur where arable land was sharply limited by mountains, deserts, or seas. Thus, I argued, conquest warfare arose first and proceeded furthest in those parts of the world where agricultural land was *circumscribed*. Once settlement became dense enough in a circumscribed area, any war fought there would leave a defeated village with no place to flee. It would be forced to remain in place and be subjugated by the victors.

In areas where arable land was *not* circumscribed, warfare would fail to have this effect. Thus, in extensive and unbounded regions like Amazonia, warfare rarely led to conquest, but rather to a process of "fight and flight". Since arable land was available almost everywhere around, the losers in war could readily move away from their more powerful enemies and thus retain their independence. As a result, local autonomy was rarely transcended in Amazonia. And, with only a few exceptions, political evolution here did not go beyond the stage of autonomous villages.

Auxilliary Hypotheses

When I first formulated the circumscription theory I was aware of a few cases that appeared to be exceptions to the rule. These exceptions were states which had developed early and autochthonously, but in areas where arable land was *not* circumscribed. The two principal exceptions, it seemed to me at the time, were North China and the Maya Lowlands. I never doubted that these states, like all others, had been forged into existence through conquest warfare. But it seemed to me that something other than environmental circumscription must have precipitated conquest warfare in those areas.

In an effort to account for the rise of states in such areas, I sought for an *auxilliary factor* to put into the equation. This auxiliary factor turned out to be *social circumscription.*

The notion of social circumscription is one I borrowed from Napoleon Chagnon (1968: 251), and it originated in the following way. It seemed to Chagnon that those Yanomamö villages located in the central or "nuclear" part of Yanomamö territory were larger, more closely spaced, had stronger headmen, fought more frequently, and formed alliances more readily than villages in the peripheral areas. Thus, in nuclear Yanomamö territory political evolution seemed to have advanced a step or two beyond what it had elsewhere. This fact Chagnon attributed to a greater crowding of villages in the nuclear area, a crowding which had led to more frequent and intense fighting among villages located there. But at the same time, this crowding made it harder for defeated villages to flee and find a safe haven elsewhere, since secure territory was not always available. The smaller and weaker villages of this nuclear area, which were usually the victims in war, were often unable to move out of range of their stronger enemies. To compensate for this, they formed defensive alliances with other villages and sometimes even temporarily amalgamated with them.

In Chagnon's notion of social circumscription I saw an important process, one by no means limited to the Yanomamö. In fact, it seemed to me that social circumscription and its political consequences were much more strikingly manifested in certain other areas of the world. The process, I thought, could be seized on and put to work to account for the rise of the state in areas such as the Maya Lowlands and North China, which appeared to me to lack environmental circumscription.

In state formation, social circumscription operates in much the same way as environmental circumscription. Under the latter, a group defeated in war is kept from moving away and thus retaining its autonomy by the existence of surrounding mountains, deserts, or seas, which force them to stay in place. In cases of *social* circumscription, it is not physical barriers that prevent defeated villages from fleeing, but the fact that all the surrounding territory is already occupied by other villages. The net effect, though, is the same. Villages beaten in war are forced to remain in place and be subjugated by the victors.

There is, however, an important difference. Social circumscription is never as tight as environmental circumscription. It doesn't restrict villages movement quite as much. Social circumscription allows a certain amount of "leakage", especially during its early stages. Defeated groups can sometimes filter their way out to unoccupied territory and thus escape losing their sovereignty. Only with the passage of time does the land fill up to the point where flight becomes impossible. Thus, as a cause of state formation, social circumscription acts more slowly. And indeed, in those areas where the state arose by this means, areas such as the Maya Lowlands, Western Europe, and West Africa, it did so substantially later than where it arose through environmental circumscription.

The reader may have noticed my omission of North China from this listing. As I mentioned earlier, I once thought that the earliest Chinese states had arisen in North China through the action of social circumscription. But I am now convinced that I was wrong. My error lay in supposing that the first Chinese state was Shang, which developed on the broad, unbounded plain of the lower Yellow River. A careful reading of K. C. Chang's *The Archæology of Ancient China* (1977) though, began to persuade me that such was not the case. The earliest Chinese state, it seems to me, arose *prior* to Shang, and it arose many miles upstream, along the big bend of the Yellow River, where it is joined from the west by the Wei. This riverine area, which is shaped like an inverted "T", is flanked by mountains that rise to 5,000 feet or more and which hem in the river valley on all sides. Thus, the birthplace of the Chinese state turns out to be environmentally circumscribed after all! And accordingly, we can move North China out of the column of cases requiring the weaker explanation in terms of social circumscription, and into the main column of state formation where the stronger version of the theory, involving environmental circumscription, prevails.

Resource Concentration

As I continued to look at state formation, certain facts led me to formulate a second auxilliary hypothesis. The facts were these. From the accounts of the early Spanish chroniclers, it was very clear that during the 16th century, chiefdoms of impressive size had existed along the Amazon River. Since this region lacked environmental circumscription, and seemed to lack social circumscription as well, some additional factor was needed to account for the development of chiefdoms here. It struck me that the additional factor was *resource concentration.* Quoting from my paper of 1970, this is how it seemed to me that resource concentration had operated along the Amazon:

> With unbroken stretches of arable land extending back from the Amazon for hundreds of miles, why were there chiefdoms here?
>
> To answer this question we must look closely at the environmental conditions afforded by the Amazon. Along the margins of the river itself, and on the islands within it, there is a type of land called *várzea.* The river floods this land every year, covering it with a layer of silt. Because of this annual replenishment, *várzea* is agricultural land of first quality which can be cultivated year after year without ever having to lie fallow. Thus, among native farmers it was highly prized and greatly coveted.
>
> The waters of the Amazon were also extraordinarily bountiful, providing fish, manatees, turtles and turtle eggs, caimans, and other riverine foods in inexhaustible amounts. By virtue of this concentration of resources, the Amazon, as a habitat, was distinctly superior to its hinterlands.
>
> Concentration of resources along the Amazon amounted almost to a kind of circumscription. While there was no sharp cleavage between productive and unproductive land, as there was in Peru, there was a steep ecological gradient. So much more rewarding was the Amazon River than adjacent areas, and so desired did it become as a habitat, that peoples were drawn to it from surrounding regions. Eventually crowding occurred along many portions of the river, leading to warfare over sections of river front. And the losers in war, in order to retain access to the river, often had no choice but to submit to the victors. By this subordination of villages to a paramount chief there arose along the Amazon chiefdoms representing a higher step in political evolution than had occurred elsewhere in the basin. (Carneiro 1970: 736–737).

In a footnote to this passage I noted that chiefdoms fostered by resource concentration had also arisen in the Mojos Plain of Bolivia (see Denevan 1966: 43–50, 104–105, 108–110)—and, I might have added, the middle Orinoco. I also pointed to the lower Mississippi valley as another region lacking environmental circumscription which had nevertheless given rise to chiefdoms, and I attributed this development to resource concentration as well.

And there the matter stood until 1980. That year saw the publication of Anna Roosevelt's book, *Parmana*, a daring and ambitious attempt to reconstruct, in broad terms, the prehistory of Amazonia. Before presenting her own theory, Roosevelt examined those of others, criticizing each in turn. In discussing my views on the subject, she seemed ready to accept the notion that Amazonian chiefdoms had arisen through conquest warfare. However, she objected to the chain of arguments I had presented to account for the onset of conquest warfare. Most particularly, she argued that my use of resource concentration was flawed because, in the manner that I had applied it here it amounted, she said, to a voluntaristic theory. Since in "A Theory of the Origin of the State" I had emphatically rejected voluntaristic theories of political development in favor of a coercive one, Roosevelt felt that my use of voluntarism in reconstructing the rise of Amazonian chiefdoms involved me in a contradiction. And she was right.

The explanation of the rise of Amazonian chiefdoms I proposed in 1970 was, without my realizing it, voluntaristic. Consider carefully this key phrase (already quoted above) from my article: "... and the losers in war, in order to retain access to the river [and its resources], *often had no choice but to submit to the victors*". But what I was really saying was that they *had* a choice. I clearly implied that defeated groups *could* move away from the riverfront and remain autonomous, but instead, *chose* not to do so. Facing the alternative between political autonomy and bountiful food resources, they had chosen the latter.

Even before Roosevelt published her critique, I had been vaguely aware of this contradiction. But only vaguely. And rather than face it squarely, I had pushed it to the back of my mind and left it there. Now I was forced to confront it and, if possible, resolve it.

A Reformulation of Resource Concentration

Over the last few years I have spent considerable time reexamining the idea of resource concentration. And I have come to see its role in political evolution in a rather different light. Resource concentration did indeed help give rise to Amazonian chiefdoms, but only by acting in concert with social circumscription. Let us look at the Amazon again and see how the joint operation of these two factors seems to have led to that result.

There is little doubt that the enormous and inexhaustible amounts of fish and other riverine food resources found in the waters of the Amazon acted as a magnet. It drew people to that river from many miles around. And since these riverine resources were *wild*, they could be effectively exploited by the relatively simple subsistence techniques already possessed by hunters, fishers, and gatherers. Thus, the first steps leading toward dense settlement along the Amazon probably occurred early, during pre-agricultural times.

As word of the river's bounties spread, more and more Indians were drawn to it, and settlement along its banks continued to increase. By the time plant cultivation arrived on the scene, settlement along the river must have been fairly dense —denser, certainly, than in other parts of Amazonia.

The introduction of agriculture made possible the use of a riverine resource which had not been previously tapped —*várzea*. This rich soil, whose fertility was replenished every year by the overflowing of the Amazon, provided an additional lure. Even more peoples were now drawn to the banks of the river. And with this new subsistence attraction added to fish and manatee, turtles and turtle eggs, settlement along the Amazon became even denser. Before long, the margins of the river and the islands within it were thickly settled. Indeed, by the 16th century, population was so dense on the Amazon that the Spanish chronicler Gaspar de Carvajal could write that over a stretch of 80 leagues (240 miles) along the river, "there was not from village to village a crossbow shot" (Medina 1934: 198).

What I have just described is not materially different from what I envisioned in 1970. But here the scenario changes. A new factor comes into the picture. I am now ready to argue that the density of settlements induced by resource concentration gave rise at many points along the Amazon to ***social circumscription***. The degree of crowding that developed along the river and back away from it not only intensified warfare, which was, no doubt, already present, but made it difficult if not impossible for those groups defeated in war to move out of range of the stronger groups. Thus they became subject to conquest.

That warfare along the Amazon was frequent and intense, and that it was directed to the taking of land, is attested to by Carvajal. Noting that there were Indians living inland, away from the river, who were enemies of those settled along the river itself, Carvajal remarked that the inland tribes "come each day to drive them [the riverfront dwellers] from their homes" (Medina 1934: 190).

And this warfare, induced primarily by a shortage of coveted river frontage, had its effects augmented by the constraining factor of social circumscription. Defeated groups could no longer readily flee, but were subject to being incorporated, along with their land, into the political unit of the victors. In this manner the Amazon gave rise to an extensive string of chiefdoms which, in the 16th century, formed so conspicuous a feature of that river.

General Considerations

With this reformulation of the circumscription theory as it operated along the Amazon, we have now exorcised the demon of voluntarism. Like chiefdoms elsewhere in the world, Amazonian chiefdoms can be seen, unequivocally, as arising through conquest warfare brought on by population pressure.

But though the role of resource concentration is thus diminished in one respect, it is augmented in another. Social circumscription may well have been the immediate and essential precursor of conquest warfare along the Amazon, but resource concentration played the key role in creating the conditions that gave rise to this social circumscription. And now we must ask a question of broad implications: if it did so along the Amazon, why not elsewhere?

We know there are areas in other parts of the world which lacked environmental circumscription but which nevertheless saw the rise of chiefdoms and states. Could resource concentration have been the factor that triggered population growth in those areas too? It looks that way to me. And if resource concentration could aggregate people in areas that *lacked* environmental circumscription, why not in areas that *had* it? I am convinced that it did.

This idea had already entered my mind by 1970, when I noted in my paper on the origin of the state that resource concentration had played a major role in promoting political development on the coast of Peru. The enormous quantities of fish and shellfish found along this coast had fostered the emergence of chiefdoms here even before the introduction of agriculture —or at least before its extensive use (Carneiro 1970: 737). But I did not see then what I do now, namely, that resource concentration was a key factor in producing chiefdoms and states in almost every environmetally circumscribed area where they arose.

With the possible exception of the Peruvian coast, nowhere on earth did full-fledged states arise before agriculture became the major contributor to subsistence. And quite possibly it always takes agriculture to bring about that degree of population pressure needed for the onset of state formation. But even before the coming of agriculture, a high concentration of wild food resources could have given an area a big push toward a correspondingly dense concentration of people. At least, enough of a concentration to bring about the early stages of population pressure.

Let me run through the general argument once more. Resource concentration, *by itself*, cannot give rise to chiefdoms or states. It acts only as a *catalyst*. It *speeds up* a reaction involving other elements. Specifically, it accelerates the onset of population pressure, which is the prime trigger of state formation.

Now, the degree of population pressure needed to begin the process of state formation *could*, in time, arise in *any* environmentally circumscribed area of arable land. It could do so even in circumscribed areas deficient in wild food resources and thus sparsely settled prior to the coming of agriculture. State formation could occur here, but it simply would take *more time*. By giving population growth a head start, resource concentration probably hastened the advent of many chiefdoms and states by centuries —perhaps even by millennia.

In the rest of this paper I would like to survey some of the principal areas of the world where chiefdoms and states arose, and to see what evidence we can find that resource concentration may have played a role in accelerating their development.

Resource Concentration in Areas of Early State Formation

Let us begin with the region where the first states emerged —Mesopotamia. Was there a concentration of wild food resources in "the land between the rivers"? Here is what V. Gordon Childe had to say on the matter. In *Man Makes Himself*, Childe (1948: 107) wrote:

> Between the main channels of the Tigris and the Euphrates was a vast tract of swamps, only recently raised by the river silt above the waters of the Persian Gulf. The swamps were covered with a tangle of gigantic reed-brakes interspersed with groves of date-palms. They were interrupted only by low ridges of rocky outcrop or banks of sandy silt. But they swarmed perpetually with animal life while on either side the steppes above the flood-level were parched and barren throughout the long blazing summer and the bitter winter. Attracted perhaps by the game, wild-fowl, and fish, and by the groves of date-palms, the proto-Sumerians tackled the stupendous task of taming the Tigris-Euphrates delta and making it fit for habitation.

Although qualified by a "perhaps", Childe has caught a glimpse of resource concentration. He clearly implies that the presence of wild food resources in Mesopotamia may have drawn people there in sizable numbers before the coming of agriculture. Thus, when agriculture arrived on the scene, the settlers would already have been in place in Mesopotamia, ready to clear and cultivate the land so it could enhance their subsistence.

Now, what about the Nile Valley? Childe says little about the wild food resources that once existed along the Nile, but other authors provide us with a good account of this. For example, the Egyptologist Francis L. Griffith (1910: 45) wrote:

> Here in ancient times were found the oryx, addax, ibex, gazelle, bubale, ostrich, hyena and porcupine, more rarely the wild ox and wild sheep (*O. tragelaphus*). All of these were considered fit for the table... Pigeons and other birds were caught in traps, and quails were netted in the fields... Enormous quantities of wild-fowl of many sorts were taken in clap-nets, to be preserved in jars with salt. Fish were taken sometimes in hand nets, but the professional fishermen with their draw-nets caught them in shoals. The fishing industry was of great importance: the annual catch in the Lake of Moeris, and its canal formed an important part of the Egyptian revenue. The fish of the Nile, which were of many kinds (including mullets, etc., which came up from the sea), were split and dried in the sun: others were salted and so preserved.

Griffith was writing about Dynastic times, but if anything, the game and fish and fowl present in and along the Nile in pre-Neolithic times must have been even more abundant.

Lake Moeris, which lay in what is now the barren Fayum depression, was once a most attractive area for settlement. Around this lake have been found the remains of many early Neolithic villages —perhaps, Childe thought, the earliest such villages in all of Egypt. And Lake Moeris, Griffith tells us, was well stocked with fish. Even during Paleolithic times the lake had begun to attract human settlement, for, as Childe (1948: 70) notes, "Round the lake that once filled the Fayum depression the number of Old Stone Age tools is certainly imposing". He is not ready to say that the Fayum was densely populated during the late Paleolithic because the stone tools found there archeologically "have to be spread over so many thousand years that the population they attest [to] may be exiguous" (Childe 1948: 70).

Still, here is what Childe sees happening around Lake Moeris with the arrival of agriculture: "...quite abruptly the shore of a somewhat shrunken lake is found to be fringed with a chain of populous hamlets, all seemingly contemporary and devoted to farming. The Nile valley from the First Cataract down to Cairo is quickly lined with a chain of flourishing peasant villages, all seeming to start about the same time and all developing steadly down to 3000 BC" (Childe 1948: 70).

The simplest explanation for this rapid deployment of Neolithic villages is this: there was already a large population of hunters, fishers, and gatherers around Lake Moeris when agriculture was introduced. And these people, who were by then at least semi-sedentary, took full advantage of what agriculture allowed and began living in settled villages.

The presence of a sizable, pre-agricultural population along the Nile would naturally have shortened the time needed to produce population pressure there. Thus, it would have accelerated the process that led eventually —and I would say *irresistibly*— to the rise to the state on the banks of the Nile.

Looking south of the Sahara, we find evidence that resource concentration may have played a role in the rise of the first native states there too. Two areas where this seems to have been true are West Africa and the Interlacustrine region of East Africa.

The earliest known states of West Africa were Ghana, Mali, Songhai, Kanem, and Bornu. Ghana, the first Negro state of which we have any knowledge, was located on or near the Niger River, while the Songhai kingdom was located right on the middle Niger, and the state of Mali higher up the Niger. The kingdom of Kanem bordered the northeastern shore of Lake Chad, while Bornu was located on the southwestern shore. The question is, then, were the Niger River and Lake Chad areas of resources concentration? And the answer seems to be yes. Let us look at the evidence.

In his *Economic History of West Africa*, A. G. Hopkins says: "Fishing was practiced...in many inland waters [of West Africa], two of the largest centers being Lake Chad and the great bend of the Niger in the Western Sudan". And, he continues: "Specialized communities of highly skilled fishermen are known to have developed in these areas at an early date" (1973: 43), chief among them being the Sorkawa and the Bozo.

Hopkins goes on to note that "the availability of fish fluctuated with the...different flood periods of the rivers, so fishermen tended to migrate in search of their catch" (1973: 43). Thus, "the Sorkawa and Bozo travelled hundreds of miles each year along the Niger" (1973: 43), looking for the best places to fish. So bountiful were their catches that during the 16th century both the Sorkawa and the Bozo "paid their taxes to the rulers of the Songhai empire exclusively in dried fish" (1973: 43).

Some idea of the magnitude of fish available in the Niger can be gained from the fact that as recently as 1954 the towns of Mopti and Segou on the middle Niger exported to the Ivory Coast no less than 10,000 to 12,000 tons of dried fish a year (Hopkins

1973: 246). During the first millennium AD, when the earliest states were arising in this region, the fish supply must have been at least as plentiful.

Lake Chad also seems to have been well provided with wild foods. Speaking of the northern and northwestern shores of the lake one source says: "This region abounds in big game and birds are plentiful... Fish abound in its waters..." (Anonymous 1910: 786–787).

It seems fair to conclude, then, that a rich substratum of wild food resources existed along the banks of the Niger and the shores of Lake Chad in early times, and that it was instrumental in drawing human populations here and allowing them to flourish and multiply. This in turn would have provided the impetus for the population pressure that led these areas to develop states before the surrounding regions had done so.

In the Interlacustrine region of East Africa, wedged between Lakes Victoria, Albert, Tanganyika, Kivu, Edward, and Kyoga, early European explorers found a surprising number of native states. And the clustering of so many states in this region has continued to impress observers down to the present day. Thus Murdock (1959: 354) says that "Ruanda and Uganda...are characterized by the extraordinary prevalence of despotic states". And Fallers (1965: 23) has made a similar observation:

> Most of the Bantu Peoples of Uganda, northern Tanganyika and Ruanda-Urundi traditionally had political systems of the centralised state type. Ethnographically most famous among these Bantu states are Buganda, Bunyoro and Ankole, but the complex is much wider, extending south-west around lake Victoria to include the Toro of Uganda... the Ruanda and Rundi of Ruandi-Urundi and the Haya, Zonza, Ha, Nyamwezi and other peoples of north-western Tanganyika.

That there was an unusual concentration of states in the Interlacustrine basin has thus not escaped attention. What has gone unperceived, though, is the basic cause for the rise of so many states here. Debate still rages over whether or not these states arose through the conquest of a weak agricultural majority by a strong pastoral minority (e. g., Fallers 1965: 27–29; Kottak 1972: 353–353). But no one, as far as I can tell, has pointed to what seems to me the underlying reason for state formation here —the attraction of people to the shores of these lakes because of the food resources available there, and the eventual population pressure thus produced.

Adding to the likelihood that resource concentration was what triggered state formation here is the fact that those peoples who lay outside of the Interlacustrine region, whether they were Bantu-speakers or Hamitic-speakers, never developed states.

The little research I have done so far on the ecology of the Great Lakes of East Africa indicates that these lakes provided significant amounts of fish and other wild foods for those who dwelled along their margins. For example, in prehistoric times, the Wilton C people who lived along the lakes left "huge shellfish middens..." (Cole 1963: 46, cited in Kottak 1972: 354). And of modern-day inhabitants of the region, Murdock (1959: 352) says: "The Interlacustrine peoples depend very little on hunting and gathering, but many of them fish extensively..." Brief as it is, the evidence gathered so far suggests that resource concentration existed in the Interlacustrine region and eventually produced the expected results.

From one point of view, this area might more appropriately be called "Circumlacustrine" rather than "Interlacustrine". The former term would highlight the lake margins and their exploitation in the ultimate political developments of the area. But there is also a good reason for retaining the term "Interlacustrine", because it points to the geographical boundedness of the region. And indeed, a glance at the map shows that it was, to a considerable degree, environmentally circumscribed.

Part of the circumscription was provided by the lakes themselves, which formed a partial ring enclosing an elliptical pocket of land. And in addition, the Mitumba Mountains to the west formed a sort of rocky spine that further bounded the area, contributing still more to its circumscription. Thus, being *impacted* between lakes and mountains, and drawing on the lakes' bountiful waters for subsistence, the earlier inhabitants of this region lived under two conditions that favored state formation. And, as we have seen, they responded to these conditions by eventually giving rise to a large number of states.

Returning to Asia, in the case of North China we again find a river offering pre-Neolithic peoples a strong inducement to settle along its banks. According to the French Sinologist Jacques Gernet (1968: 47): "At the end of the second millennium [BC], Yellow River China was quite unlike what it became at the beginning of the Empire. All the available evidence shows a country still covered by great forests and vast marshes, with an astonishing number of birds, fish, and game of every size ..." Need more be said?

Moving to the New World, let us look first at the Valley of Mexico. In early times, this large interior basin contained five lakes: Texcoco, which was brackish, and four others, Zumpango, Xaltocan, Chalco, and Xochimilco, which were sweet. Of this basin during pre-Classic times, George Vaillant (1966: 49–50) wrote: "The lakes were shallow and their marshy shores, thick with reeds, attracted a teeming abundance of wild fowl. On the wooded mountain slopes deer abounded". Vaillant (1966: 142) later listed some of the wild foods available in the Valley of Mexico: "The seasonal migrations of the birds, which still visit the lakes of Mexico, offered a profusion of geese, ducks and other wild fowl. Small fish, netted or speared with a trident, were sometimes consumed, and the eggs which a certain fly lays on the lakes were made into a paste still eaten in Mexican communities".

Muriel Porter Weaver notes that the "large lakes in the Basin of Mexico contained a variety of fish, reptiles, amphibians, aquatic birds, and insects that provided an easy daily food supply". Later, speaking of Lake Chalco in particular, Weaver (1981: 17) says: "The lake was never very deep but was ideal for fresh-water mollusks, turtles, and fish, while the swampy shoreline was covered with cattail, reeds, water lentils, and a variety of aquatic plants... Remains of white-tailed deer, rabbit, dog or coyote, small rodents, and fish and turtles have all been preserved. Between November and May, migratory birds flocked into the Chalco region to spend a more pleasant winter. The inventory includes Canadian geese, ducks, pintails, mallards, teals, and the American coot. When we visualize the magnificent...setting...combining adequate rainfall, fine alluvium soils, a wealth of faunal and floral resources, we can see that this was an optimal ecological area in which to live" (Weaver 1981: 45–46).

We can conclude, then, that as far as conditions fostering state formation were concerned, the Valley of Mexico was doubly favored. On the one hand, it was ringed by mountains which tightly circumscribed the valley, and on the other it supported a great variety and abundance of wild foods sufficient to attract people in large numbers. It is not at all surprising, then, that a state should have emerged here early in prehistory.

What *is* surprising to many archeologists, though, is that a state —or at the very least a maximal chiefdom (see Carneiro 1981: 47)— should have emerged in the area occupied by the Olmec. As Weaver (1981: 67) notes: "The Olmec heartland is too hot, too humid, and too tropical and rainy to be singled out by anyone today as a choice place to live". Nonetheless, whether we classify its polities as chiefdoms or states, the Olmec produced a culture that went far beyond the level of autonomous villages. How did it happen?

Michael Coe, the chief investigator of the Olmec, seems ready to embrace the circumscription theory in order to account for this development. He writes as follows:

From our work at the site complex of San Lorenzo Tenochtitlan, we have gathered a great deal of evidence to show that the move toward sociopolitical complexity, beginning some time before 1200 BC, followed the lines of Carneiro's "circumscription theory" (Carneiro 1970). In fact, the process must have followed the same scenario [as] for the middle Amazon chiefdoms, with restricted river levee lands of immensely high productivity being co-opted by one or more lineages through success in war. In this case, if one has to look for prime movers, then the factors invoked by Carneiro fit the case very well, indeed (Coe 1981: 165).

The Olmec region does, in fact, seem to afford a conspicuous case of resource concentration. As Weaver (1981: 67) notes: "this is a very rich agricultural area that offered an almost inexhaustible food supply..." But even before the land was tapped for agriculture, the rivers and lakes of the Olmec region offered a rich source of fish, water fowl, reptiles, and amphibians.

We can suppose, then, that at quite an early period the Olmec area began attracting significant numbers of people, and contributed to the rapid increase of those who settled there. By the second millennium BC, the population concentrated here was sufficient to permit Olmec culture, with its various imposing manifestations, to arise.

From Mesoamerica let us turn next to South America and the coast of Peru. This coast, sharply delimited by deserts, mountains, and the sea, is, from an environmental point of view, one of the most highly circumscribed regions in the world. But besides its circumscription it was also an outstanding area of resource concentration. The archeologist Michael Moseley has had much to say about the natural food resources of this area:

The coastal biomass [living offshore in Peru] is the richest in the Western Hemisphere, if not the world... (Moseley 1975: 44) ...here marine plants —such as diatoms and other phytoplankton— flourish in profusion... The multitudes of marine plants provide lush pasturage that underlies a prodigious chain. Billions of cepepods and other minute herbivores form grazing populations fed upon by multitudes of other invertebrates and small fishes, including vast schools of anchovies. These animals are, in turn, prey for a profusion of larger organisms. As the energy transfer moves to higher levels in the food chain it sustains pinnepeds and other marine animals, as well as the largest population of oceanic birds found anywhere in the world (Moseley 1975: 10).

And it was this lavish marine food base, rather than agriculture, Moseley argues, that stimulated the development here on the coast of the first high cultures of Peru:

[These] maritime resources underwrote the most significant cultural changes that transpired [in Peru] during the Preceramic Stage. They fostered a sedentary way of life and supported a marked growth in the size of the coastal population. These phenomena also resulted in the development of permanently occupied settlements housing large numbers of people, and it was in this context that complex social forms and the beginnings of civilization first arose (Moseley 1975: 47).

When agriculture finally arrived on the coast of Peru, the large, settled, and complex societies that had already arisen here were well prepared to take the next evolutionary steps. In no other part of the world, apparently, did pre-existing wild food resources permit culture to evolve so far before the full-fledged adoption of agriculture.

Finally, we come to North America. Do we find on this continent prehistoric instances of the operation of resource concentration? Indeed we do. Anyone familiar with the late prehistory of the Lower Mississippi, and especially with such features of it as Monk's Mound at Cahokia, need not be told that in pre-Columbian times this region spawned elaborate chiefdoms, and quite possibly a state. Such political developments occurred here, I think, because the Mississippi and its larger tributaries, in addition to possessing very rich alluvial soil, supported an enormous variety and abundance of wild food that was readily available for human sustenance. In 1970 I recognized this fact and alluded to it, though briefly, in a footnote to my article on the origin of the state. During the intervening 16 years, the evidence for this has increased dramatically. Let us examine some of it.

First, though, I will quote the following statement by Bruce Smith about the location of Mississippian settlements generally:

> It was within the meander-belt zone of the lower alluvial valley of the Mississippi River that the Mississippian cultural adaptation first appeared, and where the major development of Mississippian communities took place. Subsequent expansion of...Mississippian [culture]... was primarily along the major floodplain valley corridors formed by the tributaries of the Mississippi River. Even those Mississippian settlements established outside the draininge of the Mississippi River were situated, with few exceptions, within the valley of a major river. It has been recognized for a number of years that...Mississippian [culture]...was largely restricted to the meander-belt zones of the river valleys of the Eastern United States. What has not been generally recognized is that this restriction of Mississippian populations to floodplain situations was not simply because of the availablity of easily tilled alluvial soils. Rather, [it]...was a function of the specific, complex adaptation by Mississippian populations to this habitat zone composed of linear bands of circumscribed agricultural land *and concentrated biotic resources* (Smith 1978: 481; emphasis mine).

Before looking more closely at these "linear bands of...concentrated biotic resources", it should be noted that even outside the valley bottoms game was plentiful. The surrounding forests harbored bears, deer, elk, turkeys, raccoons, woodchucks, beavers, muskrats, rabbits, and squirrels in very large number (Struever and Holton 1979: 154–155). However, it was the waters of the meander-belts which contributed in greatest measure to the native diet of wild foods. Speaking of the Illinois River in particular, but having in mind a considerably larger area, Robert E. Warren (1982: 65–68) writes:

> In general terms, fish represented a massive, predictable, and highly renewable resource that was relatively easy to collect. This was especially true during spring spawning periods in flooded backwater lakes and also during summer and fall when lakes evaporated and trapped dense populations in small, shallow, readily harvestable locales... Migratory waterfowl also were abundant seasonally and included a tremendous variety of ducks, geese, swans, and wading birds. Many of these are characterized by a relatively high meat yield and large clutch size, and huge flocks routinely made stopovers along the Mississippi and its major tributaries.

Looking more closely at fish, which were probably the major single component of the wild food diet of the middle Mississippi Valley, here is what Stuart Struever has concluded on the basis of his findings at the Koster site:

> Fish were a very important part of their [the Archaic peoples] diet, as were freshwater mussels. Among the fish they ate were bass, buffalo fish, bowfin, catfish, sunfish, and freshwater drumfish. The amount of fish remains [found at the site] came as a surprise. Apparently Early and Middle Archaic people included a great deal of fish in their diet. It had previously been assumed that during this long period (8000–2500 BC) the aborigines had lived primarily on land mammals and a few plants. But the Koster evidence tells a different story. As early as 6500 BC they were eating enormous quantities of freshwater fish and mussels. The most fascinating revelation from the fish remains was that as early as 3900 BC, when the first of a series of large villages was established at Horizon 6, Koster people were harvesting fish like we harvest plant crops, taking vast numbers at a time... We also found fish bones that were smoke-blackened. Apparently Horizon 6 people were smoking some of their fish harvest and storing the dried fish for use during the lean winter months, when fresh foods would have been scarce (Struever and Holton 1979: 155–156).

Struever then cites an estimate by a fish biologist that "Koster people may have been able to take from three hundred to six hundred pounds of fish per acre of water in the backwater lakes. When you consider that some of these lakes covered as much as several hundred acres, you can see that harvesting the crop from just one lake would have been a very productive venture for even a small group of people. The fish populations in the Illinois and Mississippi River Valley floodplains may well have made these two areas among the most productive for thousands of square miles" (Struever and Holton 1979: 157).

We may now ask, what effect did this mode of subsistence have on settlement patterns? Let us listen to Emerson and McElrath (1983: 231) on the subject: "The small, constantly shifting, mobile bands of hunters and gatherers still dominate many scholars' research on the Archaic..." However, "there is little evidence to support this latter model for the Terminal Late Archaic cultures of the American Bottom. Data indicate that these cultural groups had adopted a sedentary life-style as a result of their exploitation of the rich environment of the river valley".

All of this evidence makes two things very clear. First, the Mississippi River and its larger tributaries afforded the Indian population which occupied it an incalculable wealth of wild food. And second, this bountiful subsistence had already given rise to relatively dense and settled populations here, even before the Archaic Period came to an end. Thus, these areas of resource concentration bequeathed to the Woodland and Mississippian peoples who followed, a broad and firm subsistence base. And this subsistence base helped to generate the conditions which eventually made it possible for these areas to give rise to impressive chiefdoms and perhaps even states.

Summary

And now it is time to summarize. Where does the circumscription theory stand today? Still very much where it did before, at least with regard to the *conditions* and the *process* of state formation. I continue to see conquest warfare as the only means by which communities could be made to surrender their individual sovereignties and unite into larger political units. And I still see population pressure as *the* propulsive force needed to engender conquest warfare. Where the theory has been modified, elaborated, and refined is in the role assigned to resource concentration.

I have altered, and in some ways lessened, the role of resource concentration. Yet this change entails no loss of explanatory power for the theory as a whole, only a gain in consistency. A disguised voluntaristic element has been unmasked and expelled, leaving the theory entirely coercive.

After looking at evidence from many parts of the world, I now see resource concentration playing (all things considered) a much larger role than I previously envisioned. I see it as attracting peoples to ecologically favored areas and there giving rise to the dense populations needed to trigger conquest warfare —the *sine qua non* of the circumscription theory. In area after area —Mesopotamia, the Nile Valley, North China, the Valley of Mexico, the Olmec area, coastal Peru, the Amazon, and the Mississippi— we have found nature bountifully offering prodigious quantities of wild food to those who would come and live there. And this wild food —primarily fish— could be taken with the simple technology of Mesolithic or Archaic hunters and gatherers —whom we really should rechristen "*fishers* and gatherers". As a result, even before the coming of agriculture, not only had resource concentration drawn people to choice areas of settlement, it had allowed them to become partly or even fully sedentary.

It is even possible that at the wild food stage in these choice areas, the first steps were taken in transcending local autonomy and forming chiefdoms and perhaps states. But even if no society on earth ever became a state until after it had acquired agriculture, it cannot be denied that resource concentration brought many of them a good deal closer to that stage before they had adopted agriculture.

At what rate, or over what period of time, a given region went from the level of autonomous villages to that of a state depended on many factors. These factors included the size of the basal population, the abundance and types of wild food, the date of the introduction of agriculture, the fertility of the soil, the size of the area, its degree of environmental circumscription, and so on. A delicate equation was at work in each area, and the answer it yielded depended on the magnitude of the particular factors involved.

But if we cannot predict exactly when a valley, peopled initially by fisher-gatherers, would eventually become a state, we can certainly issue a general forecast. Thus, if a large Mesolithic population, drawn by the fish and fowl that abounded in the surrounding waters, was in place early, if agriculture soon made its appearance and found fertile soil, if the valley was of moderate size and tightly ringed by high mountains, we have every reason to expect much more rapid state formation than if these conditions were entirely absent or present only to a moderate degree.

Of course, the process of state formation is complicated. The variables may act independently of one another. A high magnitude of one factor may compensate for, or even outweigh, the low magnitude of another. Details aside, though, the net effect of all this should be quite clear:

The rise of chiefdoms and states is not adventitious. It is not a matter of happenstance but of *circumstance.* It is a determinate process with identifiable causes. These causes, moreover, are neither innumerable nor unentangleable. They are few in number and parallel in operation. To be sure, in actual cases they occur combined with all manner of extraneous and inessential conditions and events. And it takes a practiced eye —more, it takes a disposed mind— to separate the one from the other, keeping the underlying and necessary factors, and setting aside the irrelevant ones. But the quicker we are in recognizing the few key factors at work, and the readier we are in applying them, the surer and truer will be our reconstruction of the major events of the past.

REFERENCES

ANONYMOUS

1910. "Chad", *Encyclopædia Britannica*, 11th edition vol. 5: 786–787.

BLANTON, R.E., *et al.*

1981. *Ancient Mesoamerica*, Cambridge University Press, Cambridge.

CARNEIRO, R.L.

1970. "A Theory of the Origin of the State", *Science* vol. 169: 733–738.

1972. "From Autonomous Villages to the State, a Numerical Estimation", in: *Population Growth: Anthropological Implications*, edited by Brian Spooner, The MIT Press, Cambridge: 64-77.

1981. "The Chiefdom: Precursor of the State", in: *The Transition to Statehood in the New World*, edited by Grant D. Jones, and Robert R. Kautz, Cambridge University Press, New York: 37–79.

CHAGNON, N.A.

1968. "The Culture–Ecology of Shifting (Pioneering) Cultivation among the Yanomamö Indians", *Proceedings of the VIIIth International Congress of Anthropological and Ethnological Sciences* (Tokyo and Kyoto), vol. 3, Ethnology and Archæology: 249–255.

CHANG, K.C.

1977. *The Archæology of Ancient China*, third edition, Yale University Press, New Haven.

CHILDE, V.G.

1948. *Man Makes Himself*, Watts and Company, London.

COE, M.D.

1981. "San Lorenzo Tenochtitlan", in: *Supplement to the Handbook of Middle American Indians, vol. 1, Archæology*, edited by Jeremy A. Sabloff, University of Texas Press, Austin: 117–146.

DENEVAN, W.M.

1966. *The Aboriginal Cultural Geography of the Llanos de Mojos of Bolivia, Ibero-Americana* vol. 48.

EMERSON, T.E., and D.L. McELRATH

1983. "A Settlement-Subsistence Model of the Terminal Late Archaic Adaptation in the American Bottom, Illinois", in: *Archaic Hunters and Gatherers in the American Midwest*, edited by James L. Phillips and James A. Brown, Academic Press, New York: 219–241.

FALLERS, L.

1965. *Bantu Bureaucracy, A Century of Political Evolution Among the Basoga of Uganda*, University of Chicago Press, Chicago.

GERNET, J.

1968. *Ancient China*, Faber and Faber, London.

GRIFFITH, F.L.

1910. "Egypt: Ancient", *Encyclopædia Britannica*, 11th edition, vol. 9: 39–48.

HOPKINS, A.G.

1973. *An Economic History of West Africa*, Longmans, London.

KOTTAK, C.P.

1972. "Ecological Variables in the Origin and Evolution of African States: The Buganda Example", *Comparative Studies in Society and History* vol. 14: 351–380.

McEVEDY, C.

1980. *The Penguin Atlas of African History*, Penguin Books, London.

MOSELY, M.E.

1975. *The Maritime Foundation of Andean Civilization*, Cummings Publishing Company, Menlo Park, California.

MURDOCK, G.P.

1959. *Africa; Its People and Their Culture History*, McGraw-Hill Book Company, New York.

SMITH, B.D.

1978. "Variation in Mississippian Settlement Patterns", in: *Mississippian Settlement Patterns*, edited by Bruce D. Smith, Academic Press, New York: 479–503.

STRUEVER, S., and F.A. HOLTON

1979. *Koster; Americans in Search of Their Prehistoric Past*, Anchor Press/Doubleday, Garden City, New York.

VAILLANT, G.C.

1966. *The Aztecs of Mexico*, Revised edition, Penguin Books, Harmondsworth.

WARREN, R.E.

1982. "The Historical Setting", in: *The Cannon Reservoir Human Ecology Project*, edited by Michael J. O'Brien, Robert E. Warren, and Dennis E. Lewarch, Academic Press, New York: 29–70.

WEAVER, M.P.

1981. *The Aztecs, Mayas, and Their Predecessors*, Second edition, Academic Press, New York.

WEBB, M.C.

1984. "The State of the Art on State Origins?", *Reviews in Anthropology* vol. 11: 270–281.

COMMON SENSE VS. OLDFASHIONED THEORY IN THE INTERPRETATION OF THE CULTURAL DEVELOPMENT OF THE ANCIENT NEAR EAST

JAMES MELLAART
London University

It is probably fair to say that most of us field archæologists as were trained in the arts with a predominantly classical education followed by historically oriented approach to prehistory, feel ill-equipped when confronted by such philosophical concepts as urbanism, the city, civilisation and the like, on which our views are asked, as it is us that have uncovered them.

Nobody now doubts that the views of Gordon Childe have greatly influenced theoretical thinking about the development of civilisation; every schoolboy knows about the "Neolithic revolution" and the "Urban revolution". In this context it is often forgotten that Gordon Childe was a prehistorian who did his pioneering studies on the Danube —Central and East European—, before he became interested in British archæology. He never dug in the Near East, which was definitely not his field of competence, yet he was fascinated with the idea of Ex Oriente Lux and was its most powerful exponent. Such is the power of the pen, that his thoughts; decidedly controversial —to put it mildly— , have triumphed over the less articulate, but better founded views and statements of practitioners in the field. He synthesized from book knowledge and wove the scraps into a magnificent theory; a parachute covering the whole Near East, Egypt included; impressive on paper, faulty in detail, and marred by anachronistic ideology. In modern parlance it was a "model" and a daring one.

It was soon to be put to the test by excavations outside the two supposed homelands of civilisation: Egypt and Lower Mesopotamia. In the decades after the Second World War, one excavation after the other showed conclusively that the geographical basis of Gordon Childe's theories was too narrow, the environmental studies misjudged, and the timescale grossly underestimated. One assumption after the other was demolished; the Neolithic did not take place in Egypt or Lower Mesopotamia *c.* 4750 or 5000 BC, a habitat devoid of the wild progenitors of wheat and barley, but in such peripheral areas as Palestine, Syria, Anatolia, Iran, Baluchistan, etc., all previously regarded as barbarous. Neolithic cultures did not spread from Egypt or Lower Mesopotamia to less fortunate neighbours, arriving there much later. The very reverse is the case; wild wheat and barley were brought down from the hills, mountains and plateaus and were gradually introduced into the new environment of the lowlands; and where rainfall was insufficient to ensure their growth, it was not until irrigation techniques had developed that they finally colonised Egypt and Lower Mesopotamia, the Indus valley etc. by *c.* 5000 BC or a little earlier respectively in Southern Mesopotamia. The entire process, from the earliest use and cultivation of crops in the Natufian in the 10th millennium, had already evolved over 5000 years or so, something which defies the term "revolution".

Creatio ex nihilo, already condemned by Aristotle, is not a very useful concept; evolution and continuity are. The ancestors of our neolithic farmers are the hunters and fishers of an earlier age —the Upper Paleolithic—, whose real accomplishments are still very badly known in the Near East, but who were certainly there, and provided the impetus to what was to develop later. For invasions or immigrations of newcomers there is at present no scrap of evidence; population pressure, invoked by some to account for the transition from food gathering to food production, lacks a solid base of observable facts. Overpopulation as the result of succesful food production, can however be demonstrated at Çatal Hüyük.

Childe's second revolution, his urban one, is equally tenuous. The rise of cities in the Protoliterate period (Late Uruk in Mesopotamia) and Protodynastic Egypt (when exactly: Nagada II or under the First Dynasty?) seems rather sudden and inexplicable if it had no predecessors. They would inevitably have come to light long before if they had offered any inducements, in the hope of finding even earlier texts. Having concluded that they did not, interest in "prehistory" evaporated rapidly, both in Egypt and Mesopotamia. Such a timid approach has probably robbed us of much that remains to be discovered and views that Egyptian civilisation owed its existence to Mesopotamian contacts had an adverse effect on further research in Egypt, where the idea of a foreign contribution to its culture has for long been unthinkable and anathema. In this respect Egyptologists are undoubtedly right, but the best way to prove it is by excavating "Protodynastic" cities and reveal their level of culture in the Nagada II, (Gerzean) period —something that is long overdue, and alone can remove the pro-Mesopotamian bias, of H. Frankfort and G. Childe's theories. Comparing the contents of Egyptian cemeteries with the building remains of Mesopotamian (and Syrian) cities (without cemeteries) in the fourth millennium BC can hardly be regarded as the ideal way to evaluate cultural contacts. As a basis for far reaching reconstructions of cultural development vis à vis the other, the evidence is clearly unsatisfactory in the extreme.

To this ought to be added the simple fact that urbanisation in Egypt is another of these assumptions that dog the Childe theory —the evidence is not there; no Egyptian city of this period has been excavated; and a wall or gate does not prove the existence of a city. Not surprisingly perhaps Wilson described ancient Egypt as "a civilisation without cities", appropriate enough before the Middle Kingdom, *c.* 2000 BC. To argue about urbanism in Egypt in the fourth (or third) millennium is wholly premature. It is worse, for it is misleading. What we have got is monumental tombs, mostly at Saqqara, though those at Abydos and Naqadeh also fall in this class —but none before the First Dynasty (*c.* 3200–3000 BC) so far. These in turn are followed by the monumental Step Pyramid complex of Netjerykhet (later known as Zoser), which is justly hailed as a translation into stone of buildings previously constructed of reeds, mats, wood and mudbrick. Interesting as they are, they appear to represent a royal fortress with the appurtenances of the offices of state, replica temples as well as a royal residence and its workshops and storerooms. What they are clearly not, is a city, like e.g. Kahun in the XIIth dynasty. Monumentality in architecture seems confined to the buildings of impressive tombs and associated funerary and valley temples during the Old Kingdom. What the palaces were like we do not know, just as we do not know what mansions, private houses of the nobles were like. A row of houses outside the mastaba tomb of Queen Khentkawes (early Fifth dynasty) built for her funerary priests is almost the only evidence for domestic architecture; it is functional.

One would imagine that Egyptian royal palaces, the nobles' mansions etc. were distinctly non-urban (in the Mesopotamian sense) and something much more like a garden suburb; spacious and shady, with pools and fish, birds and trees. Only the artisans, merchants and the poor probably lived in huddled quarters with narrow lanes and crooked passages, still to be seen today. Egypt after all had no serious enemies and our ideas of urbanism conjure up heavily walled cities, overpopulated in crowded quarters with a minimum of gardens, except perhaps around temples and palaces, something typical of a land with many and mutually hostile city states, —Lower Mesopotamia, the ancient lands of Sumer and Akkad. I said "our ideas", but a closer look at the three best preserved examples of urbanism of the fourth millennium: Uruk itself, Habuba Kebira and Jebel Aruda do not conform to such ideas. At Uruk within a vast enclosure there is plenty of space around the temples on the Anu Ziggurat and the unexcavated huge mound nearby topped by a Seleucid temple. In the E-anna complex also a few temples and numerous other public structures (all called temples) but probably palaces and buildings of state alternate with vast open areas and areas of domestic dwellings, as yet hardly touched by excavation. At Habuba Kebira a walled town abuts on an administrative quarter (in T. Qannas) fringed by an area of gardens and a southern suburb. J. Aruda to the north, may contain the small temples and houses for the cult personnel; high up on a hill with a fine view of the Euphrates. The comparatively spacious layout of these places comes much closer to what one might deduce for Egypt at the same period than what was to appear later, and what we think of, rightly or wrongly, as typical, when overpopulation and defence created the predecessors of slum conditions and the social unrest that arose as a result, as told of Lagash in the days of Urukagina and of Memphis during the First Intermediate period, when central government broke down.

It is also necessary to bear in mind that the first settlers at Habuba Kebira did not feel the need for fortifications; probably the Syrian natives were not yet hostile, and the sheer number of settlers may have been sufficient to overawe them. The best early defense is not walls, but men, in sufficient numbers to deter.

How old is urbanism then and I am thinking here in terms of Uruk, Habuba and similar sites, like Tell Jawa in the Black Desert of Jordan. Does it really only appear in the Uruk period, the fourth millennium BC? The question should perhaps be put in a different way: what makes an *urbs*? and what distinguishes it from a village? The confusion this question creates very strongly suggests to me that we are asking the wrong question, which ought to be rephrased in ancient terms. How did the ancient Egyptians or the Sumerians distinguish between settlements? How much tax could each afford to pay? The answer is disappointing; there are settlements, forts and in Egypt, vineyards/estates, specifically mentioned, but there are no separate words for city, town and village. If the ancients were not interested in such distinctions, should we be? Gordon Childe's attempt to link civilisation with the appearance of cities seems flawed to me with one of his two main contenders falling out of the race. Nobody could deny that e. g. Habuba Kebira shows all the signs of urbanisation; it is wellplanned, neatly parcelled out with streets at right angles to a main thoroughfare, complete with a system of drains, quays, a city wall, an administrative centre, what resembles a park and standard type housing. What one easily forgets if one takes such a site as typical, is that Habuba Kebira is a new foundation of the mid-fourth millennium, mostly build on virgin soil. Most cities, however were not new but grew on top of earlier settlements or within an enclosed terrain (with fields around e. g. or within a defensive wall) or around a temple, well, sanctuary: a rather higgledy piggledy arrangement unplanned except for a couple of main streets crossing at the centre, near some important building. Many cities sat on rocks, taking advantage of natural strength, with a fort, or temple on top and a vaguely radiating plan of quarters all around, enclosed by a wall and not infrequently built over the remains of earlier occupation. Some planning is always involved; how long it remains clear and undisturbed is another matter. When settlements grow they not only frequently subdivide space, block off lanes and streets but eventually spread beyond their walls, rebuilt further out.

It is only short-lived settlements, especially new foundations that have preserved neat and clear evidence for planning, subsequently undisturbed, which impress architects and archæologists alike.

Expansionist policies, like the unification of Egypt, or the Uruk occupation of the upper stretch of the Syrian Euphrates undoubtedly led to much new building by aggressors and resisters alike. Whether we call this urbanisation or not seems to me irrelevant; far more important is the fact that it was not new, and did not mean a universal change from village to town and city. Town and city life created problems of food supply, cramped quarters facilitated the spread of disease and social unrest. It had of course advantages: an assured labour force for manufacture of goods that could be exchanged for raw materials. The marked increase in international trading networks during the fourth, third and early second millennium was undoubtedly facilitated by an increase in towns and cities and a greater concentration of wealth. All this is common knowledge; what concerns us here is the conditions before the fourth millennium.

For many years now it has been known that the Ubaid period in Mesopotamia was probably of much greater importance than the meagre archæological evidence recovered suggested. Ubaid remains occur below nearly every important Sumerian city and temples, ancestral to the Uruk ones have been encountered at Eridu, Uruk, and Tepe Gawra in Assyria. Excavations in the Hamrin area have now produced, at T. Abada, an entire settlement plan of Ubaid 3 with architecture of a sophisticated type, hitherto unsuspected, and matched by a number of soundings at neighbouring mounds. Although T. Abada is small, it may represent the seat of the local administration of the district; the principal building been architecturally emphasized, with buttresses and recesses. The site conspicuously lacks a temple, unless the central building (in plan no different from the others) had in fact such a function.

The question which arises, is, whether these Ubaid sites like Abada and T. Gawra are "Ubaid cities" or their more humble counterparts, the "towns" pending the excavation of a large Ubaid "City". It seems inherently unlikely that cities of the Uruk period were not preceded by Ubaid cities, but they were mere "villages", on the revolutionary principle. I am suspicious of such evolutionary jumps, when the archæological evidence is scantly in the extreme. Striking contrasts, used by theorists to prove their point, on reflection and further evidence, tend to be less extreme. Those sudden bursts of intelligence, like the invention of writing, have recently been shown to be illusory by D. Schmandt-Besserat. Writing had many predecessors, merging back into the 8th millennium. Hardly a revolutionary event! Similarly the bevelled rim bowls of the Uruk period are preceded by the flint scraped or brushed Coba bowls of the Ubaid 4 period and with a similar vast distribution from Cilicia and North Syria to Susiana, suggesting widespread trade and contact. What Gordon Childe thought were innovations in the late fourth millennium, now turn out in the light of more advanced knowledge to be a culmination of processes which had started long before. Even Braidwood was aware of what he called a "*koine*" during the Ubaid period, and it requires no mental effort to recognise at least two predecessors of the same process in the Halaf period of the sixth millennium and the "PPNB" of the seventh, not to mention vestigial traces in the 8th millennium in PPNA/Early PPNB in the Levant, North Syria and North Mesopotamia, engaged in the obsidian trade.

These pre-Ubaid cultures already exhibit a considerable range of size in settlement; size evidently has nothing to do with time, but depends on opportunity, favourable circumstance and human endeavour. There are much larger Halaf sites than anyone has so far excavated and to assume that pre-Ubaid sites all consist of villages is most unreasonable as the size of e. g. Jericho PPNA (4 ha), Tell Abu Hureyra (11 ha) or Çatal

Hüyük East (16 ha) demonstrates. Great disparities in size in fact appear as early as agriculture is established; if a site can successfully feed itself, its location is pretty well assured; commercial relations develop and arts and crafts flourish. Subsidiary villages along trade routes to the sources of raw materials are already there in the Neolithic; Umm Dabaghiyah being a good example, so too probably was Arpachiyah.

In other words, there is now good evidence for a line of ancestors for Childe's urban cities in the later fourth millennium BC, linked to the appearance of civilisation. A somewhat less revolutionary explanation of Near Eastern cultural achievements is clearly called for, as all the features of material culture assembled as "civilisation" (cities, irrigation, monumental structures, art, writing, etc.) are already present in some form or another in the earlier stages; none of them (writing excepted) being demonstrably a new invention. Archæological research has overtaken the premises on which Gordon Childe's urban revolution was built. Instead we now see a much larger vista of cultural and socio-economic evolution that is clearly demonstrable from the Natufian period onwards and almost certainly has roots that extend even further back. The new stress is on evolution and continuity, with suitable embellishments and regional adjustments that eventually reach a climax in the third millennium BC and in an area of far greater extent than Mesopotamia and Egypt, extending from the Aegean to India. The widening of horizons has had some very salutary effects; chief among them the realisation that Lower Mesopotamia was not the centre of the world and source of all culture as H. Frankfort would have us believe. Nor the source of Indian civilisation, last suggested by Mortimer Wheeler. Anatolia, Iran, western Turkestan, and Baluchistan have all come into their own as more or less independent centres of culture owing preciously little to Mesopotamia before the fourth millennium.

The contribution of the Levant, also, has come as a surprise, with Syria looming large vis à vis Sumer and Akkad. The teeming population in Chalcolithic Palestine and Sinai and its extraordinary metalwork (N. Mishmar) contemporary with the Ubaid period can also no longer be ignored. One could easily go on; what matters is that our knowledge has been augmented to such an extent that G. Childe's theories have lost much of their voice. Were he still with us, he would have been the first to admit it.

Çatal Hüyük: Problems of Interpretation.

That a number of early sites do not easily fit into schemes of cultural development conceived before their presence was known is one of the risks taken by all who theorise. The intelligent way out of the dilemma is to modify the theory to incorporate the new evidence or information. It is unhelpful to ignore it and pretend it does not exist, regard it as a freak, or treat it as a white elephant. If Jericho presents problems to accepted wisdom, Çatal Hüyük is clearly a nightmare to old-fashioned thought. Had it been found not in Anatolia, but in Egypt, and had it been dated to the period immediately preceding the Badarian, it would have been hailed as the earliest evidence for Egyptian civilisation. It is large enough to represent an ancient capital, was neatly built of mudbrick, wood and plaster. Its inhabitants practised irrigation agriculture and had domesticated cattle and dog, and perhaps some sheep. They wore garments of animal skin, including leopard, but also wove linen and woollen cloth. They were armed with bows and arrows and spears, tipped with finely made points of obsidian and flint, practised pressure flaking on daggers, mounted in wooden, stone, or carved bone hilts. They used maces for fighting and killing prey, and used the mace as a symbol of authority. They carved bone and stone, made copper and lead trinkets, used shell for beads; fished with hooks and nets; played drums and flutes, danced, hunted deer, wild boar, onager and wild ass and fowled birds in the marshes. They made very simple pottery, but better wooden vessels, modelled clay and carved stone statuettes; put up clay reliefs of a goddess and numerous heads of bulls;

engraved in the thick plaster levels of their buildings and above all they painted. In those scenes they depicted their gods and goddesses in human form with their animal attributes; early landscapes, celebration with animals, deerhunting, fishing nets and a boat with an steering oar, enormous bulls, vultures, and headless corpses and a number of other clearly religious subjects, but no scenes of fighting or war. Writing was still unknown.

Roof entry into the houses and secondary burial were unfamiliar, Archaic features perhaps, derived from a more distant past.

As an ancestor of the Badarian culture Çatal Hüyük would have fitted admirably, allowing for a few, but not very substantial changes, in its material culture.

Here the Egyptologist's dream ends; Çatal Hüyük for all its resemblances to later predynastic cultures of Egypt, lies in Anatolia and its date is not the sixth, but the seventh millennium BC after calibration. To some theorists Çatal Hüyük must seem out of place and out of time. Before the establishment of C^{14} dating (now reinforced by dendrochronology) this would have presented few difficulties. "In our opinion the excavator has dated his site too high" was a familiar argument. Only a desperado would now ignore one of the best dated sites in Western Asia. Nor can the fertile Konya Plain in which the site is situated, be regarded as a peripheral backwater, another favourite argument in the nineteen-fifties. Uniqueness is a term that has been used for the site — something peculiar, out of the ordinary; macabre, uncanny. But wallpaintings of the same period are now also known from Bouqras, Umm Dabaghiyah, Zaghe, and earlier ones from Can Hasan III, Aşikli (painted plaster only), Mureybit III. Animal heads at Mureybet Ib, Ganj-Dareh and Zaghe and the peculiar burial habits of Çatal Hüyük are akin to the Natufian to Halaf sequence of the Levant and North Mesopotamia, as are the use of red plaster floors, houses entered from the roof (Mureybet, Cheikh Hasan, Ganj-Dareh, Umm Dabaghiyah) as well as at Can Hasan and Erbaba. The use of copper is attested even earlier at Çayönü, the same type of very early pottery is also found at Tell Assouad on the Balikh river in north Syria, unbaked pottery as early as Mureybet III *c.* 8000 BC. The claim to uniqueness is dissipating fast and with it the excuse for ignoring the site as untypical. The unpalatable fact remains that everybody had underestimated the real potential of the Near Eastern Neolithic and some scholars have not yet recovered from the shock. They feel outraged at the thought of ***neolithic cities and civilisation***, which is my interpretation of the phenomenon, and which is clearly out of step with methodological and anthropological thinking, especially in the U.S.A. Archæology without interpretation is fairly sterile and little more than a boring catalogue. As Sir Mortimer Wheeler once said: "archæology is about people, not objects". How much thought and imagination one must bring to bear in archæological interpretation leads to many tedious debates; a few bright insights to me are more important than mountains of trivia. Real archæologists are born, not trained; common sense in particular is essential, and so is a sense of history. Without these, one does not go very far in Near Eastern archæology.

To me, it seems inconceivable that there is still a school of thought that expects and demands that new discoveries should be trimmed to fit in a preconceived straight-jacket of theoretical thought about cultural evolution, be it Childe's or Braidwood's, Redman's or anybody else's. Surely theory should adept to the discovery of new facts (evidence) and not vice versa.

In his *Threshold to Civilisation* (New York, 1975: 140–207), W.A. Fairservis Jr. attempts to fit Çatal Hüyük into an anthropological framework under threat by new discoveries. In his analysis, the emphasis is on playing everything down by modern anthropological parallels. The real point is missed; we are not dealing with Eskimos or the Konyak Naga, Andaman Islanders or Papua-New Guinea, still around today and no doubt very accomplished in making tools, music art, all in their spare time; but with a culture flourished some 8500 years ago and which did not necessarily responded in the same way. It is important, as far as that is possible, to remove the anachronism, put ourselves in their place and try to find out how they saw things. After all they did not have the hindsight, which affects us all. When as I have tried to point out that such concepts as urbanisation, cities, towns, villages or civilisation are modern constructs, conceptually unimportant to literate Egyptians and Sumerians, how valid or useful are such ideas as kingship, class society, slave owning and the rest to ancient man? As far as I am concerned the terminology is unimportant and when I use *neolithic* city I do not mean Mesopotamian city, Greek city, Roman or modern city, but something which I imagine is considerably more complex than simpler forms, "towns" or "villages". Similarly I use civilisation not in the retricted Anglo-Saxon sense, but in the French one, on a par with "Les civilisations du Paléolithique". It is nor only a question of semantics, but a question of interpretation; I would object to such a term as "the civilisation of the Mongols", not to quote more recent examples. There is more involved here than purely material culture, technology, economics; there is also the moral attitude towards life, in my, ***humanistic***, opinion, far more important than materialistic or technological achievements, in the definition of civilisation. Civilisation without civilised behaviour, i. e. tolerance and respect, moral virtue and religion, is a ***contradictio in terminis.*** It seems to me that the aspects of religion (i. e. divine ordinance for proper human behaviour) has got lost in the definition of civilisation. A splendid example of what I mentioned before, is our supercilious view or theirs; atheistic interpretations are unlikely to contribute much (if anything) to our understanding of early civilisations; they entirely miss *the* point; *man's beliefs*, which are far more potent than socio-economic modern fantasies about him. Whether at Çatal Hüyük, Old Kingdom Egypt or fourth and third millennium Mesopotamia, the religious emphasis is not on technological improvements in plant on animal husbandry, but on continuity of life and the established order, granted or ordained by the gods; on prosperity, fecundity, children and wellbeing as the result of ritual obedience. It is not kings that are to be obeyed, but the god himself (in Egypt) or his priest or governor (Sumerian *EN.ensi*, not *lugal*, the warleader). Once again, the modern interpretation got it all wrong. "Civilisation" as narrowly defined is not linked with monarchy, but with God or his representatives. This is no innovation, although it has been represented as such. To this day, Mesopotamian scholars stress the role of the priest and temple versus the idea of a secular monarch.

A claim to leadership is based on special relation to the deities; power without religious sanction seems inconceivable in the Ancient Near East. What we innocently call history, is in fact a ruler's account and justification of his stewardship to his deity. Admission of guilt, incompetence and failure have no place in it and historians are well aware that one has to read between the lines. One of the disadvantages of writing as compared to oral tradition is the ease of falsification of the records, accessible to few. What is common knowledge is hard to falsify; the historian's dilemma is to weigh the evidence and try to get at the truth.

It is a fair guess that the earliest texts containing references to society in the later fourth millennium BC record something already current, and not newly devised. Egyptian and Sumerian king-lists go back far beyond the earliest writing; starting with a series of gods, followed by semi-legendary or legendary figures and ending up with individuals, some of whom happen to be known from contemporary inscriptions. Osiris and Horus are portrayed as primeval kings of Egypt; Gilgamesh and Enmerkar as kings of Uruk.

Where literary evidence gives out archæologists, arguing that as later kings lived in palaces, their predecessors must have done the same, look for palaces and when they do not find them, proclaim a pre-dynastic period. Good classical logic, no doubt, but what happened if early kings did not live in palaces, but in temples (as a Sumerian text preclaims)? and before that in something even more simple like a house of the same plan as their subjects? There are other ways in which its special nature could be indicated beyond a simple ground plan, all that we find. So-called negative evidence is never definitely positive. The ancient Egyptians never thought of having had rulers other than god-kings, not mere humans. It is not for us to contradict them in their belief that Egypt had always been a "kingdom". Why doubt this? On a pot of the Naqada I (Amratian) period, dated to the second half of the fifth millennium, there is a clear representation of the Red Crown of Lower Egypt (Ashmolean Museum, Oxford). The evidence is pretty unequivocal; why depict a crown, symbol of kingship, if the latter does not yet exist? This is not the only example of what appear to be "regalia", scepters, crowns, etc. from the phenominal Nahal Mishmar arsenical copper hoard in Israel, dated to *c.* 4000 BC from the reed matting in which it was wrapped, when ***hidden*** in the Ghassulian period (*c.* 5000–4000 BC). They could have been manufactured at any time during this period. That these are status symbols is beyond much doubt and with *cire perdue* casting and elaborate symbolism, the chances that such superb objects once belonged to village head men or desert sheikhs seems rather remote. It is exactly the same sort of insignia that a thousand years later we see in the hands of Egyptian kings, but not in those of their subjects. Kings in Egypt and Palestine in the fifth millennium BC? Why on earth not? Because of G. Childe's theory that they did not appear until much later? One can be sure that he would have pushed kingship back much further had these discoveries been known in his lifetime. Most Mesopotamian archæologists now agree that the Uruk period, in the later half of which, urbanism, kingship, writing, etc. the basis for the idea of "civilisation" become evident, go back to the preceeding Ubaid period of the fifth millennium, that is, roughly contemporary with the Egyptian and Palestinian evidence we have been discussing. It is not the degree of power —small or big kings, or even bigger ones— that matter, but the idea of kinship, god-given power, that concern us, irrespective of whether it was primitive, developed or whatever one might like to call it. In archæological context, the fifth millennium, be it in Egypt (Badarian, Amratian), Palestine (Ghassulian-Beersheba), Syria and Mesopotamia (Ubaid period) see the first flourish of irrigation agriculture, agricultural surplus, incipient urbanism, widespread trading contacts, technological improvements, locally spectacular metallurgy (as at N. Mishmar), prospecting for resources (Sinai), the cumulative effects of which, in the fourth millennium Uruk period, were once heralded as the advent of civilisation (the basis for G. Childe's thesis).

If, on the basis of new discoveries, one can add a good millennium of development to such a climax, one is equally entitled to ask the question: and where did all that come from? The answer is obvious: from previous developments during the Neolithic period, and so on, back into the Upper Palæolithic. One can also turn the question around and ask; what evidence is there for a "pre-kingship" stage? Very little indeed in memories of ancient literate cultures; none in Egypt and precious little in Sumer, ably summarised by Thorkild Jacobson on the basis of Sumerian religious texts, as a sort of primitve democracy. Much as I respect his judgement, I feel unconvinced, and though the wish for consultation in affairs may be conceded, one very much doubts —in a Near Eastern context— that it bore much weight, among those who made the decisions. S. W. Asia has given us many things, but I would venture to suggest that democracy, primitive or not, was not among them.

I have dared to suggest that some form of kinship may go back as far as 5000 BC. What preceeded it, if anything different during the sixth and seventh millennia? Fairservis classifies Çatal Hüyük as a "chiefdom", a *Clan System*, something deemed to have existed at a lower level than a "kingdom" —a "civilised state" (assigned to the late fourth millennium and later). The essence of clanship —the tribe and its *soil*— is the intense affection for its inheritance felt by all its descendants, but with the subdivision into still more individual, but equally sacred holdings. The very meaning of "clannishness", and a source of immense pride and loyalty. The family, descended from him or her, who "first raised fire or boiled water upon these lands" was the basic unit, not a larger territory over which the head of the clan may have gained sway. Family interest transcended any other; the soil belonged to the chief, but the clan would prevent him from misusing it; chiefs for all their hereditary (mostly patriarchal) power could be and were deposed when acting against the interest of the clan. The social structure was simple; a common descent from the founder united all; the chief, his family, their descendants, his officials and those who had chosen to follow him, former members of other clans, refugees etc. Class distinction did not exist; even the poorest clansman was or regarded himself as superior to any foreigner. All revelled in the fame and exploits of their real or adopted ancestors, and looked down upon the rest.

The strength of this system is such that it has survived —in attitude— even to the present day; natural pugnacity and clannishness happily survive, —among the Scots— wherever they may be. Whether such a clan system can be transferred back in time to Çatal Hüyük, as Fairservis thinks, I leave to your judgement. He deduces a classless society with a chief (or priest) in authority over a small population of under 3000 souls, living in a village. The society is egalitarian, there is no craft specialisation, no public buildings and only domestic cults. Land is owned by the society, not the chief, and trade is of little importance. The basis of the economy is successful agriculture. Art is iconic, not symbolic.

Such a barren label would actually fit any neolithic settlement in the Near East, Jericho, Çayönü, Jarmo, etc., *except* Çatal Hüyük. If this is all anthropology can contribute to the interpretation of my site, its services are no longer required. It is as useful as including Man among the Mammals, without any further details, of how he is different.

It is easy to see why so many archæologists in the Near East prefer a historical approach to interpretation, shake off outmoded theories and go their own way.

What Dr. Fairservis probably did not know is how ordinary clansmen in Scotland and Ireland regarded their society; he would be distressed to know that they called it a kingdom, like e. g. the MacDonald Kingdom of the Isles, never subject to the King of Scots.

THE BEGINNINGS OF URBAN SOCIETY AND THE FORMATION OF THE STATE: TEMPLE AND PALACE AS BASIC INDICATORS

LINDA MANZANILLA
Universidad Nacional Autónoma de México

Introduction

The various models of pre-feudal societies that have been proposed for the past have emerged from studies based on historical sources and ethnographic analogies. Thus, Service's classical division into bands, tribes, chiefdoms, and states, or Fried's egalitarian, ranked, stratified and class societies seem to have organized the ethnographic and archæological universe in discrete classes. Yet, in spite of multiple efforts carried out in this sense, the process of change to the most complex forms has not been clarified. In my opinion, one of the problems that have biassed archæology recently is the fact that particular data has been forced inside these models, rather than producing new theoretical formulations that account for the present state of the facts.

With respect to the formation of the first urban societies and archaic states, the first fact is that we do not have ethnographic correlates for these societies. I will define an urban society as one with complex division of labor, that is, the existence of specialists in activities different from the production of subsistence goods; this type of society also presents institutions that coordinate economic processes and have authority over common people; and finally, specialists and authority live in an urban center that provides specific services to the surrounding region, such as, for example, the distribution of a large variety of goods.

On the other hand, an archaic state society would have a class social organization, with the first clear indications of accummulation of wealth, a relatively precise demarcation of frontiers, conquest as a form of territorial appropriation, and tribute as an instance of economic subjugation.

If we observe the difference between both definitions, we will see that they refer to phenomena which are not necessarily contemporaneous, but sometimes succesive in time. Moreover, the basic indicators of each of these stages lie precisely in the scope of the institutions that function as the seat of "urban" authority: temple and palace.

Our strategy will be to focus on the indicators of change inside de organization of these two institutions. The goal of the study of archæological indicators is to establish a relationship between empirical data and the phenomena stipulated in the conceptual framework. The importance of considering the archæological context as a functional and spatial–temporal reference is vital. Let us take, for example, the presence of a complex division of labor as one of the key elements for detecting the presence of early urban societies. At the level of handicrafts, we would look for traces of specialization in the archæological record, such as the presence of crafts quarters, workshops, specialists' graves with their instruments, artisans' marks in their production (specially notorious in the Samarra pottery of Mesopotamia), elements that would indicate a particular ability or complexity in the manufacture (such as that implied in the elaboration of Halaf polychrome pottery, in which each piece is a unique masterpiece, or in the manufacture of bronze objects), and finally mass–production that can be observed, for example, in mold- or fast wheel-made pottery (Manzanilla 1979, 1986).

One factor that should be specified is the degree of specialization. In dealing with lithic workshops, examples can be cited where a variety of instruments are made, whereas, in other cases, only a single type of instrument is elaborated.

Another point to consider is that these indicators are not deprived of relations to the general contexts of use-consumption where we should evaluate what social necessities they are satisfying. In 1979 I proposed that the only way to correctly evaluate the social impact of goods is to correlate the information coming from production contexts —for subsistence elements, manufactures or construction— with those derived from use and consumption contexts, in the branches of immediate and productive individual consumption, in distribution or exchange, and in political or ideological instances (Manzanilla 1979, 1986). Thus, for example, it is not the same to find a workshop where prismatic blades, destined to be used as domestic instruments, are made, than it is to find a similar workshop under the auspice of the temple, whose products serve for auto-sacrifice. Therefore, it is necessary to analyze not only production sites, but also the distribution of these same products in primary contexts that denote a particular function.

At the manufacturing level, one can think of two types of specialists:

a) Those who work independently, on their own, where we would have two modalities:

   1. The houses and workshops of those who work individually would be randomely distributed in the settlement.

   2. Those who are involved in corporate activities inside a particular process might be assembled in wards, though other factors could also be working in this decision.

b) The other type of specialist is the crafstman who works exlusively for the temple or the palace. In this case, we would expect to find their houses and workshops in the sector contiguous to the ceremonial or palace precinct.

Another element that serves to define the first urban societies is the existence of an institution that coordinates various economic processes. As I mentioned in 1983, the earliest pertinent indicators are related to the temple as an institution that controls the specialized production of certain goods, the storage of others, distribution and long-distance exchange. This subject will be developed further.

Finally, as a part of the definition I pointed out the existence of an urban center, that it houses the majority of the specialists that are not devoted to subsistence and offers specialized services to the rural communities that occupy the immediate surrounding territory. At the level of regional settlement patterns, with data coming from excavated and dated contexts, we would not only hope to find a differential distribution of activities but also a greater variety of goods that circulate in urban centers, compared with dependent communities. The basic control institutions —temple and palace— would obviously be found inside the urban center.

I. The Mesopotamian Case

The Mesopotamian example is of special importance in the problem we are dealing with because it represents the earliest and better known case regarding the formation of urban society and the state. Many scholars have proposed different hypotheses to explain the transformation process towards a "complex society". I will now consider some of these ideas that were developed more thoroughly in 1979.

1. *The Hydraulic Model*

One of the most widely discussed propositions is that of Karl Wittfogel, which emerged from the study of some periods of Chinese history, and was generalized by him and his followers at a world-wide level. The central concept is that of "hydraulic society", an agrarian social organization in which irrigation works (for productive and defensive purposes) and other constructions (communication, defense, service, etcetera) are administered by a strong government (Wittfogel 1974: 17), whose efficiency would derive from its capacity to organize and control man-power for their construction and maintenance, as well as the distribution of water and irrigated land. A hydraulic society would tend to acquire a state conformation because the state would be the most efficient political system able to integrate the formal patterns of authority that are required by the aforementioned tasks (Sanders and Price 1968: 177). However, Wittfogel conceives hydraulic societies at the level of chiefdoms, and also non-urban hydraulic societies.

This is an example of a proposition which has been generalized to a number of regions which have not been thouroughly studied as the one that served as the basis for the original ideas: the Chinese case. One of the criticisms that should be leveled at this proposition is that, even though it is seemingly presented within a dynamic temporal framework, the form the explanation takes is clearly ahistorical. In most cases, regions are cited without precision with respect to the particular period to which the argument is referred; in others, the process is not followed in a diachronical way. The data is fragmented and, thus, there is no complete exposition of any particular case.

In his analysis of seven contemporary small-scale irrigation systems, Millon (1962:56) has concluded that there is no clear relationship between degree of centralization of authority and the dimensions of the irrigation system or the size of the population it maintains. Moreover, the dependence on a common irrigation system results not only in antagonisms, but also in cooperation.

Due to the fact that Wittfogel and Steward included Mesopotamia in the category of "compact, simple, state, hydraulic society" (Steward 1955: 2), other scholars have insistently searched for the indicator the defines this type of society, reducing the problem to that of irrigation canals. Extensive surface reconnaissance projects were planned, particularly in the central and southern Mesopotamian plain (Diyala, Akkad, the center of Sumer and the area of Uruk-Warka). The main goal of these projects was to detect the correspondence between the presence of these hydraulic elements and the moment of emergence of the "state" or of "urban society".

After years of research, one of these scholars, Robert McC. Adams, has come to the conclusion that, for Mesopotamia, the state preceeds the vast hydraulic networks that could generate political structures of the type Wittfogel defines as hydraulic by a millennium or more (Adams 1965: 41). The archæological evidences of irrigation before the middle of the third millennium BC are so scarse and controversial that they do not have any weight.

Finally the administrative registers of the third millennium BC indicate that people depended primarily on extensive cultivation under rotation, with alternate fallow years and leguminous crops. Small-scale irrigation constituted a subsidiary part of an interdependent network of subsistence techniques and economic relations, and it cannot be isolated as a causal agent (Adams 1969; Hole 1974: 271).

2. *The Demographic and Warfare Models*

These two propositions will be discussed together because they are intimately related, explicitly or implicitly. A great defender of them, in order to explain the emergence of the state, is Robert Carneiro (1970), who proposes that the progressive increase of population provokes constant conflicts due to the competition for agricultural land or recollection, hunting, gathering, or fishing sectors (thus creating pressure over limited resources). The corollary of constant friction would be the conquest of some groups by others, and tributary relationships between conquerors and conquered. Through these mecanisms, the size of political units would progressively increase, as would the degree of complexity and centralization.

For Mesopotamia there are three models that would fit these propositions. The most orthodox, proposed by Cuyler Young (1972), was intented to apply Boserup's (1965) and Caneiro's (*op. cit.*) propositions to Mesopotamia, an area considered as a relatively circumscribed geographical unit. From 6000 to 4500 BC, the population would increase thirteen times, and this would provoke an increasing pressure on resources, which in turn would result in the intensification of agriculture and in the migration of one sector of the population to marginal zones. This would happen in the Ubaid Period.

Yet, the colonization of new territories and the intensification would have limits. When these were reached, conflicts over cultivation lands would worsen. During the Uruk Period, marginal lands would have been abandoned for the sake of creating settlement clusters that could be more easily defended, constituting "buffer zones" between populated sectors. Thus, urbanism would be the means of organizing and controlling increasing population, the intensified economic structure, and man-power that made this intensification possible, but only in ranked social forms.

On the other side, McGuire Gibson (1973: 458–460) proposes a model which specifically refers to the Uruk, Nippur, and Kish sectors, presenting some variants with respect to the previous one. In the first place, population increase would be a function of land productivity, and not an independent variable, as in Boserup's model.

Demographic increase would continue to be the principal indicator, although a new factor is introduced, considered as crucial in the urbanization process of Mesopotamia: at a particular moment, the eastern riverbed of the Euphrates and the population move to the west, near the new riverbed. This displacement agglutinates people even more, reducing available land. Again, the only way out is to intensify agriculture, the economic network and social organization. Yet, the system of large settlement units is no more efficient, and, as a result, large sites are disintegrated in favor of small villages throughout the territory. This allows a new demographic increase, and new attempts to intensify exchange and craftsmanship. Competition for land and goods is now so intense that war is a common and a much better organized phenomenon. Instead of migrating, now it is convenient to face conflicts directly.

The last model derives from Robert McC. Adams' (1972: 62–63) ideas which share the central structure of the aforementioned models, but proposes that the effects of these processes should be analized not only on the intercommunity level (with an increased emphasis in hostilities, whose effect would be the appearance of walled city-states), but also at the intracommunity level, that is, its effects on social stratification, which would provoke the emergence of state superstructures.

Following Hole (1974: 277), it should be mentioned that there is historical evidence of war until 2500 BC, which is some centuries after the culmination of the process. Service (1975: 304–308) adds that evidences of violent actions are found sporadically at any moment of the process. Furthermore we should not confuse two types of conflicts: friction between rival neighbors, where one center defeats others after a dispute over borders, and conflict between sedentary and nomadic people. On a very different scale is the systematic militarism of an expansive state, which would be present from Akkadian times onwards. Taking into consideration the original model, only from these periods onwards could we speak of a State, thus having the problem of explaining the Sumerian "city-state".

Let us return to the basic problem: demography. One of the weakest points of these hypotheses is the archæological strategy adopted. By means of the study of archæological material found on the surface, particularly the distribution of sherds, it is pretended not only to assign the site to a particular period, but also to determine the type of settlement and to calculate its population. Also it is important for the model to explain changes in "settlement patterns" at a regional level, where demography is in terms of the number of sites in each period. Yet, due to the inexistence of extensive excavation in these strategies, there are no precise chronological controls and, thus, there is no confirmation that the sites assigned to a particular period are really contemporary. On the other hand, it is well known that Near Eastern sites are characterized by the fact that different occupations are superposed to form huge mounds called *tell*, in which the majority of the occupations lie many meters below tons of fill and debris. The same is true for settlements buried under tons of lime. Thus, some of the factors that would decisively have influence on the extension and density of surface materials are the depth at which the different levels of occupation lie, the means by which the materials belonging to the underlying levels reach the surface, the degree of erosion of the site and the number of superposed levels, to mention only some (see Manzanilla 1979, chapters 7 and 9).

I propose that the problem be considered as follows. In the first place, the demographic estimate must not be an imperative goal in a regional research design. Not only this approach produces quantitative results that are afterwards taken as definite without further discussion, but also great endeavours are wasted on goals not yet accesible. In any case, the most accurate estimate would come from the number of houses in the site, specially when we have data from excavation contexts and when the function of the structures has been determined. In the second place, I doubt the fact that the demographic increase should be considered an independent variable. As Childe once pointed out, one of the effects of the improvement of the production system was demographic increase, and this improvement would certainly be evident in the increased efficiency of the redistributive system, as we will see afterwards.

3. *The Administrative Hierarchy Model*

Wright and Johnson have proposed a model to explain the origin of the state in the Khuzistan. In this model the emergence of centralized government institutions, with specialized administrative functions in decision making, would be involved, and they would imply three or more levels in the control hierarchy, in order to be considered a state. The indicators would be the supremacy of some settlements over others and the technology used in administration.

The mecanisms that are proposed to explain the emergence of such levels are the following. On the one hand, the fact that different variables are changing simultaneously and, thus, the capacity of the decision-making institutions to process information would be surpassed. On the other hand, there are two mechanisms that take place: the change from diversified functions to specialized and complementary functions for the administrative assistants, which would prevent conflicts that could emerge out of their position in the different levels of decision-making, and the interaction between variables (Wright and Johnson 1975: 267–268, 285).

The problem with this hypothesis is that these authors never determine what type of variables are in gear, and neither is there a comprehension of how a complex administrative system emerges in the Near East, or how it functions. I will devote some consideration to this subject later on. The administrative system would be an effect, not a cause, of a broader system.

4. *The Exchange Model*

Colin Renfrew (1975) has developed a series of models by means of which the role of exchange as an indicator for understanding the degree of organization and complexity of a "civilization" and the emergence of a state would be explained. The basic concepts that are mentioned are the "central place", as the specific site where exchange takes place, and the "early state module", as the autonomous territorial unit where these sites are embedded. In the first place, he establishes a difference between chiefdom and state in relation to the criteria of permanence of central places. The origins of civilization would be analyzed through the emergence of these permanent central nodes that would correspond to the principal demographic concentrations of the state modules. An example would be the Sumerian "city-states".

In the second place, Renfrew establishes a model of economic relations that would take place in three different scales: internal exchange, within each module, whose prevailing type would be redistribution; intermediate exchange, which would maintain uniformity between state modules by reciprocity; and finally, long-distance exchange, between the modules and foreign regions.

Some archæologists have begun to discuss the subject of pre-literate Near Eastern exchange. The slowness with which the outline is being produced is due to the complexity of intervening factors, and to the difficulty of evaluating, only with archæological data, certain modes and means. Generally a particular raw material is discussed, as for example chlorite, lapis lazuli or copper, as indicators of interaction between the Sumerian or Iranian influence spheres, during the third millennium BC. With respect to earlier periods, one should mention the obsidian case, that Gary Wright (1969) has accurately described.

Yet, the subject of exchange cannot be treated only under the raw material approach, but we should also discuss the use to which these are destined, the forms of exchange, the routes, who participate and who are benefitted. From the tabulation of objects found in north Mesopotamian Chalcolithic sites a very complex scheme emerges. Some raw-materials, that in a particular period have been destined to the manufacture of domestic goods, are later transformed into prestige goods (for example, alabaster, obsidian, and limestone). Others, such as serpentine, basalt or diorite, with which working instruments were elaborated, afterwards are represented in weapons or in administrative objects. Others begin as elements of personal decoration and eventually are the material basis of instruments and weapons (copper, for example). Some others penetrate in bulk and suddenly appear as prestige goods, representative of status (lapis lazuli, gold, and silver). One should ask what phenomena underlie these facts. What emerges from the analysis of the transformations that have occurred with time in the exchange networks is the following. With the Ubaid Period (Perkins 1949; Tobler 1950; Mellaart 1975) we can observe, for the first time, the complete occupation of the Mesopotamian plain and, thus, vast interregional contact. We therefore find obsidian from Eastern Anatolia in the Persian Gulf sites.

Settlements such as Tepe Gawra, in the northern plain, seem to have controlled the provisioning of turquoise, copper, lapis lazuli, and alabaster through sites such as Tepe Giyan, Sialk and Tepe Hissar. On the other hand, the relationship with Eastern Anatolia continues, for the supply of obsidian and, maybe, copper and gold.

For the Uruk and Gawra Periods (Herrmann 1968; Caldwell 1976; Wright, H.T. 1972; Dyson 1965; Beale 1973; Tosi 1973; Lamberg-Karlovsky 1973, 1974), there is a vaster integration of the aformentioned pattern. With respect to long-distance exchange, one could mention two circuits that perhaps intersect at Susa (Manzanilla 1979, 1986):

a) The northern route that includes the line that unites Tepe Gawra, Tepe Giyan, Sialk and Tepe Hissar, a route that in later millennia received the name of "the great road of Khurassan". Through this route, lapis lazuli and turquoise penetrated as far as Tepe Giyan, and probably from there to Susa, Uruk, and Tepe Gawra.

b) The southern network related the Persian Gulf (for sites such as Bakun A and Jagin) to the Kerman (Tepe Yahya and Tal-i Iblis) and the Khuzistan (with Tepe Farukhabad and maybe Susa). The majority of these sites were dedicated to the elaboration of specific handicrafts.

For Susa, it has been mentioned that, at the beginnings of the Uruk Period, the articulation of exchange networks is related to the periodic appearance of vast groups of specialized nomads in search of goods (Wright and Johnson 1975: 279).

An important fact is Moorey's (1976: 106) observation in the sense that the distribution of bevelled-rim bowls outside Mesopotamia is directly related to sources of raw-materials that interested this area. One might think of a chain that would connect different redistributive centers in charge of the provision of foreign raw materials for their respective regions. As we will see afterwards, exchange would thus be one of the institutions fostered by redistribution.

For the Jemdet Nasr Period (Tusa 1977; Weiss and Young 1975; Young 1972; Lamberg-Karlovsky and Tosi 1973; Frankfort 1971), we observe the establishment, sometimes by force, of proto-Elamite colonies at various points implied in the provisioning of Iranian raw-materials. As Beale (1973) mentions, the political control of the exchange networks and of the deposits was the dominant factor that determined the most important fluctuations.

According to Caldwell (1976), Susa captured the lapis lazuli exchange, and sites such as Tepe Giyan and Sialk present a hiatus in their sequence, or evidence of destructions and sacking, followed by the implantation of proto-Elamite outposts.

On the other hand, Lamberg-Karlovsky and Tosi observe two vast interaction spheres during this period: one from Sumer to the Persian Gulf and the southwest Iranian plateau, and the other that would comprise the southern sector of Turkmenia and the northeast Iranian plateau.

According to Herrmann (1968: 47), during the Early Dynastic Period, there seems to be a suspension in the affluence of foreign elements to Sumer by way of the northern route, due to the scarse presence of lapis lazuli. During Phase II, an important network between Kerman and Sumer would exist, maybe with Susa as an intermediary. The southern route that is mentioned in historical texts belonging to the end of the Early Dynastic is the maritime route that linked Sumer with Dilmun, and from here goods went to Makran and to the northwestern coast of India. In this respect, there has been the suspicion that the opening of a maritime route was may have been a response to the Elamite barrier in the Khuzistan.

In relation to the northern network, Lamberg-Karlovsky and Tosi (1973: 44, 50) propose that there was a minor macro-regional and a greater zonal integration. In the Iranian plateau, the former pattern of cultural interrelations is disintegrated, and this provoked later on the isolation and, thus, the collapse of urban life, accelerated by internal social tensions and by population movements from the north (Tosi 1973: 445).

A well-known fact is that in a region without rocks, minerals, and metals, such as Mesopotamia, long-distance exchange is very important for the provisioning of raw-materials used in the manufacture of the most basic domestic implements, weapons, and prestige items. Yet, as I suggested in 1979, long-distance exchange was only one of the by-products of an organization centered in the temple, in which redistribution assured the availability of goods. Earle and D'Altroy (1982: 267) agree with this idea, in view of their study of redistribution in Hawaiian societies. Let us pass to the last hypothesis considered, in order to integrate exchange in a more complete framework.

5. *The Control of Production and Redistribution Model*

This model has its origin in V. Gordon Childe's analysis of urban organizations. Childe mentioned that the city is a by-product of progressive demographic increase, which could only be produced by the accumulation of surplus. The population of the first cities not only represented a new order of magnitude, but also a different character in the presence of full-time specialists (Childe 1964: 29–30). In the Mesopotamian case, surplus production beyond the domestic needs of the village inhabitants would serve to maintain new economic classes that would have special access to the basic productive resources, stimulating social stratification. On the other hand, this surplus would be devoted to exchange and redistributive activities. There would then be an implicit need to create institutionalized forms to concentrate and redistribute such surplus: a central authority (Childe 1968: 275–276).

Service (1975: 207–210) has mentioned the fact that in the Uruk Period, the temple administrators concentrated and rationed foodstuffs, stored and redistributed raw-materials and products, and also were in charge of exchange with foreign groups. For the Early Dynastic, redistribution is transformed when power moves to secular political spheres.

Following a similar line of thought, Frank Hole (1974) proposes that when man becomes sedentary, he sacrifices diet variety and mobility for security in production. In this process, a development of restricted–specter patterns of exploitation is implicit and, thus, man is obliged to exchange his products for those goods that are not locally produced; the result would be the loss of self-sufficiency. The fact that the Near East is characterized by an unequal distribution of resources would stimulate intercommunity specialization. Therefore, there is the need of an organization that controls production and redistribution, either in the hands of one person —the chief—, an institution —the temple—, or a settlement.

Furthermore, in a redistributive economy, more surplus would be correlated with more stratification, that is, a significant portion of the surplus would be channeled to the temple and to the activities related to the elite, such as the provision of raw-materials, the maintenance of craftsmen that process them, etcetera.

My proposal goes in the sense of this last model, but it distinguishes between two stages with different organizations: that of the temple and that of the palace.

II. *The Temple Sphere*

As mentioned earlier, the area in which the first indicators of centralization can be better tracked is naturally Mesopotamia, where we have an abundance of excavated and analyzed sites. From the Neolithic onwards there is evidence of communal activities, particularly storage. Due to the fact that we do not have data that distinguishes one social group from the rest, either in residential practices or in graves, one comes to the conclusion that, as in actual "lineage societies", the elders' council would have served as the coördinating institution of common activities, specially the distribution of stored production in the center of the villages, in a circuit I and others have called "circular redistribution" (Manzanilla 1983). Furthermore, there could be a communal property sphere beyond the family sphere.

With the appearance of the tripartite temple, from the Ubaid Period onwards, we observe clear evidence of specialization in different tasks. The priesthood takes the place of the elders' council as a coördinating institution of economic and ideological communal tasks, but adds a higher order of complexity: even though the elders' council does not disappear at the level of particular communities, the priesthood lies at a higher hierarchical order and resides in the main settlement where it is possible to maintain craft- and exchange-specialists, as a subsidiary part of the assymetrical redistribution circuit. The stored production concentrated in the small rooms that flank the sanctuary of the tripartite temples flows in the form of foodstuff rations to the faithful and dependent personnel, and it is also channeled to long-distance exchange, to make a provision of different raw-materials which do not exist in Mesopotamia. These are brought from Anatolia, the Levant or Iran. The rations are evident in historical sources belonging to immediate later periods, but also in the mass production of truncoconial bowls (Espinosa and Manzanilla 1985) and other vessels that lie in temple storehouses of various sites, such as Arslantepe in Oriental Anatolia (Palmieri 1973).

Assymetrical redistribution is important in regions with relatively homogeneous and limited resources, such as the Mesopotamian plain or the Mayan Lowlands, due to the fact that the goods stored by the temple serve to foster specialized manufactures and to exchange them for inexistent raw-materials. On the other hand, in regions with ecological diversity due to the existence of orographic differences, the response is the "economic symbiosis" of settlements that participate in intercommunity specialization networks, as Sanders (1968:100) has proposed for the Late Formative of the Basin of Mexico, and that is also present in the Valley of Oaxaca (Flannery and Coe 1972).

An opposite alternative would be the one that pretends to achieve self-sufficiency for each community and the exploitation of ecological niches located at different altitudes; this would constitute Murra's (1975) "vertical control" model for the Central Andean Area.

When institutions that centralize a good part of the productive process, and the consequent distribution of goods appear, such as the temple, urban centers emerge where the priesthoods promote the existence of a series of specialists in different productive and exchange activities. Such centers become the axis of macro-regional circuits of concentration and distribution of goods. To assure the supply of foreign raw-materials for the elaboration of sumptuary goods, the priesthood sends emissaries that would establish a type of pacific colony inside local communities in other regions, and sent those goods back.

Examples, such as Teotihuacan, in the Basin of Mexico, and Tiwanaku, in the Titicaca Lake, would fit in this framework. As a matter of fact, it should be mentioned that in the central-northern Andean region, from Formative times onwards, a type of settlement designated as "ceremonial center" emerges, in which the central portion is occupied by the temple surrounded by storehouses (Lumbreras, personal communication, 1986).

Thus, the redistributive circuit could be detected in the archæological record particularly in the warehouses controled by the temple, in the flow of stored goods in the form of food rations or communal banquets, using mass-made pottery and, in the Near Eastern case, in an elaborate administrative system that implied accounting and control procedures of the goods that entered and departed from the warehouses. The corollaries to this system were the emergence of specialists detached from food production, sustained with the stored foodstuffs, and the supply of foreign raw-materials by means of temple emissaries. These elements constitute important characteristics of what we could call "pristine urban societies".

Two special cases are China and Egypt, where from early times onwards, we have evidences of divine kingship. This is the case of China during the Shang Dynasty (Chang 1974), in which the palace and the ceremonial area of the capitals were articulated in one sector; on the other hand, groups of craftsmen, and the residential and storage areas were located in clusters surrounding the central sector. This detachment of the storehouses from the central institutions could indicate that redistribution occurred at the lineage or clan level. Chang (1984) has emphasized that the monopoly of shamanism was the key the ruler had to accede to ancestral and divine wisdom, and this was the basis of his political authority.

Another notorious case is that of Egypt from the Late Predynastic through the Ancient Kingdom, in which the pharaoh was the head of all important functions, either civil or religious. The warehouses of Upper and Lower Egypt lay in the palace sphere and redistribution depended on him.

In the Chinese and the Egyptian examples, the articulation of the ruler' s authority with his divine functions is so intrincate that the priesthood stands in a second place, generally devoted to the funerary ceremonies of past rulers. It also seems that the Valley of Oaxaca, during Classic times, could have had a similar organization.

III. *The Palace Sphere*

During the Early Dynastic, at Sumer, the temple gradually collapses as the economic axis of society and this fact is contemporary with the appearance of the palace as a rival economic institution, concerning land, cattle, products, and centralization. The palace acquired the administrative functions of the temple, and although the goods in storage did not reach common people, as in the former sphere, they did form part of a closed circuit.

In the Early Dynastic Period the two institutions coexisted, but in the following Akkadian Period, a new kind of organization becomes clearly outlined, with the palace as the axis: the expansive and tributary state (Manzanilla 1979).

The defining characteristics and organization of this moment are the following ones:

1) Tribute substitutes redistribution as the form of centralization of surplus production. Conquest assures the continuous flow of goods to the palace.

2) Another consequence of conquest is the emergence of the concept of wealth accumulation, in the form of war spoils, and in the adscription of marginal land. It can be said that, even though there are some cases of landlords rewarded with land, the basic property cell is the community (*calpulli, ayllu, nome*) which is the one that assigns land and tasks to its members. Then, this step marks the beginning of class societies and the state.

3) A third element, in close relationship with the palace and its organization, is the market and the emergence of the merchant. The market is an institution that comes late in the process; it is created in close articulation with the palace interests, as has been pointed out by Carrasco (1983). Since the Early Dynastic II in Mesopotamia, merchant groups associated with the palace stimulated popular demand for bronze weapons and luxury goods (Adams 1973). In the Basin of Mexico, the *pochteca* substitute the older temple emmisaries.

4) As a consequence of the appearance of the market, in addition to the temple or palace craftsmen, one can also observe independent craftsmen, generally grouped in wards, exchanging their production freely.

Examples such as the Akkadian Empire and the successive Babylonian and Assyrian states, the New Egyptian Kingdom, the Central Mexican Postclassic, and the Chimu and Inca Empires belong to this stage.

In the majority of these cases there are still relicts of the former redistributive organization. For example, Carrasco (1982) and Broda (1976) have written on different cases in which the Aztec *tlatoani* opened his warehouses to give foodstuff to common people, or insignia to the warriors. Another example is that of the Inca State, where we can better study the relicts of redistribution in the palace organization. Scholars such as Morris (1978), Earle and D'Altroy (1982) and others have described storehouses in different Peruvian centers, particularly in the region of the provincial capital of Huánuco Pampa and in the Mantaro Basin.

These warehouses, that Murra (1975) mentions are vital when there are frosts or drought, served also to maintain state personnel, the army, and all the people that participated in state activities. The deposits were located either inside the settlements or on the mountain slopes, depending on the permanent or temporal character of the population (Earle and D'Altroy *op. cit.*). On the other hand, Morris (*op. cit.*) mentions the fact that at Huanuco Pampa, the two open sectors or *plazas* were devoted to the elaboration and redistribution of *chicha* and other foodstuffs.

Due to the persistence of the basic elements of the former stage, we could consider the Inca case as a transitional example. An element that would support this idea is the inexistence of markets and merchants inside the Inca Empire.

We would like to conclude, thus, saying that the temple organization, centered in an intrincate redistributive circuit, was the basis for the "urban revolution", and that only in later times, a state centered in the palace as the axis of a tributary circuit and the top of a class society emerges.

REFERENCES

ADAMS, R. McC.

1965. *Land behind Baghdad. A History of Settlement on the Diyala Plains*, The University of Chicago Press, Chicago.

1966. *The Evolution of Urban Society. Early Mesopotamia and Prehispanic Mexico*, Aldine Publishing Company, Chicago.

1969. "The Study of Ancient Mesopotamian Settlement Patterns and the Problem of Urban Origins", *Sumer* vol. XXV, 1–2, Baghdad: 111–124.

1972. "Demography and the 'Urban Revolution' in Lowland Mesopotamia", in Spooner, Brian (ed): *Population Growth. Anthropological Implications*, The MIT Press, Cambridge: 60–63.

1973. "Some Hypothesis on the Development of Early Civilizations", in Leone, Mark P. (ed): *Contemporary Archæology. A Guide on Theory and Contributions*; Southern Illinois University Press, Carbondale: 359–364.

BARNETT, R.D.

1958. "Early Shipping in the Near East", *Antiquity* vol. 32, no. 128, December: 220–230.

BEALE, T.W.

1973. "Early Trade in Highland Iran: A View from a Source Area", *World Archæology* vol. 5, no. 2: 133–148.

BOSERUP, E.

1965. *The Conditions of Agricultural Growth. The Economics of Agrarian Change under Population Pressure*, Aldine Publishing Company, Chicago.

BRODA, J.

1976. "Los estamentos en el ceremonial mexica", in Carrasco, Pedro *et al.*: *Estratificación social en la Mesoamérica Prehispánica*, SEP-INAH, México: 37–66.

CALDWELL, D.H.

1976. "The Early Gliptic at Gawra, Giyan and Susa, and the Development of Long Distance Trade", *Orientalia* vol. 45, n.s., Rome: 227–250.

CARNEIRO, R.L.

1970. "A Theory of the Origin of the State", *Science* vol. 169, 21 August: 733–738.

CARRASCO, P.

1982. "1. La economía del México prehispánico", in Carraso, Pedro and Johanna Broda (eds): *Economía política e ideología en el México prehispánico*, Editorial Nueva Imagen, México: 13–76.

1983. "IV. Some Theoretical Considerations about the Role of the Market in Ancient Mexico", in Ortiz, Sutti (ed): *Economic Anthropology. Topics and Theories*, Society for Economic Anthropology, New York: 67–82.

CHANG, K.C.

1974. "Urbanism and the King in Ancient China", *World Archæology* vol. 6, no. 1, June: 1–14.

1984. "Ancient China and its Anthropological Significance", *Symbols* spring/fall, Harvard University, Cambridge: 2–4, 20–22.

CHILDE, V. G.

1964. *Evolución Social*, Problemas Científicos y Filosóficos 29, UNAM, México.

1968. *Nacimiento de las Civilizaciones Orientales*, Ediciones Península, Barcelona.

DURING CASPERS, E.C.L.

1971. "New Archæological Evidence for Maritime Trade in the Persian Gulf during the Protoliterate Period", *East and West* n.s., vol. 21, no. 1–2, March-June, IsMEO, Rome: 21–44.

DYSON, R.H.

1965. "Problems in the Relative Chronology of Iran, 6000–2000 BC", in Ehrich, Robert H. (ed.): *Chronologies in Old World Archæology*, The University of Chicago Press, Chicago: 215–256.

EARLE, T.K. and T.N. D'ALTROY

1982. "12. Storage Facilities and State Finance in the Upper Mantaro Valley, Peru", in Jonathon E. Ericson and Timothy K. Earle (eds): *Contexts for Prehistoric Exchange*, Academic Press, New York: 265–290.

ESPINOSA, G., and L. MANZANILLA

1985. "Consideraciones en torno a la capacidad de los cuencos troncocónicos de Arslantepé, Turquía Oriental (Período VIA, Bronce Antiguo I)", *Quaderni de 'La Ricerca Scientifica'* no. 112, Rome: 64–85.

FLANNERY, K.V., and M.D. COE

1972. "Social and Economic Systems in Formative Mesoamerica", in Binford, Sally R., and Lewis R. Binford (eds): *New Perspectives in Archeology*, Aldine Publishing Co., Chicago: 267–283.

FRANKFORT, H.

1971. "Chapter XII. The Last Predynastic Period in Babylonia", *The Cambridge Ancient History* vol. I, Part 2, Cambridge University Press Cambridge: 71–92.

FRIED, M.H.

1967. *The Evolution of Political Society: An Essay in Political Anthropology*, Random House, New York.

GIBSON, McG.

1973. "Population Shift and the Rise of Mesopotamian Civilisation", in Renfrew, Colin (ed): *The Explanation of Culture Change. Models in Prehistory*, Duckworth, Gloucester Crescent: 447–463.

HERRMANN, G.

1968. "Lapis Lazuli: the Early Phases of its Trade", *Iraq* vol. 30, London: 21–57.

HOLE, F.

1974. "Investigating the Origins of Mesopotamian Civilization", en Lamberg-Karlovsky, C.C., and Jeremy A. Sabloff (eds): *The Rise and Fall of Civilizations. Modern Archæological Approaches to Ancient Cultures*, Selected Readings, Cummings Publishing Co., Menlo Park: 269–281.

JOHNSON, G.A.

1972. "A Test of the Utility of Central Place Theory in Archæology", in Ucko, Peter J.; Ruth Tringham, and G.W. Dimbleby (eds): *Man, Settlement and Urbanism*, Gerald Duckworth and Co., Hertfordshire: 769–785.

LAMBERG-KARLOVSKY, C.C.

1974. "Trade Mechanisms in Indus-Mesopotamian Interrelations", in Lamberg-Karlovsky, C.C., and Jeremy A. Sabloff (eds): *The Rise and Fall of Civilizations. Modern Archæological Approaches to Ancient Cultures*, Selected Readings, Cummings Publishing Co., Menlo Park: 302–313.

1975. "8. Third Millenium Modes of Exchange and Modes of Production", in Sabloff, Jeremy A., and C.C. Lamberg-Karlovsky (eds): *Ancient Civilization and Trade*, (A School of American Research Book), University of New Mexico, Albuquerque: 341–368.

LAMBERG-KARLOVSKY, C.C., and M. LAMBERG-KARLOVSKY

1973. "An Early City in Iran", in: *Cities. Their Origin, Growth and Human Impact*, (Readings from Scientific American), W.H. Freeman and Co., San Francisco: 28–37.

LAMBERG-KARLOVSKY, C.C., and M. TOSI

1973. "Shahr-i Sokhta and Tepe Yahya: Tracks on the Earliest History of the Iranian Plateau", *East and West* n.s., vol. 23, no. 1 and 2, March–June, IsMEO, Rome: 21–57.

LUMBRERAS, L.G.

1968. Course: "Los fundamentos económicos de la sociedad andina", Centro de Investigaciones y Estudios Superiores en Antropología Social, México.

MANZANILLA, L.

1979. Comentarios en torno a un proceso histórico: la constitución de la sociedad urbana en Mesopotamia (cuarto milenio aC), tesis de maestría, ENAH, México.

1983. "La redistribución como proceso de centralización de la producción y circulación de bienes. Análisis de dos casos", *Boletín de Antropología Americana* no. 7, julio, Instituto Panamericano de Geografía e Historia, México: 5–18.

1985. "Templo y palacio: proposiciones sobre el surgimiento de la sociedad urbana y el estado", *Anales de Antropología* vol. XXII, UNAM, México.

1986. *La constitución de la sociedad urbana en Mesopotamia. Un proceso en la historia*, (Serie Antropológica no. 80), Instituto de Investigaciones Antropológicas, UNAM, México.

MELLAART, J.

1975. *The Neolithic of the Near East*, Thames and Hudson, London.

MILLON, R.

1962. "Variations in Social Responses to the Practice of Irrigation Agriculture", in Woodbury, Richard B. (ed): *Civilizations in Desert Lands*, (Anthropological Papers no. 62, December), University of Utah Press, Salt Lake City: 56–88.

MOOREY, P.R.S.

1976. "The Late Prehistoric Administrative Building at Jamdat Nasr", *Iraq* vol. 38, part 2, autumn, London: 95–106.

MORRIS, C.

1978. "Chapter 13. The Archeological Study of Andean Exchange Systems", in Redman, Charles L. *et al.* (eds): *Social Archeology. Beyond Subsistence and Dating*, Academic Press: 315–327.

MURRA, J.V.

1975. *Formaciones económicas y políticas del mundo andino*, Instituto de Estudios Peruanos, Lima.

PALMIERI, A.

1973. "Scavi nell'area sud-occidentale di Arslantepé. Ritrovamento di una struttura templare dell'antica etá del Bronzo", *Origini* vol. VII, Roma: 55–228.

PERKINS, A.L.

1949. *The Comparative Archeology of Early Mesopotamia*, (Studies in Ancient Oriental Civilization no. 25), The Oriental Institute of the University of Chicago, Chicago.

RENFREW, C.

1975. "1. Trade as Action at a Distance: Questions of Integration and Communication", in Sabloff, Jeremy A., and C.C. Lamberg-Karlovsky (eds): *Ancient Civilization and Trade*, (A School of American Research Book), University of New Mexico Press, Albuquerque: 3–59.

SANDERS, W.T.

1968. "Hydraulic Agriculture, Economic Symbiosis and the Evolution of the State in Central Mexico", in Meggers, Betty (ed): *Anthropological Archeology in the Americas*, The Anthropological Society of Washington, Brooklyn: 88–107.

SANDERS, W.T., and B.J. PRICE

1968. *Mesoamerica. The Evolution of a Civilization*, Random House, New York.

SERVICE, E.R.

1975. *Origins of the State and Civilization. The Process of Cultural Evolution*, W.W. Norton and Co., New York.

STEWARD, J.H.

1955. "Introducción. El symposium sobre las civilizaciones de regadío", *Las Civilizaciones Antiguas del Viejo Mundo y de América*, (Estudios Monográficos no. 1), Unión Panamericana, Washington: 1–5.

TOBLER, A.J.

1950. *Excavations at Tepe Gawra* Vol. II. Leveles IX–XX, University of Pennsylvania Press, Philadelphia.

TOSI, M.

1973. "Early Urban Evolution and Settlement Patterns in the Indo-Iranian Borderland", in Renfrew, Colin (ed): *The Explanation of Culture Change: Models in Prehistory*, Duckworth, Gloucester Crescent: 429–446.

TUSA, S.

1977. "I sigilli e le impronte", in *La Cittá Bruciata del Deserto Salato*, Erizzo Editrice, Venezia: 251–261.

WEISS, H., and T.C. YOUNG Jr.

1975. "The Merchants of Susa. Godin V and Plateau-Lowland Relations in the Late Fourth Millenium BC", *Iran* vol. 13, London: 1–17.

WITTFOGEL, K.A.

1974. "Developmental Aspects of Hydraulic Societies", in Lamberg-Karlovsky, C.C., and Jeremy A. Sabloff (eds.): *The Rise and Fall of Civilizations. Modern Archæological Approaches to Ancient Cultures*, Cummings Publishing Co., Menlo Park: 15–25.

WRIGHT, G.A.

1969. *Obsidian Analyses and Prehistoric Near Eastern Trade: 7500 to 3500 BC*, (Anthropological Papers no. 37), Museum of Anthropology, University of Michigan, Ann Arbor.

WRIGHT, H.T.

1972. "A Consideration of Interregional Exchange in Greater Mesopotamia: 4000–3000 BC", in Wilmsen, Edwin N. (ed): *Social Exchange and Interaction*, (Anthropological Papers no. 46), Museum of Anthropology, University of Michigan, Ann Arbor.

WRIGHT, H.T., and G.A. JOHNSON

1975. "Population, Exchange and Early State Formation in Southwestern Iran", *American Anthropologist* vol. 77, no. 2, June, Washington: 267–289.

YOUNG Jr., T.C.

1972. "Population Densities and Early Mesopotamian Urbanism", in Ucko, Peter J.; Ruth Tringham, and G.W. Dimbleby (eds): *Man, Settlement and Urbanism*, Duckworth, Hertfordshire: 827–842.

THE URBAN REVOLUTION OF MESOPOTAMIA —RECONSIDERED

HANS J. NISSEN
Freie Universität Berlin

I am afraid, it would be quite easy to dismiss Childe's concept of an "Urban Revolution" into the history of our field, as over many years it had been dismantled by all concerned with that question. I would like to show, however, that seen from the Mesopotamian scene the term still can be used profitably to designate an important intellectual change. It should be understood, however, that the term "revolution" is retained only in its very loose meaning: the building up towards a change which then comes rather quickly and results in an irreversible situation.

I do not want to spend much time on discussing chapter seven of his "New Light on the Most Ancient East" where Childe (Childe 1952) most explicity applied this concept to Mesopotamia, and at the same time derived it from Mesopotamia. I only want to mention one point where in addition to all criticism of his criteria listed there, and most concisely, in his article "The Urban Revolution" (Childe 1950) there is an inconsistency in his basic reasoning. This happens to be a point where moreover recent research most drastically has altered the basis. This concerns his ultimate argument for calling the Uruk-period in Mesopotamia "urban":

> Writing is used by sociologists as a convenient criterion to distinguish civilization from barbarism. So we can assert that in Messopotamia civilization was created in the Uruk period. We can now term the larger settlements —Eridu, Erech (Uruk), Ur, Lagash, Uqair— cities and compare the great temples to Cathedrals (Childe 1952: 128).

Obviously, he considered the most decisive criterion the emergence of writing. It seems strange that he should place so much emphasis on this fact because elsewhere he himself attacks the concept of drawing a sharp line between "historic" and "pre-historic" periods, a concept originating with Hegel with the term "historic" being tied to the presence of a script. Hegel's concept has been contested on the ground that as a matter of course the earlier periods are part of the history of mankind as well, a reason why in Germany we use the term "Urgeschichte" ("primeval history") instead of "Vorgeschichte" ("pre-history"). But there are other arguments why both Hegel's concept and Childe's opinion on the emergence of writing marking the advent of urbanism lost their basis. As I will mention later in context writing in Babylonia has not been used to render fully the spoken language for more than 500 years after its appearance. Its emergence at first is nothing else but a refinement of existing controlling devices for economy (Nissen 1983b; 1985; 1986). The *emergence* of writing thus is not fit to serve as a historic dividing line. If one wants to think in these categories at all, it is the *time of the first full use of the potential of writing* for rendering the full flow of the spoken language, and subsequently, for writing historical and literary texts what should be taken to mark the beginning of a new phase.

Rather than taking up the issue of discussing one of Childe's criteria for the "Urban Revolution" after the other I would like to sketch my own view of the early developments in Mesopotamia which I have given in greater detail in other places (Nissen 1983a; 1983b; 1987).

Let me start with the assumption that at a time and because of reasons we shall not discuss here settlements which lie close together will develop differently from each other. One of the reasons is the emergence of so-called "central functions" in one settlement, i. e. of aspects which gain importance for a larger area than presented by one settlement alone.

Another assumption is that from a certain level of organization on, division of labor and specialization have the tendency to develop more complex forms if the outer conditions permit.

It should be noted that I shall limit the following discussion to settlements the inhabitants of which derive their main subsistence from cultivating the surrounding area. Settlements relying heavily on trade ("ports of trade") or any other means of income thus will not be covered.

The thesis is that such a development of specialization finds its limits when a special activity needs a larger customer potential than can be provided by the population of one settlement, or of those settlements, which by force of the central functions are tied together to form a settlement system. The degree of the development of special activities and of central functions thus turns out to depend on the size of the settlement, or of the settlement system.

Each settlement under consideration is surrounded by agriculturally used land of a size which is necessary to furnish the inhabitants of the settlement with their necessary supplies. The number of inhabitants thus is tied to the size of the surrounding agriculturally usable area. While theoretically is should be possible to extent this area up to its natural limits in order to accomodate as many people as possible, this meets an obstacle. The maximun radius of the area cannot be larger than half the daily traffic capacity of man, which in early societies means the marching capacity. Size of the area, number of inhabitans and thus customer potential find certain limits which cannot be surpassed.

There is a possibility, however, to surpass this border-line in the case of the customer's potential, if instead of one settlement we think of the enlarged potential of a settlement system, assuming that special activities may attract their customers also from the other settlements of a system. Theoretically, such supra-local activities may settle in any settlement of a system but in reality, they will tend to cluster in one of the settlements. Normally, this will be the central place of such a system.

The size of the area of a settlement system again theoretically may be extended to some natural borders, but in reality, this will not be so. First, the territories of the constituting settlements cannot be extended as we saw; secondly systematic close contacts between settlements as are necessary to qualify for the term system can be maintained only through constant daily contacts, which in turn, require short traffic connections. Hence we assume that only one circle of settlements will surround each central place. Should the available area suited for settlement be larger than covered by a system consisting of a central place and surrounding settlements, the remainding area will be taken up by the formation of another such system rather than by mere expansion of the first one. Again, settlements of such additional systems may serve as a reservoir for enlarging the customer potential for any special activities who might form in a central place serving all settlements of such a coherent settlement area.

In the case described before the formation of special activities will be possible which need an even larger customer potential than before. On the basic assumption that the size of the total population from which the number of customers can be recruited sufficient to support a special activity has to be the larger the more specialized the activity is, we find again that the degree of specialization grows with the complexity of settlement systems.

The situation discussed before rested on the assumption of several settlement systems consisting of a central place and surrounding settlements being in that close contact with each other that another level of superimposition was reached by the formation of a centre on a level higher than the central places existing previously. This problem can be approached also from another angle, by accounting for the necessity of conflict regulation. After a long time of interpersonal conflict regulation size and density of settlements will expedite the formation of permanent agencies for such regulation, which at the same time will be entrusted with the formulation and execution of rules. As the extent of conflicts does not grow linearily with the size of the groups involved but shows peaks at various points, we may assume that at certain points of the extension of settlement systems the foundations and mechanisms of conflict regulation will have to assume another quality. From a certain level on in the development of settlement systems this will require specialists and thus be one reason of the formation of a ruling or administrative group.

Neither the formation of a permanent ruling or administrative group nor the ongoing process of a refinement of the division of labor can have passed without major impacts on the social organization of the normal population. In fact, it may be assumed that these processes were part of and enhanced a change of the social situation, particularly a trend to substitute family divisions by functional divisions. And again, we can tie this development to the development of settlement systems without being able to name the point within that development.

Of one thing we can be sure, however: the inter-dependant developments in the spheres of the division of labor, of the formation of ruling classes, of functional divisions and of settlement systems all combine in a probably slow but nevertheless substancial change in the structure of the economic, social and political situation which did not take place in areas where outer conditions were such as not to allow the development of settlement systems. Thus the periods of the formation of more complex forms of settlement systems at the same time must have seen a differentiation in the level of organization between settlements according to their location.

At this point, I should clarify what I mean by the term of a "coherent settlement area" before I am going to use it again. I maintained before that in order to qualify for a settlement system the various settlements should be able to have regular, daily contacts. Though exceptions are possible, we can continue by saying that opportunities are optimal, if there are no natural obstacles impeding such traffic. Also in other respects will such outer conditions be optimal for the formation of settlement systems as in large plains the area around settlements may assume its optimal size and at the same time, the connecting routes between the settlements may remain as short as possible. Areas which in this sense allow settlement systems to develop optimally I call "coherent settlement areas". It is furthermore understood that normally such coherent settlement areas would be separated from each other by natural boundaries.

Within a natural landscape there will be any number of such coherent settlement areas of various sizes. According to our contention that the degree of hierarchization of settlement systems depends on the opportunity to establish parallel systems within the same area and to develop superimpositions we can postulate that in any such larger region with different coherent settlement areas there will be systems on different levels of complexity. That means, there will be differences in the degree of the division of labor, in the specialization of activities and functions, and the general social and economic conditions.

Let us now consider the evidence. There is one region in particular in the Near East where some of the ideas proposed can be checked against both the outer conditions and the information available. The area between and including the Babylonian plain to the West and the Iranian plateau to the East provides the outer frame since within a relatively small area it contains many differently sized coherent settlement areas, and on the other hand, we have enough evidence to know that they were inhabited during various early periods, partly during the same periods. As far as can be derived from archæological evidence suggests that there is a relation between the size and time of full occupation in the sense that the larger areas were last to be settled to their full capacity (Nissen 1983a).

In addition, the area mentioned lends itself to a discussion of the phenomena of Early Urbanism because it includes the Babylonian plain and therefore among others all those places mentioned by Childe in the quotation cited above as early cities. Apart from the terminology used, there is no doubt that this area saw the formation of large, complex settlements at an early moment. This applies at least to those large centers of the time around the middle of the third millennium, the period which in our terminology we are used to call Early Dynastic III. These settlements cover an area of several hundred hectares with a high density of the built-up areas; a high degree of specialization in crafts and administrative functions is accompanied by the existence of permanent ruling groups. Although the term can be a matter of dispute —we call these settlements "city-states"—, at the top of these settlements we find hereditary rulers, supported by a thoroughly organized body of administrators.

The fact that excavations have been conducted almost exclusively in the central areas of ancient Babylonian cities plus the assumption that the use of the script was restricted to official institutions both made us believe for a long while that the entire life within such a settlement was controlled by the central institution of the city. This notion has changed to some extent as we find strong hints as to the presence of several large "households" within a city. The form of the "temple-economy" tied to the concept of the "temple-city" is not considered anymore the only, or even the most important one. However, this does not change the general concept according to which all inhabitants were employees of one or the other kind of household. We have no evidence for the existence of private entrepreneurs in the middle of the 3rd millennium BC (Diakonoff 1974; Gelb 1969).

Another piece of relevant evidence is provided by the observation of the distribution of settlement sizes in the survey area of the hinterland of Uruk (Adams and Nissen 1972). Of the total aggregate settled area formed by computing the area of all contemporary settlements more than 80% can be attributed to the category of cities. Short of entering the dubious calculations of the number of inhabitants from the size of the settled area we nevertheless are safe in assuming that around 80% of the total population lived in cities by the middle of the 3rd millennium in Babylonia. The percentage could be even higher if we account for the probability that population density was higher in larger places than in smaller ones (Nissen 1984b, 1987).

To arrive at a date for the beginning of urbanization we should simply try to trace back in time the features mentioned above to their first occurrence. Unfortunately, however, such approach is met by a number of difficulties. The first is that the Early Dynastic III period is the first when writing is used to its full potential of rendering all details of spoken language. This immediately led to the composition of royal and literary texts which allow us to get a glimpse of the historical texture of that period. Earlier, the script had only been used to write more or less detailled catch-words, and its is quite understandable that this should cause many problems for our understanding these texts. In spite of all these problems, we have been able over the past years to decode a number of these texts, of the so called "Archaic Texts from Uruk", of the time of shortly before 3000 BC. In some cases we are even able to be more specific about the economic and social conditions of the period we call the Late Uruk period (Green 1980; Nissen 1985, 1986).

According to this and related information the conditions during this time of around 3000 BC seem to have been very close to those mentioned before. Our best evidence comes from Uruk itself. This place covered an area of more than 100 ha and had extensive public areas in the centre. The best single piece of evidence is a list of titles of officials and craftsmen, apparently enumerating them according to their ranking. There is a fixed system of hierarchies both in the offices and the crafts. The list starts with a title which otherwise is unknown from later tradition except that in a much later vocabulary it is translated by the later word for "king". The following titles again are composed with an element standing for "leader" or the like, containing in addition words like "barley", "city", "troops" which we take to indicate their area of responsibility. There is even one title which later means the "chairman of the assembly". Unfortunately, our studies have not progressed to the point that we should have any notion of the interrelations of these officials (Nissen 1974, 1985, 1986).

I mentioned already that no private quarters have been excavated. However, we have evidence from one settlement on the Middle Euphrates, Habuba Kabira, in present-day Syria which is spite of its distance from Babylonia shows such striking similarities in all details to Babylonian inventories of the Late Uruk period that also its plan must be accepted as a substitute of the missing evidence from Babylonia (Strommenger 1980). If so, we can assume a high density of the built-up area in Babylonian centres, particularly of the private quarters. As a cautionary note, however, I should recall the passage of the Gilgamesh-epic, reading that 1/3 of the city area was covered with houses, 1/3 with gardens and 1/3 with pits, presumably the source of the building material.

The hinterland of Uruk was occupied during this period by a multitude of large and small sites which obviously were centered around Uruk. Applying models from settlement geography resulted in the observation of a multi-tiered settlement system with Uruk at the top (Adams and Nissen 1972; Nissen 1983b, 1987).

This short sketch has revealed the close relationship in organization of those central places to those settlements of *c.* 600 years later which we had no difficulties in calling cities. If we go one step back in time from the Late Uruk period we come across a strange fact: immediately before the Late Uruk period we encounter a totally different situation in Babylonia. This older situation is marked by the presence of isolated settlements separated by wide distances without any chances of forming settlement systems. We have no idea about the internal structure of these settlements; writing or anything else indicating some degree of complexity are absent, and an occasional central temple complex does not suffice to postulate the existence of a ruling group.

Though to its full extent this situation became known only in the past years, there nevertheless was enough information available many years ago to incorporate the notion of a certain suddenness of the emergence of the Late Uruk culture into the common belief. I have no doubt that this notion influenced Childe's terminology to quite some extent.

The perspective changes, however, once we consider the larger region mentioned above, because the situation shows more consistency once we look at the adjacent area to the East. Large centers like Chogha Mish, Susa and others on the one hand display the same Late Uruk features which we saw in Babylonia, like indications for a complex economy or large public areas; but differently from Babylonia, the immediately preceding situation, in our terminology Susa A, presents itself as just an older stage within the same development (Nissen 1983a; Wright and Johnson 1975).

These earlier places obviously were central places too, in particular they can be shown to form the top of three-level settlement systems. Again we find large central installations and the extensive use of stamp-seals points to a complex economy. To be sure, we have only vague hints for a developed stage of the division of labor and specialization, as well as for the existence of a permanent ruling group, but all points to the existence of those phenomena.

With this stage we have reached the earliest range supplying us with actual information for our purposes. As for still earlier stages we never find enough parts of the entire assemblage. If nothing else, this excursion into the Khuzestan area at least has supported the idea that the Late Uruk situation in Babylonia by no means was the result of a sudden development, and certainly was not the first phase to show those features which judging from the later developments we took as indicators for urban centers.

The attempt to trace back in time the various features thus has not had striking results, since nowhere did we come across such breaks in the development which would signalize a deep qualitative change. As I mentioned, for still earlier periods the evidence is too scarce; on the other hand, our earlier deliberations suggest that some of the phenomena were present already at an earlier stage, like the existence of a permanent ruling group or a high degree of the division of labor, since these features probably should not be separated from the formation of settlement systems.

Does this mean that the process of urbanization was a totally continuous one? Yes and no. If we consider only the archæological evidence the answer must be affirmative because obviously in our record there is nothing to suggest such a change in quality as to warrant the designation of an "Urban Revolution". If there ever was such a moment then it must have been before the periods for which we have sufficient archæological evidence. The answer may be different if in addition to the features listed above for the Early Dynastic III cities we introduce a phenomenon which, however, leads us into the field of speculation: I mean the aspect of an urban consciousness. Obviously, there is hardly any way imaginable how to use archæological evidence in this respect, and even using textual evidence will not help too much. So we must take refuge to some general considerations.

At another occasion (Nissen 1980) I have assembled hints for the presence of a fixed group-consciousness during the Early Dynastic period, used to set oneself apart from other groups. In particular, this can be shown for the dichotomy between settled and non-fully-settled but also for the differences between town and country (Nissen 1980). Most probably, one line of development of such group-consciousness is tied to the development of central places; that would mean that also this feature originated quite early. However, I have the feeling that a radicalization can be tied to a specific historic situation: the rapid changes in settlement patterns during the time from Late Uruk through Early Dynastic III, or from 3300 to 2400 BC, with the trend of decreasing the number but increasing the size of settlements. I mentioned already the final outcome: by the Early Dynastic III period about 80% of the settled area is taken up by large centers while there are only few small settlements left. This quite obviously means a polarization between towns-people and inhabitants of the countryside whether they lived in small settlements or in semi-permanent settlements.

This process of changing settlement configurations can be followed up through the Early Dynastic period with the tendency becoming visible already during the preceding period. However, it would be difficult to reach at the conclusion mentioned before if we had not some poor support from textual evidence. In particular I am referring to the literary genre of dialogues of which the ones between Dumuzi and Enkimdu (i. e. between farmer and shepherd) or between barley and sheep (Kramer 1961) are the most interesting ones in our context, as they display a certain arrogance from the side of the town people. Even the errection of city walls, what in Babylonia starts in Early Dynastic times to my mind should be placed in this context. As much as they were meant for protective purposes against attacks from other cities they became an absolute demarcation line between the inhabitants of the city and everyone living outside. I am inclined to take the errection of city walls not only as the final answer after too many attacks but also as a deliberate move towards the building up of a group's identity. Whereas in spite of all differences in earlier times an interchange between town and countryfolk was possible this new development resulted in an irreversible situation.

Returning to the term "Urban Revolution" and considering the archæological evidence, we find none of the essential criteria appearing that rapidly that the term "revolution" would be justified, except if we consider the development of an "urban consciousness". Forgetting for a moment all the squabble about the choice and validity of single indicators for identifying a settlement as urban the general direction of the concept of the "Urban Revolution" was anyway more on the urban consciousness than on single traits. Certainly much of the material used by Childe is outdated by now and of course some of his conclusions cannot be accepted anymore. But the question of proving or disproving the concept cannot be tied to the question whether his datings still hold good. In fact, in my mind, the term still is applicable if talking about the emergence of an urban consciousness, and it is this aspect a consolidation of which in fact justifies that one starts a new phase within the development of settled life.

REFERENCES

ADAMS, R.M.

1981. *The Heartland of Cities*, Chicago.

ADAMS, R.M., and H.J. NISSEN

1972. *The Uruk Countryside*, Chicago.

CHILDE, V.G.

1950. "The Urban Revolution", *Town Planning Review* 21.

1952. *New Light on the Most Ancient East*, 4th Edition, London.

DIAKONOFF, I.M.

1974. *Structure of Society and State in Early Dynastic Sumer*, Monographs of the History of the Ancient Near East 1,3, Malibu.

FALKENSTEIN, A.

1974. *The Sumerian Temple-City*, Monographs of the History of the Ancient Near East 1,1, Malibu.

FINKBEINER, U., and W. RÖLLIG (eds)

1986. *Jamdet Nasr: Period or Regional Style?*, Wiesbaden.

GELB, I.J.

1969. "On the Alleged Temple and State Economies in Ancient Mesopotamia", *Studi in Onore de E. Volterra*, Rome.

GREEN, M.W.

1980. "Animal Husbandry at Uruk in the Archaic Period", *Journal of Near Eastern Studies* 39.

JOHNSON, G.A.

1977. "Aspects of Regional Analysis in Archæology", *Annual Review of Anthropology* 6.

1982. "Organizational Structure and Scalar Stress", in C. Renfrew *et al.* (eds), *Theory and Explanation in Archæology*, London.

KRAMER, S.N.

1961. *Sumerian Mythology*, New York.

NISSEN, H.J.

1974. "Zur Frage der Arbeitsorganisation in Babylonien während der Späturuk-Zeit", *Acta Ant. Hung.* 22.

1980. "The Mobility Between Settled and Non-Settled in Early Babylonia: Theory and Evidence", in: T. Barrelet (ed), *L'archéologie de l'Iraq du début de l'époque Néolithique á 333 avant notre ére*, Colloques du CNRS no. 580, Paris.

1982. "Die 'Tempelstadt': Regierungsform der früdynastischen Zeit in Babylonien? (The 'Temple City': form of organization of the Early Dynastic period in Babylonia?)", in H. Klengel (ed), *Gesellschaft und Kultur im alten Vorderasien*, Schriften zur Geschichte und Kultur des Alten Orients 15.

1983a. "Political Organization and Settled Zone", in Young, Smith and Mortensen 1983.

1983b. *Grundzüge einer Geschichte der Frühzeit des Vorderen Orients*, Darmstadt.

1985. "The Emergence of Writing in the Near East", *Interdisciplinary Science Reviews* 10.

1986. "The Archaic Texts from Uruk", *World Archæology* vol. 17, 3.

1987. *Outline of a History of the Early Periods of the Near East*, Chicago.

STROMMENGER, E.

1980. *Habuba Kabira, eine Stadt vor 5000 Jahren*, Mainz.

WRIGHT, H.T., and G.A. JOHNSON

1975. "Population, Exchange and Early State Formation in Southwestern Iran", *American Anthropologist* 77.

YOUNG, T.C., P.E.L. SMITH, and P. MORTENSEN (eds)

1983. *The Hilly Flanks and Beyond. Essays on the Prehistory of Southwestern Asia*, presented to R.J. Braidwood, Studies in Ancient Oriental Civilization no. 36, Chicago.

URBANIZATION IN PERIMESOPOTAMIAN AREAS: THE CASE OF EASTERN ANATOLIA

MARCELLA FRANGIPANE and ALBA PALMIERI
Rome University

The propagation model of Near Eastern urban civilization outlined by Childe, being based on the internal expansive characters of the new economy, stresses on the need for such a diffusion. "In fact not only the products of urban industry, but the new economy that produced them spread and were bound to spread. To persuade their possessors to exchange the needed raw materials for manufactures, they must be induced not only to demand the latter, but also to adjust their economy to absorb them" (Childe 1942: 140).

In Childe's perspective, this model, although referred to specific regions, held a general value and concerned a vast chronological and geographical scale, since it intended to account for the eastern influence on the entire development of European prehistory. The mode of diffusion appeared as a corollary of the very nature of urban civilization, the advent of which was seen as one of the revolutions which marked the overall trend of historical development.

Both definitions of "urban revolution" and "neolithic revolution" are still basic, on a general level, particularly in that they recognize the fundamental importance of the development of productive forces as well as of the social relationships of production, therefore qualifying a conceptual and methodological perspective.

Childe's diffusion model has been variously reconsidered, either through re-adaptations to special situations, or through criticisms based on different points of view.

An approach which stresses the relevance of local processes instead of the diffusion of "influences", has led, also on the basis of new data, to a different interpretation of the western Mediterranean cultural developments, focused on their being independent of the urban civilizations of the Near East (Renfrew 1970); on the other hand, the weight of Near Eastern relations on the Aegean urban developments still appears to be a very important element to be properly assessed.

It has recently been underlined that the choice of the geographic scale in the study of innovation processes is very closely related to the type of explicative model adopted (Renfrew 1986). With respect to Childe's model, the new data at our disposal allow us to consider the diffusion of the southern Mesopotamian urban civilization in a greater detail both chronological and geographical. The interpretative models can therefore be more circumstantiated and concern the specific mechanisms in turn involved. Within the early Mesopotamian urbanization, in fact, one can presently recognize phases with different characteristics as regards both the local organizations (Johnson 1973; Adams 1981) and the interregional exchange systems (Palmieri 1985b).

At the first full establishment of the urban processes, in the Late Uruk period, remarkable cultural connections join together the alluvial plain with some peri–Mesopotamian regions, so that these must be included in the general analysis of the urbanization phenomenon. In this paper we will consider the explanatory possibilities offered by the phenomenon of the rise, in northern peri–Mesopotamian areas along the Syro–Anatolian Euphrates, of such cultural aspects strongly connected with the south. This phenomenon mainly concerns the Late Uruk period, even if it is difficult to find in the north the exact equivalent of the chronological and cultural divisions in the south. The chronological limits of the Late Uruk horizon in Syria and eastern Anatolia might include at least part of the time span of Jemdet Nasr in the south, an aspect which is not recognizable in northern areas by the presence of peculiar traits. In fact, the horizon characterized in eastern Anatolia by Early Reserved Slip and Uruk-like wares is followed here by a Late Reserved Slip ware horizon, which appears to be roughly contemporary with the archaic phases of the Early Dynastic, as it is indicated, for example, by the stylistic connections shown by the glyptic of Hassek Hüyük (Behm-Blancke 1981). Moreover, within the horizon with Uruk-like wares, the northern sites up to now investigated might also have been spread out over a certain period of time. The radiocarbon datings show that, apart from a single date for Habuba Kabira[1] which appears to be fairly early, there is in general a good correspondance among the dates from other sites repesenting the southern expansion in central-western Iran (Godin)[2], in the Middle Euphrates (Jebel Aruda)[3] and in Anatolia (Hassek Hüyük[4] and Arslantepe[5]), which altogether fall between 3400 and 3000 BC, using MASCA correction[6]. It can be noticed that the Arslantepe dates overlap only in part with those from Jebel Aruda and Hassek, also covering the final part of this period, between 3200 and 3000 BC.

Here we shall deal globally with the Late Uruk horizon of the northern Syro–Anatolian areas, with the purpose of making some suggestions about the possible mode of diffusion from the southern Mesopotamian regions. A strong cultural affinity relates areas which were previously heterogeneous and which subsequently reconfirm their heterogeneity. This should indicate not a pre-Uruk absence and post-Uruk discontinuance of trade, but probably a change in the form and intensity of trade relations.

Within the universe of ideas which was given scientific form by Childe, hints for reflections range from the field of the general interpretative reconstructions to that of the definitions of conceptual parameters. The basic Childean concept of archæological culture is referred to here. In fact, focusing on the wide geographical extension of the Uruk "culture", the identification of the cultural diffusion in terms of either colonization or acculturation is seen as fundamental.

The debate which arose around Childe's concept of archæological culture (Childe 1936, 1956: 34, 111–124) led to link more closely archæological studies with the development of the general theory of culture. Attempts have been made to delve into the taxonomic and interpretative problems of this concept

---

1 5085 ± 65 BP (Oates 1983: 272).

2 4474 ± 100 BP (Young 1969: 48, note 21).

3 Three dates between 4490 and 4410 BP (Oates 1983: 272).

4 Three dates between 4470 and 4390 BP (Behm-Blancke 1985: 93–94).

5 Sixteen dates ranging from 4460 to 4240 BP (Alessio *et al.* in press).

6 A good correspondence is also visible with two dates from Sharafabad (4380-4331 BP) in the Susiana plain (Bovington *et al.* 1973).

(Clarke 1968; Klejn 1982), to question certain aspects such as the normative approach and the direct correlation between culture and people or society (Binford 1965; Binford, Sabloff 1982); some authors have even chosen to reject the utilization of "culture" as a useful concept in archæology in order to better show up the social nature of the problems involved (Renfrew 1978, 1982).

In fact the interpretation of the patterns of formal similarities as they emerge in different situations from the archæological data, rather than being based on general equivalences with an overall validity, should be founded case by case on the investigation of the identifiable structural elements, such as the social organization of activities. Even the spatial delimitation of the cultural *facies*, often indistinct, sometimes clear, poses specific interpretative problems which go beyond the question of their recognition and precise definition.

The different degree of formal similarities recognizable in the Late Uruk pottery of the Syrian–Anatolian areas must be investigated paying attention to the global characterization of the individual sites and to their possible function in the network of interregional relations in the period concerned.

Habuba Kabira/Tell Kannas is a settlement which well represents a Late Uruk presence on the Middle Euphrates, together with Jebel Aruda and Tell Hadidi (Sürenhagen 1974–75; Strommenger 1980; Finet 1979; Van Driel, Van Driel-Murray 1979, 1983). Here the so-called "colonial" settlement pattern is characterized by a kind of urban centres which are peculiar both in their internal structure and in their relationship with the cultural environment in which they are found. Their appearance as uniformly planned settlements is quite sudden on sites which have not revealed immediate pre-Uruk occupation, as is also sudden their abandonment at the end of the period. Their temple areas include structures with the same architectural characteristics of the Mesopotamian buildings (fig. 1a).

Going upstream along the Euphrates into eastern Anatolia, other centres appear to be connected with these. Some of them are of remarkable proportions —Samsat and Lidar, south of the Taurus (Özdogan 1977)— others are smaller such as Hassek Hüyük (Behm-Blancke 1981, 1984, 1985), again south of the Taurus, and Tepecik (Esin 1982, a, b), to the north. The excavation of the small mound of Hassek has uncovered a walled settlement, also without immediate precedents, which seems to be largely made up, besides a "temple", by a big building whose plan is a faithful reproduction of the model known at Habuba Kabira and Jebel Aruda (fig. 2).

Arslantepe, in the Malatya plain north of the Taurus, shows different characteristics. Here the excavation of the levels belonging to the end of the fourth millennium (period VIA) has until now unearthed exclusively buildings of a public character following one another (Palmieri 1973, 1981, 1985a, c; Frangipane and Palmieri in press). The area destined to the central institutions seems already to be quite extensive. The biggest sector is taken up by a structure which could be considered a palace–like building (building IV) as it is furnished with a storeroom complex that does not seem to be directly connected to cult activities. In other two buildings the religious and ceremonial function appears prevalent. One in particular (building I) is recognizable as a temple (fig. 1b).

The Late Uruk pottery production shows certain general unifying characteristics, such as the manufacturing techniques with the total or partial use of the potter's wheel and the diffusion of reserved-slip decoration. Generally speaking this pottery, which includes mass–produced categories of vessels, appears to have been made by skilled specialized potters and probably derives from a centralized activity. The repertoire of pottery shapes is particularly rich and homogeneous in the south, whereas in the Syro–Anatolian areas it shows a variable range of types and different degrees of affinity with the southern sites. This is also one of the reasons why some of the settlements in the north have been interpreted as "colonies" (Nissen 1974) and others have appeared to be the result of an acculturation process (Palmieri 1981, 1985 b; Palmieri and Frangipane 1986).

A number of particular aspects could be investigated in this field (Howard and Morris 1980), but the presently available data are far from sufficient. It was therefore decided to use pottery shapes to measure the reciprocal affinity of northern sites, selected on the basis of the available documentation, as well as their relations with southern Mesopotamia taken as a whole.

The Habuba Kabira pottery production appears to be linked to that of southern Mesopotamia and Susa by an extremely high percentage of common profile types (fig. 7a). The southern vessel categories are all represented in the Middle Euphrates outpost and reproduce almost the entire typological range recognizable in the south (figs. 3, 4). There are only a few shapes peculiar to Habuba Kabira, and some of these are probably linked to the local Syro–Palestinian environment (Sürenhagen 1974/75: tab. 19, 161).

Moving on to the sites on the Turkish Euphrates, the closest connections with Habuba Kabira are shown by Hassek Hüyük, where the typological repertoire includes for the most part pottery shapes found both at Habuba and in the southern area (figs. 7b, 4). In this scale of southern affinities Tepecik is a close second (fig. 8a), while Arslantepe differs for the autonomy that its pottery production shows with respect both to the southern areas and especially to Habuba Kabira —very low number of actually similar forms— (fig. 8b), though fully fitting in with the general Late Uruk characterization. The intensity of southern relations does not therefore seem to depend on the geographical location of the sites, as Tepecik is the northermost.

The three Turkish sites, besides their different connections with Habuba Kabira and with the south, show significant reciprocal relations consisting also in the sharing of some classes of black burnished pottery with central Anatolian similarities (however exceptional at Hassek) and some kitchen pots and *pithoi* (fig. 6, 9a). Another noteworthy link is represented by the characteristic wheel–made reserved-slip jars with expanded shoulder they have in common (fig. 5: *10, 11, 12*). On the whole the typological repertoire of all three sites is simplified, less rich in the number of categories and profile types than that of both the southern regions and the Middle Euphrates Late Uruk settlements (fig. 9b).

Despite the substantial similarity of the three Upper Euphrates sites, the links between Tepecik and Hassek seem stronger than those shown by these two sites with Arslantepe, essentially due to their greater affinity with Habuba Kabira and with the south.

To the large outpost of Habuba Kabira we may therefore much more closely relate the two sites of more limited dimensions, while Arslantepe, which seems to have been of larger size, is remarkably autonomous. From a chronological point of view we have seen that on the basis of radiocarbon dating Arslantepe could cover the final phase of the Late Uruk Syro–Anatolian development, partially overlapping with Jebel Aruda and Hassek. Anyhow the site shows very sound typological connections with Hassek (e. g. virtually identical reserved-slip jars), while among the material which is considered imported for the unusual characteristics of the paste there is a spouted bottle with a reserved-slip decoration highly reminiscent of some Habuba examples (fig. 3: *10*). The remarkable autonomy shown by the Arslantepe pottery production hence does not seem to be accounted for by an eventual partial chronological displacement.

Even in the rich series of seals reconstructed at Arslantepe from over two thousand clay–sealings, both iconography and style indicate prevalently autonomous elaborations, though they are somehow related to the Syrian and northern Mesopotamian traditions (Frangipane and Palmieri in press). There is in any case a clear difference from the Mesopotamian-influenced glyptics of Habuba Kabira and Jebel Aruda (Töpperwein 1973; van Driel 1983). A marked influence of the classic southern Late Uruk culture, however, is recognizable in the use of symbols of particular strenght and diffusion, and especially in pictures regarding the central authority and the activities of the burocratic élite. Animals with intertwined tails and representations of porters can be mentioned. The most interesting case is the scene of the transporting of a regal figure on a sledge-car depicted on a cylinder-seal of local style, which reproduces even details of the design engraved on a southern plaque (Palmieri 1981: Pl.XVa; Sürenhagen 1985: figs. 1, 2). On the plaque the transported figure represents the iconographic type of the king well-known at Uruk.

From an organizational point of view it can be recognized at Arslantepe the operation of a centralized redistributive system clearly inspired by the Mesopotamian model (Palmieri and Frangipane 1986). The organization of a storehouse system with a complex administrative control on the circulation of goods is largely witnessed: besides a number of clay-sealings found *in situ* in one of the storerooms of the palace-like building (Palmieri 1985c), a great concentration of sealings was found in particular dumping places located inside the two main public buildings of the period. In the stratified deposits of the dump in the "palace" three main stratigraphic phases could be identified, each characterized by mostly different seals and by prevailing concentrations of certain kinds of containers, as they are shown by the clay-sealings (Ferioli and Fiandra, in press; Frangipane and Palmieri in press). It seems that the various groups of rubbish layers may refer to diversified series of administrative operations, possibly relating to the functioning of warehouses of different types.

As in other Syrian and Mesopotamian contexts, a mass-production of bowls, all wheel-turned in the case of Arslantepe, is associated to the evidence of centralized storing and administration. As it has been hypothized for the Mesopotamian bevel-rim bowls (Nissen 1970: 137-38; Johnson 1973: 129-39; contra Beale 1978), the concentration of their capacities around volume groups, even though these are quite wide (Espinosa and Manzanilla 1985; Frangipane and Palmieri in press), suggests the possibility they were used for the distribution of food rations to the employees of the central institution, probably not as actual measuring-vessels, but simply as ration-containers.

The redistribution system seems to have been based, at Arslantepe, on an agricultural production of a certain complexity, including also vine, besides barley and wheat. It is to be stressed, in this respect, the central position of the site in the Malatya plain, in an area full of springs, while the smaller sites are generally situated along the bank of the Euphrates (Marcolongo and Palmieri in press).

In the Late Uruk period it can be observed at Arslantepe also a marked change in the general pattern of animal breeding with respect to the preceding local Late Chalcolithic. The most outstanding trait of such a change is represented by the remarkable increase in the number of sheep and by the contemporaneous marked reduction of pigs, the latter being characteristic of a domestic rearing (Bökönyi in press). This pattern seems to be tied up with the advent of a centrally-governed economy, inspired by the Mesopotamian model, in which redistribution elements such as milk and especially wool must have played an important role[7].

---

[7] There is a good evidence in the Uruk archaic tablets both of the great weight of sheep-rearing within the central administration (Green 1980) and of the important role held by the distribution of wool and textiles (Nissen 1985).

Unlike Arslantepe, the smaller site of Hassek H. shows pigs as the most common animal reared (Boessneck and von den Driesch 1981). Besides characterizing the local dwellers as fully sedentary agriculturalists, such a prevalence indicates a specialized breeding which was possibly based on the external exchange even of elements of primary production. It is probable on the contrary that the strong incidence of sheep at Arslantepe may reflect the central–place character of this site.

Considering on the one hand the specific characteristics of the organizational structures of the sites under examination, and on the other hand formal aspects, such as in particular those of the pottery production considered above, we see that Arslantepe, central-place with a Mesopotamian–inspired organization, has very few formal affinities with the southern areas, while smaller sites with less evident proofs of a marked centralization, like Hassek H. and Tepecik, show strong southern links as regards formal similarities. We may thus note the association of greater structural affinities with fewer formal similarities, and vice-versa.

The "colonial" configuration of the two smaller sites appears stronger, though they do not completely lack local elements. On the other hand it must be pointed out that the pottery production of Arslantepe, of which we stressed the local typological nature, owes little to the preceding Late Chalcolithic tradition. This capacity for an autonomous reorganization of the pottery typological repertoire is associated possibly not by chance with the strong centralization structure acquired, as it seems, under external stimuli by the local society. The new specific needs brought into play by the changed conditions must have been reflected in the manufacture of new functional categories as well as in the working up of distinguishing stylistic features, largely limiting the adoption of the formal southern repertoire.

Reconsidering the general Childean model of diffusion of urban civilization in the light of the Late Uruk expansive phenomenon, the assumption concerning the necessity for Mesopotamian centres to obtain raw materials seems well supported. In this same period the obvious importance of acquiring metal in Mesopotamian societies is attested to by the diffuse and diversified use they made of copper as well as of other metals (Adams 1981: 80; Nissen 1985: 358; Frangipane 1985; Moorey 1982). Concerning the ways of procurement, it has to be considered the possibility of conditioning from a distance non-urbanized or limitedly urban–oriented groups, obtaining from them adequate answers to the stimuli. The acquisition of fundamental raw materials seems to have taken place in this period through a cultural assimilation of the areas of interest, probably realized through colonization and subsequent acculturation.

Childe's model appears to be valid in general terms as far as the hypothized acculturation process is concerned, if in sites such as Arslantepe one may recognize the emergence of a local élite or, as it seems less probable, the strong local conditioning of an alien élite. Effective interactions must anyway have taken place within geographically limited areas, as is suggested by the probable mediatory function of colonies.

It would seem that the Late Uruk interregional exchange system had the character of an intermediate trade (Renfrew 1975) resting upon a network of "central places" with cultural assimilation playing an important role.

The fall of such a system in the subsequent period, in connection with the economic–organizational growth of the Early Dynastic communities in Mesopotamia, might be due to the rise of new ways and forms of exchange, resulting in the establishing of a different kind of relationship with the northern peri–Mesopotamian regions together with the opening of new routes towards far eastern areas (Palmieri 1985b).

The cultural assimilation of the regions involved in the exchange network during the formative Late Uruk phase might have been stimulated just by the difficulty for the Mesopotamian centres in managing direct forms of raw material procurement from a long distance, as they probably carried out in the successive Early Dynastic period.

List of the Pottery Types Here Considered

Type 1: Habuba Kabira (Sürenhagen 1974/75, Pl. 20, *10-11*) (*fig. 4, 3*); Tepecik (Esin 1982, Pl. III, *6-7*).

2a: Habuba K. (1974, Pl. 20, *39*); Hassek (Hoh 1981, fig. 11, 7) (*fig. 4, 7*); Tepecik (1982, Pl. III, *15*); Warka (Nissen 1970, 39/*27*).

2b: Hassek (1981, fig. 11, *8*; 1984, fig. 11, *4-6*); Tepecik (1982, Pl. III, *19*); Habuba K. (1974, Pl. 22, *85*); Warka (1970, 39/*36, 114*); Susa 17 (Le Brun 1971, fig. 45, *11*).

3a: Habuba K. (1974, pl. 1, *18*; 20, *30-32*); Tepecik (1982, Pl. III, *20*); Warka (1970, 39/*22*).

3b: Habuba K. (1974, Pl. 1, *17*; PL. 20, *33*); Hassek (1984, fig. 11, *2*) (*fig. 4, 4*); Warka (Haller 1932: lev. VI, IV; 1970, 38/*9*); Susa (1976, fig. 4, *1-3*).

4a: Hassek (1981, fig. 11, *5*); Warka (1970, 34/*2*).

4b: Tepecik (1982, Pl. VII, *1*); Habuba K. (1974, Pl. 21, *52*).

5: Hassek (1984, fig. 10, *16, 18*; fig. 11, *3, 5*) (*fig. 4, 2*); Tepecik (1982, Pl. VII, *2*); Habuba K. (1974, Pl. 22. *78*).

6: Tepecik (1982, Pl. III, *16*); Warka (1970, 40/*11*); Ur (Woolley 1956, JN 1).

7: Habuba K. (1974, Pl. 2, *30-31*) (*fig. 4, 10*); Tepecik (1982, Pl. III, *13-14*); Warka (1970, 39/*18*); Susa 17B (1978, fig. 19, *23*; 23, *1*); Ur (Woolley 1956, JN 11).

8: Habuba K. (1974, Pl. 22, *67*); Hassek (1984, fig. 10, *17*); Susa 17 (Le Brun 1971, fig. 45, *5*; 1978, fig. 19, *19*).

9a: Habuba K. (1974, Pl. 3, *34*; Pl. 20, *19-20*); Warka (Haller 1932: lev. VI; Adams, Nissen 1972, fig. 56, *1*); Susa (1976, fig. 4, *9*; 1957, fig. 11, *4*); JN (Mackay 1931, 21F).

9b: Habuba K. (1974, Pl. 3, *33*); Warka (1970, 39/*25*); Nippur (Hansen 1965, fig. 32).

10: Habuba K. (1974, Pl. 20, *16*); JN (1931, Pl LXVI, *29*); Susa (1976, fig. 4, *12*).

11: Habuba K. (1974, Pl. 20, *15*) (*fig. 3, 3)*; Warka (1932: lev. VI); Susa (1976, fig. 4, *10*).

12: Habuba K. (1974, Pl. 20, *18*); Warka (1932: lev. VI); Susa (1976, fig. 4, *11*).

13: Habuba K. (1974, Pl. 1, 12); Warka (Lenzen 1958, Pl. 51 c).

14: Habuba K. (1974, Pl. 3, *32*); JN (1931, Pl. LXVII, *20*).

15: Habuba K. (pl. 1, *13*) (*fig. 3,2*); Susa 17 (1971, fig. 46, *17*).

16: Arslantepe (Frangipane and Palmieri in press, fig. 30, *3*) (fig. 5, *1*); Warka (Strommenger 1963, Pl. 39 s).

17: Habuba K. (1974, Pl. 2, *29*); Warka (1932: lev. V; 1970, 39/*29*; 1972, fig. 71, *4*); Susa 17B (1978 fig. 23, *17*).

18: Habuba K., Hassek (1981, fig. 8, *9*) (*fig. 4, 5*), Tepecik, Warka, Susa, Nippur, Tello, Jemdet Nasr, Ur.

19: Habuba K., Hassek, Arslantepe (in press, fig. 31) (*fig. 4, 1*) Warka, Susa, Nippur, Tello, Jemdet Nasr.

20: Habuba K. (1974, Pl. 2, 24); Warka (1932: lev. IV; Lenzen 1965, Pl. 26, p-s); Nippur (1965, fig. 34).

21: Habuba K. (1974, Pl. 1, *10*).

22: Habuba K. (1974, Pl. 21, *65*).

23: Habuba K. (1974, Pl. 21, *59*).

24: Arslantepe (in press, fig. 33, *8*) (*fig. 5, 9*); Warka (1970, 21/*4*); Susa 17 (1971, fig. 46, *13*; 1978, fig. 23, *12*).

25: Habuba K. (1974, Pl. 25, *50*; Pl. 27, *119*); Hassek (1984, fig. 12, *6*) (*fig. 4, 8*); Warka (1970, 37/*5*; 1972, fig. 64, *4*); Susa (1978, fig. 24, *18*).

26: Habuba K. (1974, Pl. 1, *1*).

27: Arslantepe (in press, fig. 28, *7, 9*).

28: Habuba K. (1974, Pl. 23, *8-9*); Warka (1932: lev. VI).

29: Habuba K. (1974, Pl. 23, *13*); Warka (1932: lev. VI; 1970, 39/*10*; 1972, fig. 42, *25*).

30: Habuba K. (1974, Pl. 3, *47*) (*fig. 3, 6*); Warka (1972, fig. 47, *3*); Susa 17 (1971, fig. 46, *6*); Tello (de Genouillac 1934, Pl. 22, *3*).

31: Habuba K. (1974, Pl. 3, *38, 43*; Pl. 22, *2*); Warka (1963, Pl. 38 i; 1970, 42/*1*; 1972, fig. 42, *25-26* fig. 61, *8*); JN (1931, Pl. LXVI, *27*); Susa 17 (1971, fig. 46, *5*).

32: Habuba K. (1974, Pl. 23, *11*) (*fig. 3, 5*); Warka (1932: lev. VI; 1970, 39/*15*, 31/*6*); Susa 17 (1976, fig. 3, *14*; 1978, fig. 23, *10*).

33: Habuba K. (1974, Pl. 3, *49*); Warka (1970, 39/112); Susa 1976, fig. 3, *8*).

34: Habuba K. (1974, Pl. 19, *152*).

35: Habuba K. (1974, Pl. 19, *151*); Warka (1963, Pl. 38 h).

36: Habuba K. (1974, Pl. 19, *154*).

37: Habuba K. (1974, Pl. 19, *153*).

38: Habuba K. (1974, Pl. 5, *58*).

39: Habuba K. (1974, Pl. 5, *56*); Warka (1932: lev. VI; 1970, 32/*26*); Tello (1934, Pl. VI, *4936*).

40: Habuba K. (1974, Pl. 18, *138-143*); Tello (1934, Pl. V, *4823*).

41: Habuba K. (1974, Pl. 18, *108*).

42: Habuba K. (1974, Pl. 18, *110*).

43: Habuba K. (1974, Pl. 18, *113*). (fig. 3, 4); Warka (1972, fig. 46, *3*); JN (1931, Pl. LXV, *8*).

44: Habuba K. (1974, Pl. 18, *117*).

45: Habuba H. (1974, Pl. 18, *119*).

46: Habuba K. (1974, Pl. 18, *120*).

47: Habuba K. (1974, Pl. 15, *89*); Hassek (1981, fig. 21, *3*) (*fig. 5, 4*); Tepecik (1982, Pl. VII, *5* [150]).

48: Tepecik (1982, Pl. VII, *5* [290]).

49: Hassek (1981, fig. 22, *3*); Tepecik (1982, Pl. VII, *5* [124]).

50: Habuba K. (1974, Pl. 18, *135*); Hassek (1981, fig. 22, *2*) (*fig. 5, 3*).

51: Hassek (1984, fig. 12, *3*).

52: Arslantepe (in press, fig. 28, *1-5*) (*fig. 5, 2*); Tepecik (1982, Pl. V, *13*).

53: Habuba Kabira (1974, Pl. 18, *114*); Susa (1971, fig. 50 *9*).

54: Habuba K. (1974, Pl. 18, *136*).

55: Habuba K. (1974, Pl. 18, *147*); Warka (1972, fig. 43, *39*).

56: Habuba K. (1974, Pl. 18, *124*); Hassek (1981, fig. 23, 1-2); Arslantepe (in press, fig. 33, 4) (*fig. 4, 6*); Tepecik (1982, Pl. VII, *4*); Nippur (1965, fig. *11a*); JN (1931, Pl. LXIV, *18-20*); Susa Cb (1957, fig. 11, *24*).

57: Habuba K. (1974, Pl. 18, *122*); Warka (1932: lev. VI); Tello (1934, Pl. 25, *2*); JN (1931, Pl. LXIV, *11*).

58: Habuba K. (1974, Pl. 7, *67*) (*fig. 3, 9*); Tello (1934, Pl. 21, *2*, Pl. V, 4865); JN (pl. LXIV, *12*); Susa (1957, fig. 11, *20-21*; 1971, fig. 51, *1-2*; 1978, fig. 24, *4*).

59: Habuba K. (1974, Pl. 18, *127-130*) (*fig. 3, 1*); Susa (1957, fig. 11, *30*; Steve, Gasche 1971: m. 20.20).

60: Habuba K. (1974, Pl. 18, *133*); Tello (1934, Pl. VI, *4936*); Susa Cb (1957, fig. 11, 28).

61: Habuba K. (1974, Pl. 16, *96*); Susa C (1957, fig. 11, *31*).

62: Habuba K. (1974, Pl. 18, *132, 134*); Tello (1934, Pl. VI, *5272*); Susa (1976, fig. 8, *6*).

63: Habuba K. (1974, Pl. 11, *74*); Hassek (1984, fig. 12, *4*) (*fig. 4, 9*); JN (1931, Pl. LXVI, *37*); Susa Cb (1957, fig. 11, *35*).

64: Habuba K. (1974, Pl. 7, *66*); Warka (1932: lev. IV); Susa 17 (1971, fig. 51, *9*; 1978, fig. 33, 7).

65: Habuba K. (1974, Pl. 7, *64*) (*fig. 3, 12*); Warka (1932: lev. IV; 1972, fig. 54, *11*); Nippur (1965, fig. 21); Susa Cb,c (1957, fig. 12, *9*).

66: Habuba K. (1974, Pl. 8, *68*); Susa (1957, fig. 11, *36*; 1971, fig. 51, *8, 11*; 1978, fig. 28, *1*).

67: Habuba K. (1974, Pl. 15, *90*); Tepecik (1982, Pl. IX, 7 [340]) (*fig. 4, 13*); Ur (Woolley 1955, JN 139); Susa (1976, fig. 7, 7).

68: Habuba K. (1974, Pl. 12, *76*), (*fig. 3, 8*); Nippur (1965, fig. 22); Susa 17 (1971, fig. 52, *7-8*).

69: Habuba K. (1974, Pl. 12, *75*; 9, *69*) (*fig. 3, 7*); Ur (1955: pit F, m. 9. 30); Susa 17 (1978, fig. 32, *1*).

70: Arslantepe (unpublished).

71: Arslantepe (*fig. 5, 6*); Warka (1963, Pl. 37 d).

72: Habuba K. (1974, Pl. 4, *54*); Warka (1972, fig. 59, *31*).

73: Habuba K. (1974, Pl. 4, *53*); Warka (1961, Pl. 17 a; 1932: lev. VI–V).

74: Habuba K. (1974, PL. 5, *60*); Warka (1970, 34/*14*); Nippur (1965, fig. 27); Susa (Stève 1971).

75: Habuba K. (1974, Pl. 19, *146*); JN (1931, Pl. LXIII, *10*).

76: Hassek (1984, fig. 14, *4*); Arslantepe (*fig. 5, 8*) (in press, fig. 33, 7); Warka (1970, 39/*98*); Nippur (1965, fig. 28); Susa Ca (1957, fig. 14, *2*).

77: Habuba K. (1974, Pl. 17, *101-103*), (*fig. 4, 12*); Hassek (1981, fig. 23, 7); Warka (Lenzen 1959, Pl. 21, f; 1961, Pl. 24, a-b; 1970, 37/*18*; 1972, fig. 59, *34*); Susa (1957, fig. 12, *5-7*; 1971, fig. 52, *5*; 1978, fig. 30, 13).

78: Habuba K. (1974, Pl. 17, *100*) (*fig. 3, 10*); Warka (1932: lev. VI–V); Nippur (1965, fig. 17); Arslantepe (probably imported, unpublished).

79: Tepecik (1982, Pl. IX, 7 [138]); Warka (1963, Pl. 37 f).

80a: Habuba K. (1974, Pl. 15, *91*; 16, *93*); Hassek (1984, fig. 14, *5*) (*fig. 4, 11*); Warka (1958, Pl. 49d; 50a); Nippur (1965, fig. 26); Susa (1957, fig. 11, *42*; 1978, fig. 29, *4*).

80b: Habuba K. (1974, Pl. 16, *92, 94*); Tello (1934, Pl. VI 4938).

81: Habuba K. (1974, Pl. 16, *95*); Tepecik (1982, Pl. XI, *10*); Nippur (1965, fig. 29); Susa 17 (1978, fig. 29, *5*).

82: Habuba K. (1974, Pl. 16, *98*); Nippur (1965, fig. 19).

83: Habuba K. (1974, Pl. 15, *88*); Warka (1932: lev. VI); Susa Ca (1957, fig. 11, *40*).

84: Habuba K. (1974, Pl. 13, *85*) (*fig. 3, 11*); Warka (1958, Pl. 48 d).

85a: Habuba K. (1974, Pl. 14, *83*); Warka (1958, Pl. 48 a–b, e; 1970, 39/*119*).

85b: Hassek (1984, fig. 15, *2*).

86: Habuba K. (1974, Pl. 14, *84*); Warka (1932, lev. VI).

87: Habuba K. (1974, Pl. 12, *78*); Warka (1965, Pl. 26b).

88: Habuba K. (1974, Pl. 4, *52*); Hassek (1981, fig. 18, *5*); Susa 17 (1971, fig. 50, *6*).

89: Habuba K. (1974, Pl. 13, *81*); Hassek (1981, fig. 22, *6*) (*fig. 5, 7*).

90: Habuba K. (1974, Pl. 5, *61*); Warka (1972, fig. 52, *12*).

91: Arslantepe (in press, fig. 26, *1*) (*fig. 5, 11*); Hassek (1981, fig. 18, *9*; 20, *4*).

92: Arslantepe (in press, fig. 26, *2*); Hassek (1981, fig. 19, *5*).

93: Arslantepe (in press, fig. 26, *4*) (*fig. 5, 10*); Hassek (1984, fig. 13, *6*); Tepecik (1982, Pl. IX, *6*).

94: Arslantepe (in press, fig. 26, *6*) (*fig. 5, 12*); Hassek (1984, fig. 14, *7-8*).

95: Arslantepe (in press, fig. 26, *5*); Hassek (1984, fig. 15, *4*).

96: Habuba K. (1974, Pl. 6, *63*); Hassek (1981, fig. 20, *2*).

97: Habuba K. (1974, Pl. 11, *73*); Susa 17 (1971, fig. 53, *6*).

98: Habuba K. (1974, Pl. 12, *80*).

99: Habuba K. (1974, Pl. 12, *77*; 14, *86*).

100: Habuba K. (1974, Pl. 19, *161*).

101: Habuba K. (1974, Pl. 3, *40-42*); Warka (1963, Pl. 38 m–o).

102: Habuba K. (1974, Pl. 19, *150*); JN (1931, Pl. LXV, *1*).

103: Habuba K. (1974, Pl. 4, *50*).

104: Arslantepe (in press, fig. 28, *8*) (*fig. 5, 5*).

105: Habuba K. (1974, Pl. 3, *35*).

106: Arslantepe (in press, fig. 34, *7*); Tepecik (1982, Pl. XI, *9*) (*fig. 6, 5*).

107: Arslantepe (in press, fig. 34, *8*).

108: Tepecik (1982 a, Pl. 74, *11*) (*fig. 6, 3*); Hassek (1984, fig. 12, *5*).

109: Arslantepe (in press, fig. 34, *1-2*) (*fig. 6, 1*); Tepecik (1982 a, Pl. 73, *5-6*; 74, *10, 12-13*).

110: Arslantepe (in press, fig. 34, *3*).

111: Arslantepe (in press, fig. 34, *4*) (*fig. 6, 4*).

112: Arslantepe (in press, fig. 34, *6*).

113: Arslantepe (in press, fig. 36, *4*); Tepecik (1982 a, Pl. 74, *1-2*).

114: Arslantepe (in press, fig. 36, *3*); Tepecik (1982 a, Pl. 74, *4*).

115: Arslantepe (in press, fig. 36, *1*).

116: Arslantepe (in press, fig. 37, *1*) (*fig. 6, 7*); Hassek (1984, fig. 13, *7*); Tepecik (1982 a, Pl. 73, *9-12*).

117: Arslantepe (in press, fig. 37, *2-4*) (*fig. 6, 6*); Hassek (1981, fig. 16, *7*; 1984, fig. 13, *8*); Tepecik (1982 a, Pl. 75, *4-5*).

118a: Arslantepe (in press, fig. 38, *1-2*).

118b: Hassek (1981, fig. 15, *9*).

118c: Tepecik (1982 a, Pl. 75, *3*).

119: Arslantepe (in press, fig. 39, *3*) (*fig. 6, 9*); Tepecik (1982 a, Pl. 74, *15*).

120: Arslantepe (in press, fig. 39, *1*) (*fig. 6, 8*); Hassek (1984, fig. 13, *9*).

121: Arslantepe (in press, fig. 38, *3*).

Abbreviations

Habuba Kabira: 1974 = Sürenhagen 1974/75.
Hassek Hüyük: 1981 = Hoh 1981; 1984 = Hoh 1984.
Tepecik: 1982 a = Esin 1982a ; 1982 = Esin 1982b.
Arslantepe: in press = Frangipane and Palmieri, *Origini* XII, in press.
Warka: 1932 = Haller 1932; 1958, 1959, 1961, 1965 = Lenzen 1958, 1959, 1961, 1965; 1963 = Strommenger 1963; 1970 = Nissen 1970; 1972 = Adams and Nissen 1972.
Nippur: 1965 = Hansen 1965.
Tello: 1934 = de Genouillac 1934.
Ur: 1955 = Woolley 1955.
Jemdet Nasr: JN 1931 = Mackay 1931.
Susa: 1957 = Le Breton 1957; 1971 = Le Brun 1971; Stève and Gasche 1971; 1976 = Miroscedji 1976; 1978 = Le Brun 1978.

a

b

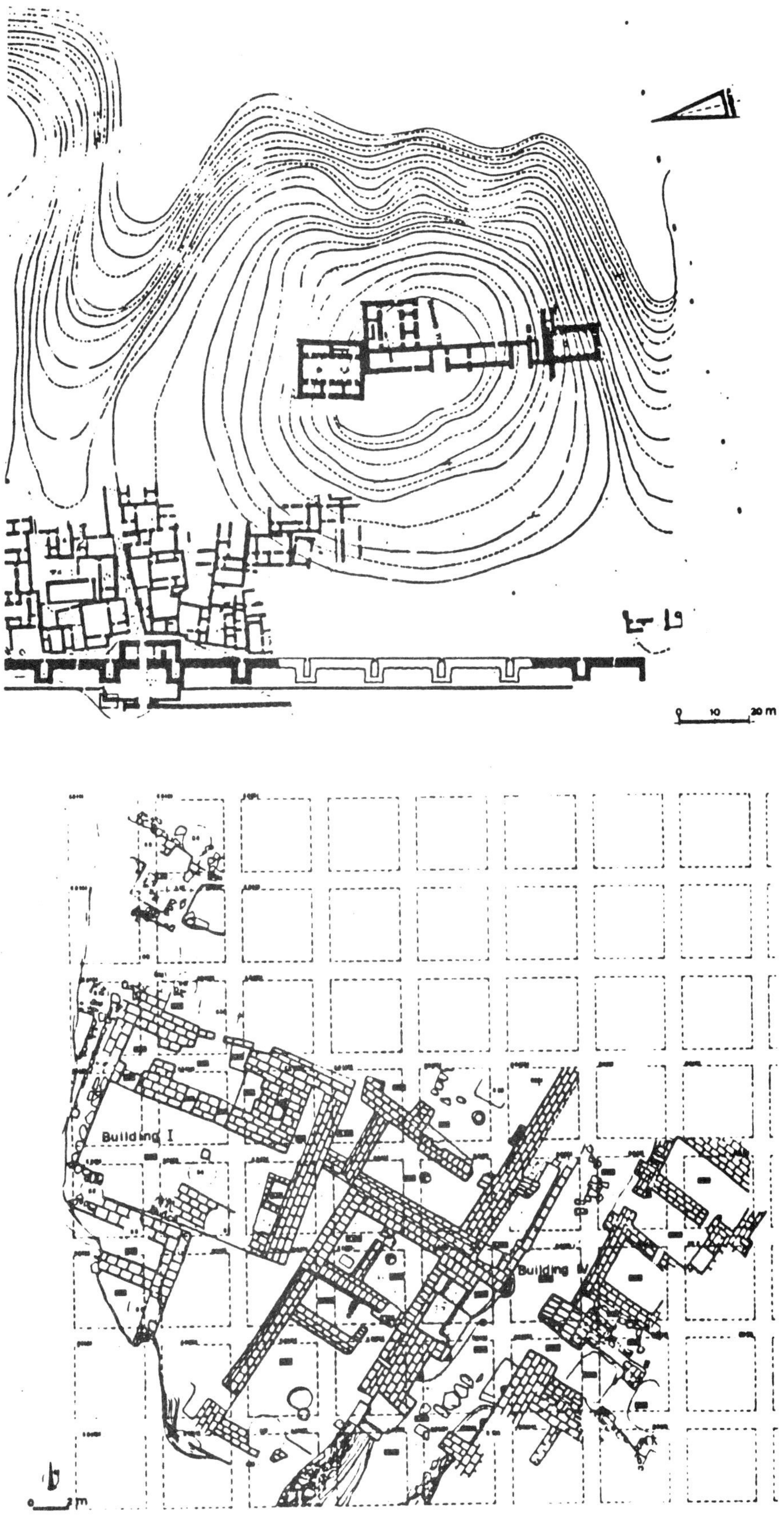

Fig. 1: *a.* The Habuba Kabira/Tell Kannas temple area; *b.* The period VIA public buildings at Arslantepe (Malatya).

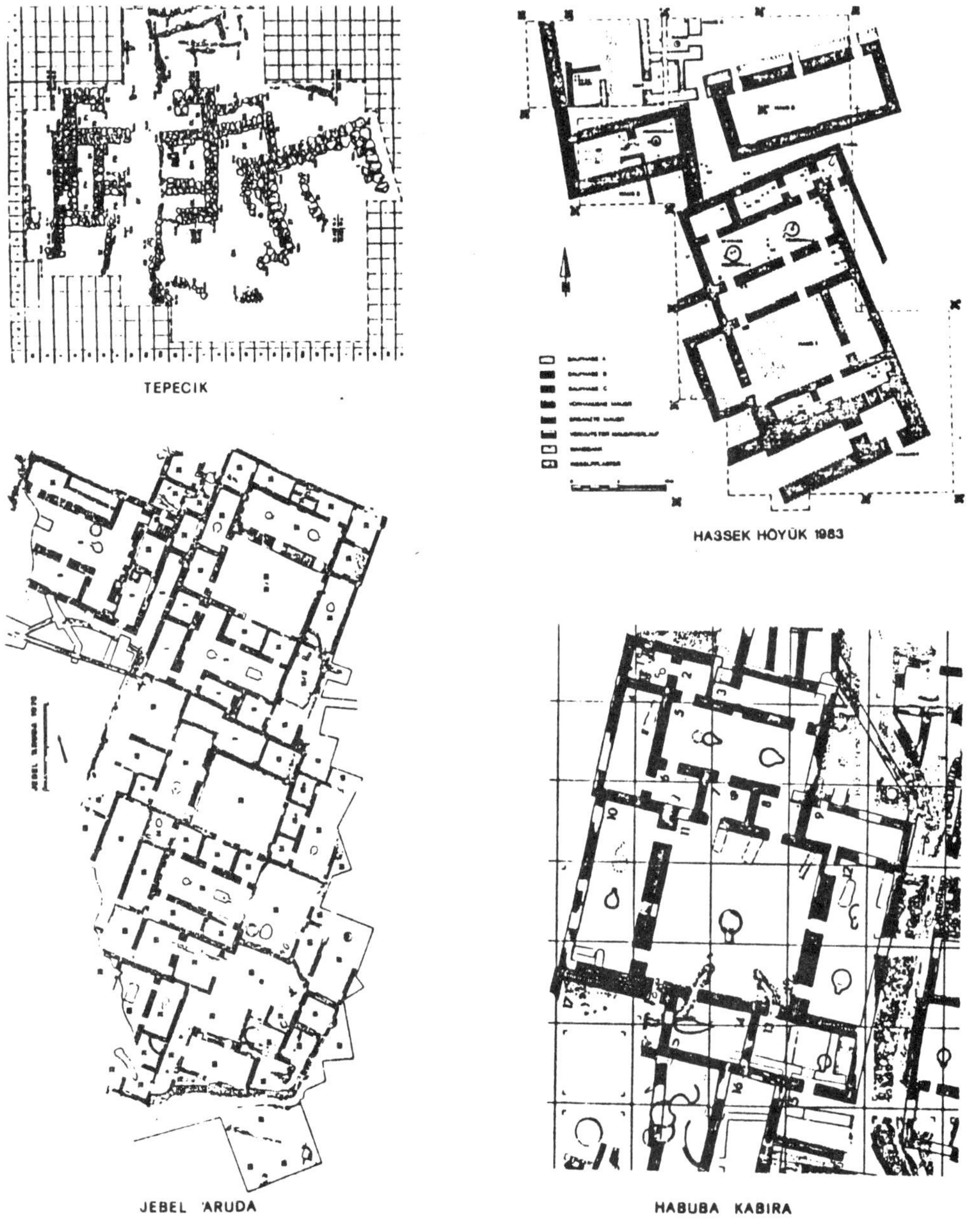

Fig. 2: Typical buildings from northern Late Uruk sites.

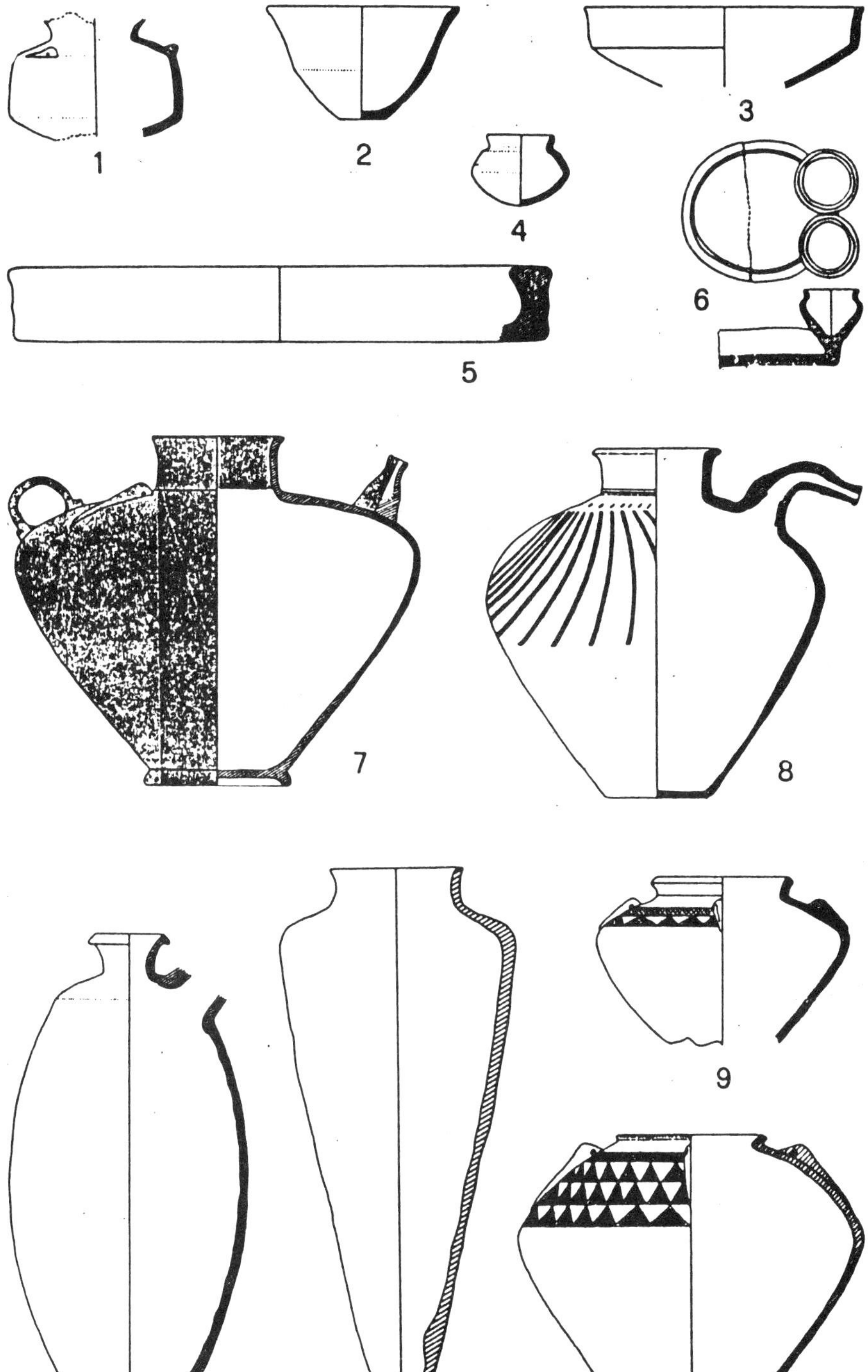

Fig. 3: Some of the most common Late Uruk pottery shapes linking Habuba Kabira with southern sites (Sürenhagen 1974/75).

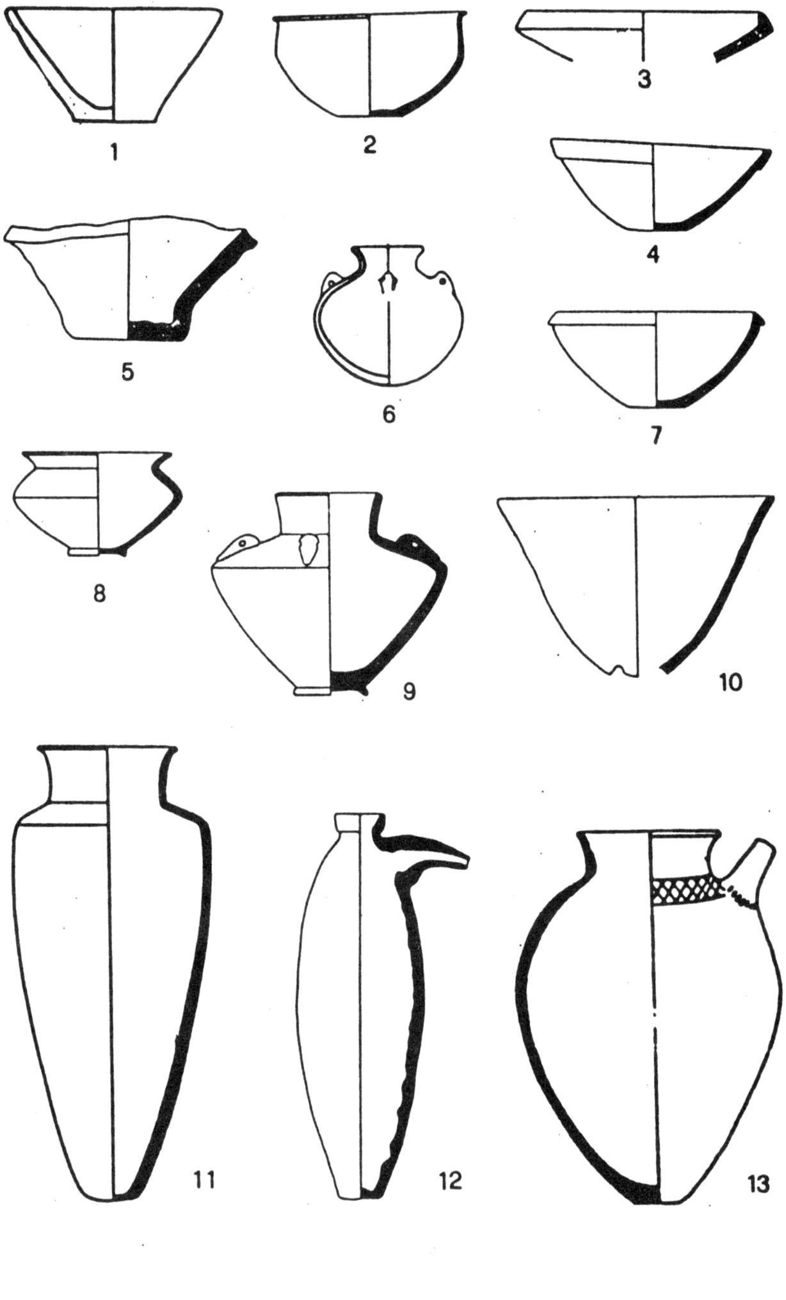

Fig. 4: The Late Uruk pottery shapes shared by the Anatolian sites and Habuba Kabira and the South. 1, 6: from Arslantepe (Frangipane and Palimieri, in press); 2, 4, 8, 9, 11: from Hassek (Hoh 1984); 5: from Hassek (Hoh 1981); 3, 7, 10, 12: from Habuba Kabira (Sürenhagen 1974/ 75); 13: from Tepecik (Esin 1982b).

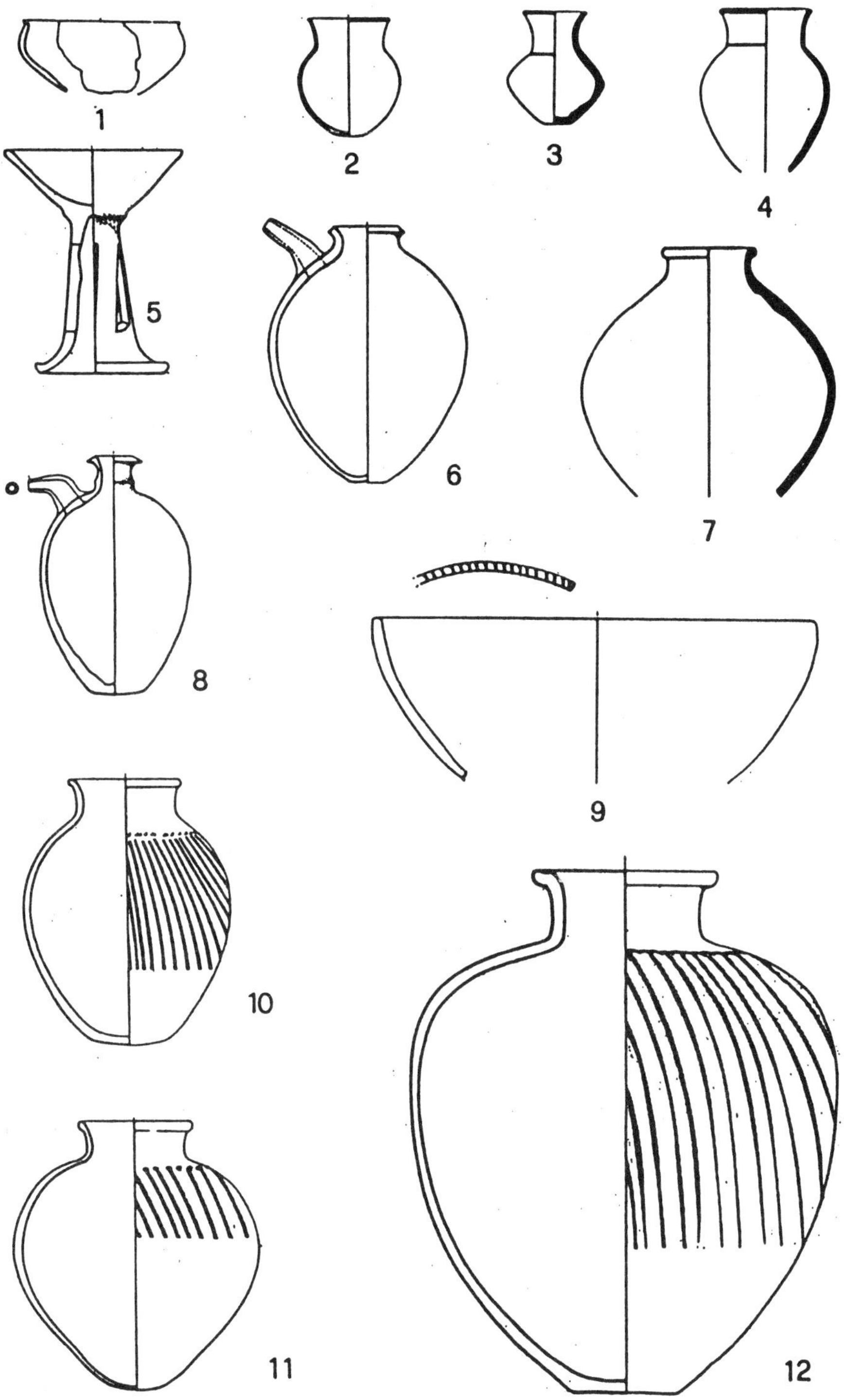

Fig. 5: *1–2, 5–6, 8–12*: typical profiles of wheel-made pottery from the Anatolian sites, in some cases (*1, 6, 8, 9*) shared with the South. *3, 4, 7*: sha- pes found at Habuba Kabira and Hassek Hüyük. 1–2, 5–6, 8–12: from Arslantepe (Frangipane and Palmieri, in press); 3, 4, 7: from Hassek (Hoh 1981).

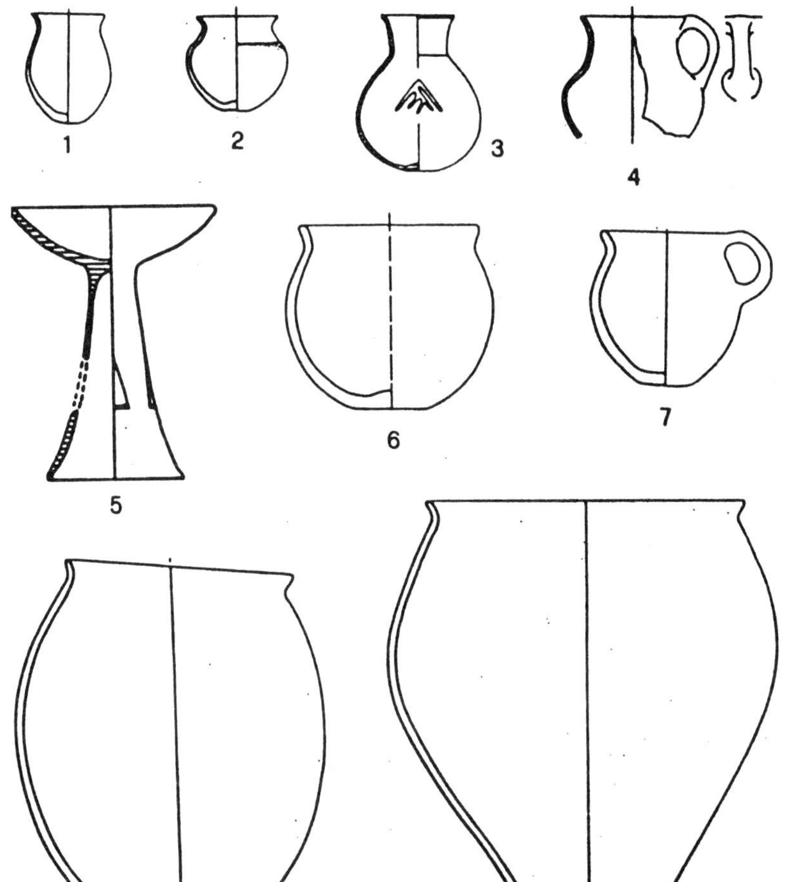

Fig. 6: Red-black (*1–5*) and Kitchen (*1–9*) pottery from the Anatolian sites in the Late Uruk/Jemdet Nasr period. 1–2, 4, 6–9: from Arslantepe (Frangipane and Palmieri, in press); 3,5: from Tepecik (Esin 1982 a).

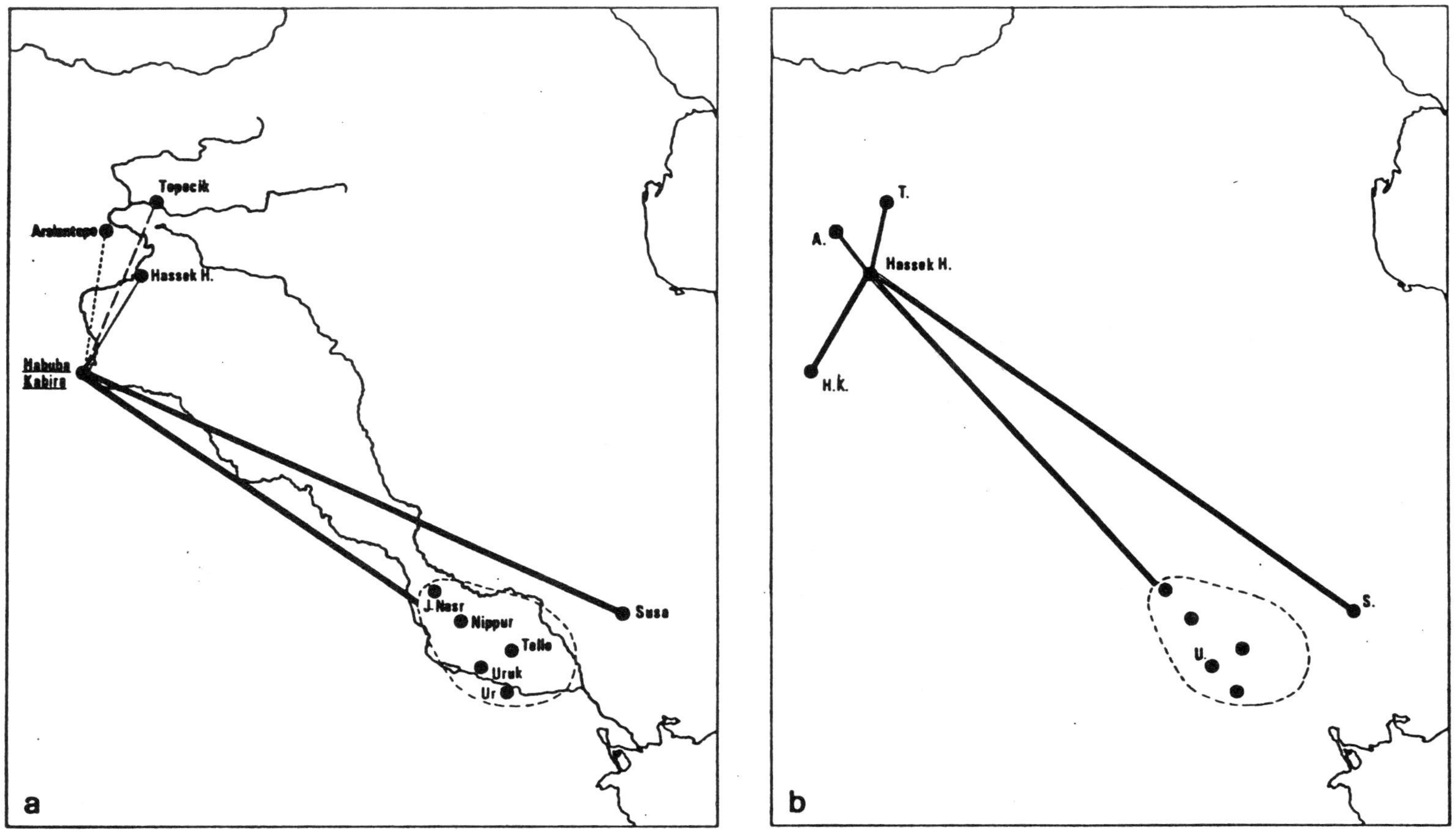

Fig. 7: Graphic representations of the different degree of affinity in pottery typology between two northern Late Uruk outposts —Habuba Kabira and Hassek Hüyük— and the other sites here concerned (southern Mesopotamia had been considered as a whole). The reciprocal relations were ob- tained for each couple of sites computing the per cent ratio of the number of common pottery shapes to the total number of the forms recognized in turn at Habuba K. (a) and Hassek H. (b). Dotted lines = from 1 to 9% of types in common; dashed lines = 10 to 19% continuous lines = 20 to 29%; double lines = 30 to 39%; triple lines = 40 to 49%; grey stripes = 50 to 60%; black stripes = more than 60%.

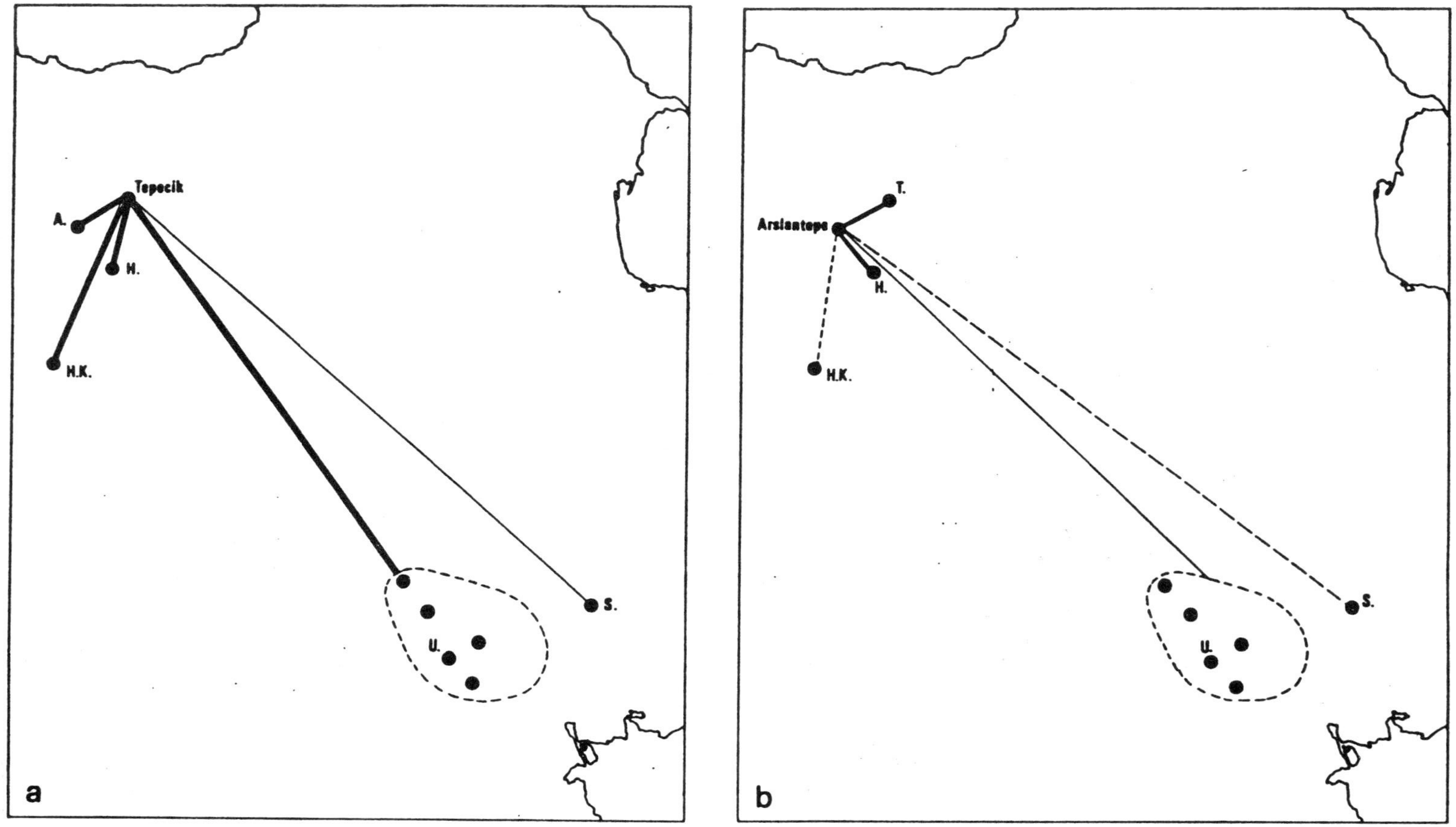

Fig. 8: Similarity relations showed by Tepecik and Arslantepe —respectively *a* and *b*— with the other sites here mentioned as far as pottery tipology measured according to the process already described in fig. 7. Symbols as in fig. 7

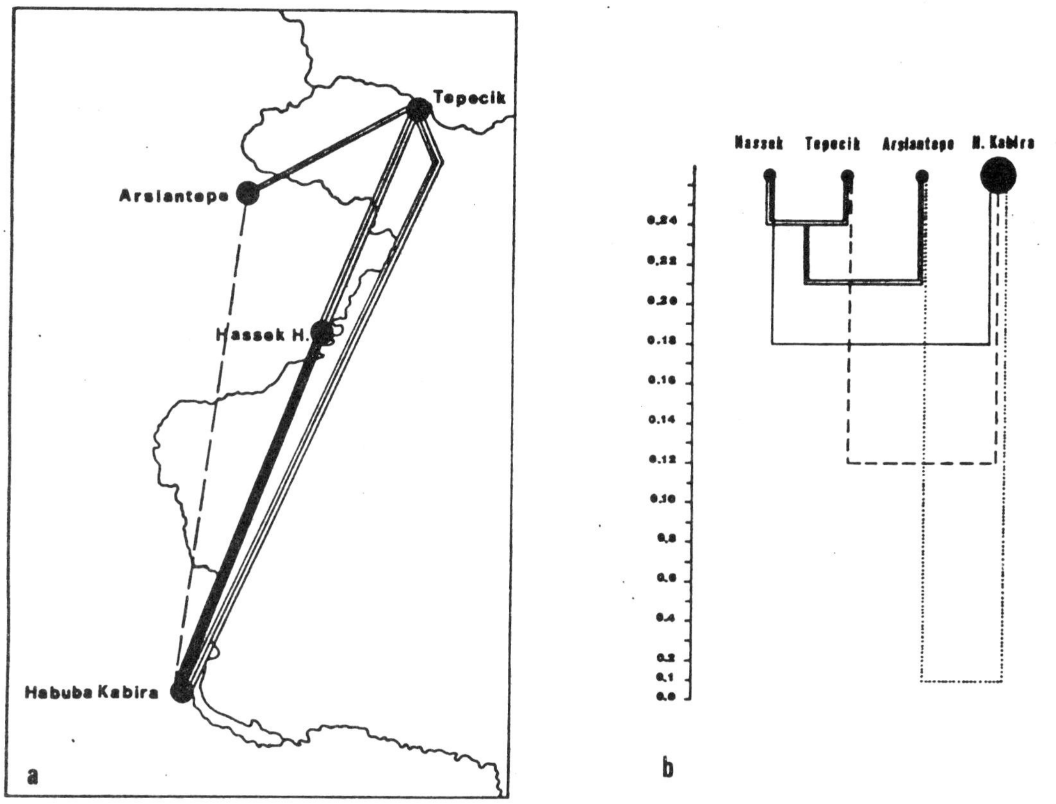

Fig. 9: Mutual affinity relations among the northern Late Uruk sites of the Euphrates region as seen from pottery production. *a*: graph obtained using the same process for measuring similarity as in figs. 7 and 8; the percentage of the shared shapes is calculated in this case with respect to the total number of forms recognized in the site with a smaller variety of types. The symbols are as in figs. 7 and 8. *b*: similarity relations resulting from the application of Jaccard's coefficient (Sokal and Sneath 1963: 133). The size of the circles stands for the number of the pottery types identified and used here: about 30 types for each of the three Anatolian sites, 92 for Habuba Kabira. It is just such a different variety in the typological range of the four sites that strongly conditions the similarity coefficient, stressing on the richness of the repertoire as an important element of differentiation between the three Anatolian sites on one hand and Habuba Kabira on the other.

REFERENCES

ADAMS, R. Mc.C.

1981. *Heartland of Cities*, Chicago.

ADAMS, R. McC., and H.J. NISSEN

1972. *The Uruk Countryside. The Natural Setting of Urban Societies*, Chicago.

ALESSIO, M. *et al.*

in press "14C Dating of Arslantepe", *Origini* XII, Rome.

BEALE, T.W.

1978. "Bevelled Rim Bowls and their Implication for Change and Economic Organization in the Later Fourth Millennium BC", *Journal of Near Eastern Studies* 37 (4): 289–313.

BEHM-BLANCKE, M.R., *et al.*

1981. "Hassek Höyük", *Istanbuler Mitteilungen* 31: 5–93.

1984. "Hassek Höyük", *Istanbuler Mitteilungen* 34: 31–149.

BEHM-BLANCKE, M.R.

1985. "Die Ausgrabungen auf dem Hassek Höyük im Jahre 1984", VII *Kazi Sonuçlari Toplantisi*, Ankara.

BINFORD, L.R.

1965. "Archæological Systematics and the Study of Culture Process", *American Antiquity* 31: 203–210.

BINFORD, L.R., and J.A. SABLOFF

1982. "Paradigms, Systematics and Archæology", *Journal of Anthropological Research* 38 (2): 137–153.

BOESSNECK J., and A. von den DRIESCH

1981. "Tierknochen vom Hassek-Höyük", *Istanbuler Mitteilungen* 31: 88–90.

BOHMER, R.M.

1972. "Die Keramikfunde im Bereich des Steingebaeudes", *U.V.B.*, XXVI–XXVII: 31–42.

BÖKÖNYI, S.

in press "Late Chalcolithic and Early Bronze I Animal Remains from Arslantepe (Malatya): a Preliminary Report", *Origini* XII.

BOVINGTON, C., *et al.*

1973. "Tehran University Nuclear Center Radiocarbon Dates II", *Radiocarbon* XV (3): 592–598.

CHILDE, V.G.

1936a. "Changing Methods and Aims in Prehistory", *Proceedings of the Prehistoric Society for 1936*: 1–15.

1936b. *Man Makes Himself*, London.

1942. *What Happened in History*, London-New York, Penguin Books.

1956. *Piecing Together the Past: The Interpretations of Archæological Data*, London.

CLARKE, D.L.

1968. *Analytical Archæology*, London.

ESIN, U.

1982a. "Tepecik Excavations, 1974", in: *Keban Project 1974–75 Activities*, I (7) Ankara: 95–118.

1982b. "Die Kulturellen Beziehungen zwichen Östanatolien und Mesopotamien sowie Syrien anhand einiger Grabungs und Oberflaechenfunde aus dem oberen Euphrattal im 4 Jt. v. Chr.", in: *Mesopotamien und seine Nachbarn*. XXV Recontre Assyriologyque Internationale, Berlin 1978, Teil 1, H.J. Nissen and J. Renger (eds), Berlin: 13–21.

ESPINOSA, G., and L. MANZANILLA

1985. "Consideraciones en torno a la capacidad de los cuencos troncocónicos de Arslantepe (Malatya)", *Quaderni de "La Ricerca Scientifica"* 112, CNR, Roma: 139–162.

FERIOLI, P., and E. FIANDRA

in press "Clay Sealings from Arslantepe VI A: Administration and Bureaucracy", *Origini* XII, Rome.

FINET, A.

1979. "Bilan provisoire des fouilles belges du Tell Kannás", in: *Archæological Reports from the Tabqa Dam Project Euphrates Valley, Syria*, D.N. Fredman (ed), Annual of the American Schools of Oriental Research 44: 79–95.

FRANGIPANE, M.

1985. "Early Developments of Metallurgy in the Near East", in: *Studi di Paletnologia in Onore di S.M. Puglisi*, M. Liverani, A. Palmieri, R. Peroni (eds), Roma: 215–228.

FRANGIPANE, M., and A. PALMIERI

in press "Perspectives on Protourbanization in Eastern Anatolia: Arslantepe (Malatya), the 1975–1982 Seasons", *Origini* XII.

GENOUILLAC, H. de

1934. *Fouilles de Telloh*, vol. I, *Epoques présargoniques*, Paris.

GREEN, M.W.

1980. "Animal Husbandry at Uruk in the Archaic Period", *Journal of Near Eastern Studies* 39: 1–35.

HALLER, A. von

1932. "Die Keramik der Archaischen Schichten von Uruk", *U.V.B.* IV: 31–47.

HANSEN, D.P.

1965. "The Relative Chronology of Mesopotamia. Part II. The Pottery Sequence at Nippur from the Middle Uruk to the End of the Old Babylonian Period (3400–1600 BC)", in *Chronologies in Old World Achaeology*, R.W. Ehrich (ed): 201–213.

HOH, M.R.

1981. "Die Keramik von Hassek Höyük", in: "Hassek Höyük", M. R. Behm-Blancke (ed), *Istanbuler Mitteilungen* 31: 31–82.

1984. "Die Keramik von Hassek Höyük", in "Hassek Höyük", M.R. Behm-Blancke (ed), *Istanbuler Mitteilungen* 34: 66–91.

HOWARD, H., and E.L. MORRIS (eds)

1980. *Production and Distribution: a Ceramic Viewpoint*, Oxford, BAR International Series 120.

JOHNSON, G.A.

1973. *Local Exchange and Early State Development in Southwestern Iran*, Anthropological Papers, Museum of Anthropology, University of Michigan 51, Ann Arbor.

KLEJN, L.S.

1982. *Archæological Typology*, Oxford, BAR International Series 153.

LE BRETON, L.

1957. "The Early Periods at Susa: Mesopotamian Relations", *Iraq* 19: 79–124.

LE BRUN, A.

1971. "Recherches Stratigraphiques à l'Acropole de Suse, 1969–1971", *Cahiers de la Délégation Archéologique Française en Iran* 1: 163–216.

1978. Le niveau 17B de l'Acropole de Suse, campagne de 1972, *Cahiers de la Délégation Archéologique Française en Iran* 9: 57–148.

LENZEN, H.J.

1958. "Mosaiktempel und Riemchengebaeude in Md-Na XV4-XVI2", *U.V.B..*, XIV: 21–35.

1959. "Kleinfunde aus dem Bezirk des Steinstift-tempels", *U.V.B.* XV: 20–21.

1961. "Die Kleinfunde", *U.V.B.* XVII: 26–28.

1965. "Die Keramik", *U.V.B.* XXI, 36–42.

MACKAY, E.

1931. "Report on Excavations at Jemdet Nasr, Iraq", *Field Museum of Natural History, Anthropology Memoirs* 1 (3): 219–203.

MARCOLONGO B., and A.M. PALMIERI

in press "Environment, Water Supplies and Cultural Developments at Arslantepe (Malatya)", *Origini* XII.

MIROSCHEDJI, P. de

1976. "Un four de potier du IV millénaire sur le Tell de l'Apadana", *Cahiers de la Délégation Archéologique Française en Iran* 6: 13–45.

MOOREY, P.R.S.

1982. "The Archæological Evidence for Metallurgy and Related Technologies in Mesopotamia, *c.* 5500–2100 BC", *Iraq* XLIV (1): 13–38.

NISSEN, H.J.

1970. "Grabung in den Quadraten K/L XII Uruk-Warka", *Baghdader Mitteilungen* 5: 101–91.

1974. "Zur Frage der Arbeitsorganization in Babylonien während der Späturukzeit", *Acta Antiqua* XXII: 5–14.

1985. "The Emergence of Writing in the Ancient Near East", *Interdisciplinary Science Reviews* 10 (4): 349–62.

OATES, J.

1983. "Ubaid Mesopotamia Reconsidered", in: *The Hilly Flanks and Beyond. Essays on the Prehistory of Southwestern Asia Presented to Robert J. Braidwood*, T.C. Young, Jr., P.E.L. Smith, and P. Mortensen (eds) Chicago: 251–282.

ÖZDOGAN, M.

1977. *Lower Euphrates Basin 1977 Survey*, Lower Euphrates Project Publications, Series 1, n. 2. Istanbul.

PALMIERI, A.

1973. "Scavi nell'area sud-occidentale di Arslantepe. Ritrovamento di una struttura templare dell'Antica Etá del Bronzo", *Origini* VII: 55–228.

1981. "Excavations at Arslantepe (Malatya)", *Anatolian Studies* 31: 101–109.

1985a. "Scavi ad Arslantepe (Malatya), 1976–79", *Quaderni de "La Ricerca Scientifica"*, 112, CNR, Roma: 75–114.

1985b. "Eastern Anatolia and Early Mesopotamian Urbanization. Remarks on Changing Relations", in: *Studi di Paletnologia in Onore di S.M. Puglisi*, M. Liverani, A. Palmieri, and R. Peroni (eds), Roma.

1985c. "1984 Excavations at Arslantepe", *VII Kazi Sonuçlari Toplantisi*, Ankara: 29–36.

PALMIERI, A., and M. FRANGIPANE

1986. "Assetto redistributivo di una societá protourbana della fine del 4° millennio", *Dialoghi d'Archeologia*.

PARROT, A.

1948. *Tello, vingt campagnes de fouilles (1877–1933)*, Paris.

RENFREW, C.

1970. "The Tree-Ring Calibration of Radiocarbon, an Archæological Evaluation", *Proceedings of the Prehistoric Society* 36: 280–311.

1975. "Trade as Action at a Distance", in: *Ancient Civilization and Trade*, J.A. Sabloff and C.C. Lamberg Karlovsky (eds), Albuquerque: 3–59.

1978. "Space, Time and Polity", in: *The Evolution of Social Systems*, J. Friedman and M.J. Rowlands (eds), London: 89–114.

1982. "Socio-Economic Change in Ranked Societies", in: *Ranking, Resource and Exchange*, C. Renfrew and S. Shennan (eds), Cambridge: 1–8.

1986. Interazione fra comunitá paritarie e formazione dello stato, *Dialoghi di Archeologia*.

SOKAL, R.R., and P.H. SNEATH

1963. *Principles of Numerical Taxonomy*, San Francisco.

STEVE, M.J., and H. GASCHE

1971. *L'Acropole de Suse*, Mémoires de la Délégation Archéologique en Iran 46, Paris.

STROMMENGER, E.

1963. "Archaische Siedlung: Keramik", *U.V.B.* XIX: 45–55.

1980. *Habuba Kabira. Eine Stadt vor 5000 Jahren*, Mainz.

SÜRENHAGEN, D.

1974/75. "Untersuchungen zur Keramikproduktion innerhalb der Spät-Urukzeitlichen Siedlung Habuba Kabira-Süd in Nordsyrien", *Acta Præhistorica et Archæologica* 5/6: 43–164.

1985. "Einige kulterelle Kontakte zwischen Arslantepe VIA und den frühsumerisch-hochprotoelamischen Stadtkulturen", in: *Studi di Paletnologia in Onore di S.M. Puglisi*, M. Liverani, A. Palmieri, and R. Peroni (eds), Roma.

TÖPPERWEIN, E.

1973. "Kleinfunde", *Mitteilungen der Deutschen Orient-Gesellschaft* 105: 20 ss.

VAN DRIEL, G.

1983. "Seals and Sealings from Jebel Aruda 1974–78", *Akkadica* 33: 34–62.

VAN DRIEL, G., and C. VAN DRIEL MURRAY

1979. "Jebel Aruda 1977–78", *Akkadica* 12: 2–28.

1983. "Jebel Aruda, the 1982 Season of Excavation, Interim Report", *Akkadica* 33: 1–26.

WEISS, H., and T.C. YOUNG, Jr.

1975. "The Merchants of Susa: Godin V and the Plateau-Lowland Relations in the Late Fourth Millennium BC", *Iran* XIII: 1–17.

WOOLLEY, C.L.

1955. *Ur Excavations*, vol. IV: *The Early Periods*, Philadelphia-London.

YOUNG, T.C., Jr.

1969. *The Godin Tepe Excavations: First Progress Report*, Toronto.

THE STATE AS AN ENTERPRISE.
THE CASE OF MARI

JORGE SILVA CASTILLO
El Colegio de México.

I must begin by justifying the choice of a subject that leads me to talk about a period at least one thousand years later than the emergence of the first Mesopotamian states in a colloquium that, since it is dedicated to V. Gordon Childe, should deal with the origins of civilization. In fact, I shall be dealing with the Kingdom of Mari, not at the moment it appeared but at the time immediately preceding its disappearance. Nevertheless, even though the city had existed long before, the state described by the texts is in some ways a new one. At that time it was in the hands of dynasts of recent nomadic origin and underneath the outward appearance of long-established state institutions, numerous elements can be perceived that show it was in fact a state in the process of formation inasfar as its social bases and political structures are concerned. To go even further —and this is what has seemed important to me to underline in this paper— in the picture that emerges from a study of it certain formative processes of the first states that emerged a thousand years earlier would seem to be repeated. This gives reason for making certain observations that may help to throw light on earlier processes.

Thanks to an exceptionally fortunate geographical position at the point where the caravan routes across the desert to cities in the center and south of Syria and the Mediterranean beyond, crossed the river routes up the Euphrates towards northern Syria and Eastern Anatolia, the city of Mari played an important role from the Early Dynastic period —the second quarter of the third millennium BC— to the time of Hammurabi (1792–1750 BC), who understanding its strategic position, tried to subjugate the city and finally destroyed it[1].

This paper deals with the last period of its splendor, during which dynasties of Amorite (Western Semitic) origin[2] were involved in intense economic, political and diplomatic activity, attested to by the majority of the twenty thousand tablets that formed the palace archives of Mari. The main interest of these records lies in the homogeneity which enables a fairly complete picture to be built up of the state structures existing at the moment when, because of the great social and political changes of the era[3], some important aspects of the dynamics of its development can be distinguished.

---

[1] J.R. Kupper in CAH vol. II, part 1, p. 14. H. Lewy, along with other authors, maintains a different opinion. See "The Chronology of the Mari Texts", in XV RAI, p. 26.

[2] Both the dynasty considered "local", whose last kings were Iahdun-Lim and Zimri-Lim, and that of the Assyrian Samasi-Adad who dethroned the former and controlled Mari until his death, which enabled Zimri-Lim to recover his kingdom, were of Amorite origin. In this paper I refer mainly to the reign of Zimri-Lim, to which most of the texts belong, although some of those mentioned date from the so called Assyrian Interregnum.

[3] P. Garelli, PDA p. 264.

Before going more deeply into the subject, some observations must be made. First, the title of this paper "The State as an Enterprise". The end of the third millennium coincided with the disappearance of the last Sumerian state, the empire of Ur, in its Third Dynasty which had brought a highly sophisticated and impersonal bureaucratic system to its perfection. This type of state contrasts with the extreme personalism characteristic of the Amorite states that emerged during the Paleo-Babylonic period immediately following Ur III, among them Mari, where the impression is that the essential economic activities of the kingdom were concentrated in the palace, which in turn seems to be nothing more than a private business of the monarch. In saying this I do not wish to deny the importance of other factors of state dynamics, especially the political control that was repressive domestically and aggressive abroad, but it does seem interesting to me to focus this study on the economic factor, with the features peculiar to the kingdom of Mari since the analysis may contribute to an understanding of the structures of a particular archaic state.

The second observation is about the nature of the Mari archives. Excavations at Tell Hariri —the modern name of the site of Mari— have become restricted to the uncovering of remains of the city, mainly the palace and ceremonial center. The palace archive touches only indirectly on matters relating to other agents active within the economy of the kingdom. Hence, the risk of the perspective of the investigation being distorted, since the image of the palace given by a study of the texts in its archives in fact occupies an important place in the overall picture that emerges.

However, some considerations which lead us into the main body of this work, make it possible to moderate the force of the two previous observations and place the subject in its true perspective.

a) In speaking of the palace we refer to the institution that embodied the government of the kingdom, its administration and its lands as a whole. The main records give the picture of a state with properties scattered over a considerable area that covered widely differing sectors of the economic activity.

b) Apart from the palace lands, the records mention other goods that some officials received from the sovereign, as part of a system of payment for services that allowed such rewards[4]. Therefore these types of possessions. although administered by the beneficiaries as personal property must be regarded as part of what the palace granted in its role as redistributor.

c) The existence of private or collective property cannot be denied; however, as regard the latter, just as in other areas when similar socio–political phenomena existed[5] there are some texts indicating a tendency to the sale of collective land and its acquisition by the members of the bureaucracy in power[6]. As far as private land, the impression given by some allusions to the poverty of the peasants in general seems to indicate that most of them, even landowners, lived at a subsistence level.

d) The same can be said of palace employees, servants or craftsmen working in their various specialized workshops, or else peasants, farmers or herdsmen, The role of the merchants, more prominent in earlier periods and in other regions is modest at Mari, where they seem to have been palace agents.

---

[4] M. Birot in ARMT IX, p. 262 and M.L. Burke in ARMT XI, p. 130.

[5] For similar situations earlier see I. Diakonoff, SLPS.

[6] Elena Cassin, "Cession Immobilière", in NSP, p. 83–84.

From the above particulars it can be inferred that the role of the Palace–State in the economy was paramount and this, with the reservations mentioned earlier, justifies us in speaking of the state as a business.

The palace, physically speaking was a vast and complex collection of buildings: two hundred and sixty rooms distributed around spacious courtyards covering an area of almost 3 hectares (7.5 acres). In addition to the rooms of the court itself —the royal family and their servants— there were many other offices; workshops producing furniture, weapons, textiles, leather, and metal articles[7].

Production was organized in what we would now call a vertical system, since every stage was dealt with, from obtaining raw material —with the exception of wood and metals that were imported— through processing to finishing. Stored in the palace were many of the products that would be distributed in the form of food or clothing rations, or else be sent to other kingdoms in an exchange that amounted to foreign trade. We also know that outisde the walls the palace owned other workshops and warehouses both in Mari and in other cities of the kingdom[8].

The palace staff and employees —officials semi-free[9] domestic servants and slaves, usually foreign— were certainly numerous. One document mentions that the number of domestics who had to be issued with clothing could be as high as four hundred[10], though it cannot be ascertained whether this included all the craftsmen from the workshops, which was probably much higher[11].

The palace, as an institution, owned agricultural lands both in the district of Mari and spread through other areas of the kingdom[12], sometimes of considerable size. It has been calculated that the most common, worked by teams of 10 to 15 men, could cover 50 to 80 hectares (approximately 124 to 198 acres), but there were doubtlessly larger ones[13]. The work done by the public administration for the upkeep of irrigation, for the harvesting and threshing of cereals needed a large labor force, frequently obtained by forced labor recruited among the inhabitants of towns and villages[14]. The same can be said about livestock, some cattle but mainly sheep, which the palace owned all over the kingdom. As in the case of water and agriculture, forced labor was used in sheep shearing[15].

Textile production[16], tanning, leather, and metal working were important activities of the palace. Letters exchanged among officials as well as many administrative documents mention all aspects of these, from the processing of the raw material to the

---

7 As an example, ARMT XIV, 14.

8 In reference to this, see ARMT V, 67, 27–34; IX, 25; 26; 97, Rev. 28 among other texts.

9 The employees —*kinātu*— did all kinds of work and although they were not exactly slaves they were not completely free as they were closely guarded (ARMT IV, 10, Rev. 13'–15') and pursued if they ran away (ARMT X, 150, 5–7).

10 ARMT VI, 39.

11 In any case, judging simply by the much higher number in a list of prisioners to be sent to the weaving workshops —almost a thousand divided into forty-five teams— it can be supposed that the four hundred servants in question, at least at a given moment, represented only part of the total palace staff (ARMT XIII, 1).

12 Lands belonging to the palace are frequently mentioned in the correspondence between officials. See for example, ARMT III, 3, 4, 31, 34, 77, VI, 3, XIV, 14, 17, 48, etc.

13 Francis Joannès in ARM XXIII, p. 111.

14 E. g., *passim* in the correspondence of the governor of Terqa the second city of the kingdom, ARMT III, 1 and 7.

15 Example in ARMT II, 140.

16 Jorge Silva Castillo in NSP, p. 109.

finishing of the products. Hundreds of ordinary garments were for the palace employees. Other, more valuable products —in addition to those set apart for the king, his court and high officials— that included metal items, better quality textiles, leather articles, weapons, and even wine, were commodities in an active exchange with other kingdoms[17]. Additionally there were workshops belonging to the palace in provincial cities, although these were less important[18]. Another extremely interesting indication of the omnipresence of the palace in the economic life of the kingdom is to be found in the fact that the temples seem to have depended on it[19]. Agricultural land dedicated to gods, that must be considered as belonging to the particular deity, is also mentioned[20]. The yield of this land probably served to maintain the respective temples.

From a more qualitative point of view, it is revealing that all this economic activity was finally controlled by the king himself. As for the concept of the palace-as-an-institution, the way in which the high administrative officials sometimes refer to it is significant "the grazing lands of my Lord"[21]...the flocks of my Lord"[22]. These same people were responsible for a series of subordinate employees organized in a hierarchy. On the other hand there was no well-established hierarchy among them, since they depended directly on the sovereign, who took the final decisions. For this reason, the messages they exchanged frequently refer to disagreements which could only settled by referring to the king himself[23]. He intervened directly in all the cases of real importance, such as supervising the reception of precious metals that were weighed with special weights "of the king" and stored in warehouses under his personal seal[24].

This is nothing more than a consequence of the nature of the sovereign's power at this time, particularly in Mari. The king's power was absolute and personal as his political and diplomatic correspondence shows. In no text do we see the slightest indication of either political, institutional or personal control of the royal power. Surprising, because it is exceptional, is the advice given by one of the highest officials in Mari, Bahdi Lim, urging prudence in politics. He reminds Zimri Lim that as king not only of the Haneans —the Amorite tribe from which he descended— but also of the Akkadians —the sedentary population— he should appear in public mounted on a donkey, not on a horse[25]. But despite his ethnic origin and the character of the state consisting of the types of society, urban and segmentary, that he ruled, there is no sign in the texts of a specially organized body at a state level strictly speaking that might be expected to exist as a vestige of the "councils" of elders that were at the head of the corporate groups characteristc of the organization of segmentary societies. There are references only to councils of elders at a village level and these dealt with purely local affairs[26].

---

[17] Denis Soubeyran, in ARM XXIII, p. 351.

[18] ARMT IX 25, R. 37; 26, R. 10; IX, 97; R. 28.

[19] Indirect proof is given in a text that says that an official has to take certain objects from a palace storeroom either of the palace or of the temple of Adad. This shows that both were regarded as his property (ARMT XIII 56, 14–15).

[20] Francis Joannes in ARMT XXIII, p. 115 follows the opinion of M. Birot, in preferring to interpret this as meaning that certain agricultural lands has shrines dedicated to these gods.

[21] ARMT XIV, 86.

[22] ARMT XIV, 20.

[23] J. Bottéro in ARMT VII, p. 353.

[24] ARMT XIII 10, 10–11.

[25] ARMT VI, 76.

[26] ARMT V 24.

This last remark leads us into a discussion of an element that is interesting from the methodological point of view if we wish to attempt to make an interpretation of the characteristic features of the state of Mari, enumerated very summarily here.

The view used to be that the kingdom of Mari was a fully urban state under pressure from tribal elements —Amorite nomads— at times allies, but most of then enemies[27]. This has been superseded by another based on the application of our knowledge of more recent segmentary societies. This new interpretation, while not denying the conflicting and always fluctuating relations between tribal and urban societies, considers them as complementary economic factors necessarily integrated into a state system —the dimorphic state— that responds ecologically to a particular geographical medium. Although this explanation that we owe to Michael Rowton[28] is so illuminating in understanding the structures of the state of Mari from a theoretical point of view, several studies are necessary of details concerning crucial points in which the relationship between the urban and segmentary societies can be clarified according to the texts. Furthermore, the economic symbiosis that is mutually complementary, reciprocal and equitative between urban and segmentary societies implied by the concept of the dimorphic state is open to discussion, unless it is further defined, when the dominant and absorbing role of the state–enterprise in the economy of the kingdom of Mari is verified.

At this point it is useful to quote a passage from Robert McC. Adams' book *The Evolution of Urban Society* in which he describes the process of stratification in the society that gave rise to the Mesopotamian states. "Standing at the pinnacle of Mesopotamian society were small numbers of princely families who seem to have been vigorously extending their control of land by purchase during the later Early Dynastic and Akkadian periods... (These) ruling families in Mesopotamia seem to have headed semi-integrated manorial estates that varied greatly in size. Their labor force, dependent in varying degrees in the distribution of rations, allotments of land, and other forms of clientage, also included a small but significant proportion of persons employed under repressive, closely controlled conditions of outright slavery. Yet in spite of the proliferation and increasing importance of these class–orientated forms of organization, there still existed kin-based communities both outside and within the manorial establishments, and, while such communities were declining in the relative amount of land under their corporate control, it is likely that they still had very large areas at their disposal"[29]. This tendency towards stratification grew during the Akkadian period with the distribution of land that the monarch granted to his political clients as a reward for services and as a means of ensuring their loyalty[30].

The analogies to be found between the formative processes of the state, unrecorded in the earliest times and rewritten in the case of Mari, only reinforce the factors present at both moments by bringing them into relief.

As we have already seen, some legal texts from the Mari archives give evidence of purchases made by the officials of the palace administration of land owned communally by groups of segmented societies[31] and point to a concentration of property on the part of the

---

[27] J.R. Kupper's work "Les nomades en Mesopotamie aux temps des Rois de Mari" is a great advance on the traditional view that regarded the migrations of the nomads as a violent invasion. In contrast he presents the nomadic populations as in a continual ebb and flow (see NMTRM); Michael Rowton, on the other hand, considers them integrated symbiotically into the state together with the sedentary populations.

[28] Michael Rowton put forward his thoughts in a series of articles listed under his name in the bibliography.

[29] Adams, EUS: 109.

[30] C.J. Gadd in CAH vol. I, part 2 A, pp. 424 and 448–449.

[31] ARMT VIII, 11–14. See also note 6.

bureacucratic stratum of society. These same officials received lands from the sovereign as a reward for their services in administration, and this doubtlessly strengthened their links of personal loyalty to their benefactor. However, at the same time it opened the way to a concentration of land in the hands of the bureaucratic class. As for the process by which the king's aristocratic control came to be established, at the time the texts of the archives were produced it is already well advanced, and was strengthened by the control of many other economic activities.

There is also evidence that there were groups of population organized on the lines of segmentary societies. In connection with this, I think it is important to point out that we have to await the results of an exhaustive study of an enigmantic but intriguing institution: the *sugagûtum*. This was a function of certain persons, the *sugâgum*, who were guided by the norms typical of segmentary societies. They were elected as a body by the "elders", represented their communities and defended their interest, at the same time acting as intermediaries before the state administration. The appointments were recognized by the admninistration of the kingdom after confirmation of their right of represenation and the payment of a kind of tribute[32]. State officials turned to these people when dealing with their communities[33]. It would seem most important to define whether such communities were ones simply made up of segmentary groups of Amorite origins (which included not only nomads, but semi-nomads, transhumants and farmers already settled in villages), or else by the villagers who were known as "Akkadians", in other words descendents of long-settled peoples, integrated into the ancient state structures inherited by the Amorite dynasties[34]. If the latter were the case, these rural communities would be comparable to those referred to by Adams in the passage quoted above and we would have a situation similar to the one he describes, but with the addition of other segmented groups of Amorite origin who did not recognize any loyalty other than that owed to the person of the sovereign, not as a head of state but as a descendent of a common putative ancestor... If this last hypothesis is the one that is confirmed, it modifies the outline suggested by Adams to explain social phenomena that were present in more remote peiods, thanks to the contribution of the dimorphic state theory put forward by Rowton.

Whatever the result of an exhaustive and more finely detailed study, the case of Mari, better documented than earlier processes by which the Mesopotamian states were born, provides elements that are likely to shed light on the dynamics of their formation. The state appears to be an enormous economic machine put in motion by the king and his bureaucratic clients, who emerge as the dominant social stratum thanks to their access to the sources of power and, through the king, to the accumulation of land. At the other extreme, and at the base of society, two types of population with different social structures are integrated within this state. The way they functioned within the state concern needs more careful study.

---

32 Jorge Silva Castillo in NSP, p. 121.

33 ARMT III, 6.

34 J. Luke (PPMP), reacting against Kupper's point of view (NMTRM), sees in the segmentary societies that appeared in Mari not elements foreign to urban society but groups of herdsmen, as a labor specialization derived from village culture. This interpretation represents an important advance, but does not take into account the undoubted presence of other segmentary societies of Amorite origin that were not natives of the region.

REFERENCES

ADAMS, R. McC.

1966. *The Evolution of Urban Society. Early Mesopotamia and Prehispanic Mexico*, New York, Aldine Publishing Company. (EUS)

ARM (T).

1950. *Archives Royales de Mari* (transcription and translation), vols. I–XXI, Paris, Paul Geuthner (ed). (ARMT)

ARM.

1954. *Archives Royales de Mari*, Editions Recherche sur les Civilisations, vols. XXI–XXIV. (ARM)

1980. *The Cambridge Ancient History*, New York, Cambridge University Press. (CAH)

DIAKONOFF, I.

n.d. *Sale of Land in Presargonic Sumer.* Papers presented by the Soviet Delegation at the XXIII International Congress of Orientalists. Assyriology Section, publishing House of the USSR Academy of Sciences. (SLPS)

GARELLI, P.

1969. *Le Proche-Orient Asiatique. Des Origines aux Invasions des Peuples de la Mer*, Paris, Presses Universitaries de France, Colección Nueva Clío. (POA)

KUPPER, J.R.

1957. *Les Nomades en Mésopotamie au Temps des Rois de Mari*, Paris, Société d'Edition "Les Belles Lettres". (LNMT)

1967. (ed) *XVe. Recontre Assyriologique Internationale "La Civilisation de Mari"*, Université de Liège, Lieja. (RAI)

LUKE, J.T.

1965. *Pastoralism and Politics in the Mari Period. A Re-examination of the Character and Political Significance of the Major West Semitic Tribal Groups on the Middle Euphrates, c. 1828–1758 BC*, Michigan University Microfilms LTD. (PPMP)

ROWTON, M.B.

1973. "Urban Autonomy in a Nomadic Environment", *Journal of Near Eastern Studies* no. 32: 201–215.

1973. "Autonomy and Nomadism in Western Asia", *Orientalia* no. 42: 247–258.

1974. "Enclosed Nomadism", *Journal of the Economic and Social History of the Orient* no. 17: 1–30.

1976. "Dimorphic Structure and the Problem of the *Apirû-Ibrî*", *Journal of Near Eastern Studies* no. 35: 13–20.

1976. "Dimorphic Structure and Typology", *Oriens Antiquus* no. 15: 17–31.

1976. "Dimorphic Structure and the Tribal Elite", *Studia Instituti Anthropos* no. 28: 219–257. [*Al-Bahit: Festschrift Joseph Henninger zum 70. Geburtstag am 12. May;* St. Augustine: Anthropos-Institut, Keilschriftbibliographie 38: 122.]

1977. "Dimorphic Structure and the Parasocial Element", *Journal of Near Eastern Studies* no. 36: 181–198.

1980. "Pastoralism and the Periphery in Evolutionary Perspective", in M.T. Barrelet (ed): *L'archéologie de l'Iraq du début de l'époque néolithique à 333 avant notre ère*, París, Editions du Centre National de la Recherche Scientifique: 291–301.

1981. "Economic and Political Factors in Ancient Nomadism", in Jorge Silva Castillo (ed): *Nomads and Sedentary Peoples*, México, El Colegio de México: 25–36.

SILVA CASTILLO, J. (ed)

1981. *Nomads and Sedentary Peoples*, México, El Colegio de México. (NSP)

CHILDE AND THE URBAN REVOLUTION: THE CENTRAL ANDEAN EXPERIENCE

LUIS GUILLERMO LUMBRERAS
Instituto Andino de Estudios Arqueológicos

When V. G. Childe stated that there had been at least two great revolutions in prehistoric times: the "neolithic" and the "urban" revolutions, he was actually —although he did not make it explicitly— setting up a scientific argument to contrast with Karl Marx's statement, formulated in the "Preface" to *A Contribution to the Critique of Political Economy.* Such a statement concerned the historical necessity of the revolutionary processes in the transition from one social form to another.

This assumption is based on the thesis that social revolutions are neither the result of individual will, nor the product of casuality, but of the maturing process of each historical formation or stage. In the famous Preface, Marx says: "At a certain stage of their development, the material productive forces of society come in conflict with the existing relations of production, or —what is but a legal expression for the same thing— with the property relations within which they have been at work hitherto. From forms of development of the productive forces, these relations turn into their fetters. Then begins an epoch of social revolution".

But he also defines the nature of the revolutionary process, when he states that:

> With the change of the economic foundation the entire immense superstructure is more or less rapidly transformed. In considering such transformations a distinction should always be made between the material transformation of the economic conditions of production, which can be determined with the precision of natural science, and the legal, political, religious, æsthetic or philosophic —in short, ideological forms in which men become conscious of this conflict and fight it out. Just as our opinion of an individual is not based on what he thinks of himself, so can we not judge of such a period of transformation by its own consciousness; on the contrary, this consciousness must be explained rather from the contradictions of material life, from the existing conflict between social productive forces and the relations of production.

Having stated its structural nature as well as its form and its consequences, Marx finally explains the historical conditions necessary to the revolutionary process:

> No social order ever perishes before all the productive forces for which there is room in it have developed; and new, higher relations of production never appear before the material conditions of their existence have matured in the womb of the old society itself. Therefore mankind always sets itself only such tasks as it can solve; since, looking at the matter more closely, it will always be found that the task itself arises only when the material conditions for its solution already exist or are at least in the process of formation.

We would like to revise this Childean concern, although we are interested in setting it up explicitly in relation to the marxist statements just cited. This will establish certain differences between Childe's description and his definition of the process, and the ones we shall propose here.

Our own comparison will be resting upon the historical experience of the Marañón Andes, an area located in the northern region of the so-called Central Andes. This area was the focus where the great Prehispanic states of South America developed, and whose final version was known in the sixteenth century AD by the name of "*Tawantinsuyo*", or Inca Empire.

The concept of Urban Revolution was set up by Childe through the study of the conditions under which the transition from a tribal form of organization —or "neolithic Barbarism"— to a class society —or "Civilization"— took place. He identified the first civilized form with the so-called "Bronze Age", based upon the historical experience of the Old World.

From this comparison between neolithic Barbarism and Bronze Civilization, Childe established that the most important indicators of change were given by the emergence of such qualitatively new elements as:

1. A specialized technology based upon the development of exact and predictive sciences, such as arithmetics, geometry and astronomy, the function of which was to regulate the success of the agricultural cycle, as well as to improve significantly the technical processes of food production, through the creation of new working tools;
2. the (complementary and consequent) creation of recording systems, such as writing and numerical notation;
3. the development (also complementary and consequent) of several forms of crafts specialization: sculpture, painting, carving, pottery, metallurgy, etc.;
4. the development (also complementary and consequent) of some sort of "foreign trade", associated with the necessity of acquiring raw materials for luxury or industrial purposes;
5. the social division into classes: specialists, "exempted from manual activities", who constitute a "ruling class", and primary workers, direct producers of consumption goods;
6. the rise of relations of production and consumption of an asymmetric nature, where primary producers are forced to give away the product of their work, as a tribute to "divine lords" or "gods", who concentrate and administer the surplus captured this way through the "ruling class". The counterpart of this relationship rested upon the planning and organization services concerning production and consumption, and naturally, the generating of new and powerful specialized means of production;
7. the generation of a production surplus (which to us seems complementary), allowing the concentration of capital in the hands of the "ruling class", and guaranteeing its social reproduction;
8. the rise of the city as a new working and living place, totally created by man, quantitatively and qualitatively different from the residential (villages or hamlets) or working (fields) places of the "neolithic Barbarism" communities. The city says Childe, must be larger than a village (in relative, and not absolute terms); it must have a population whose composition and functions are diverse; residences ought to be based upon work or occupation, rather than kinship or affinity, and their nature ought to be permanent, with the presence of public architecture, constituted by temples, workshops or warehouses, besides the residential or domestic constructions;
9. the rise of the State, associated with the existence of a Religion, which made possible the permanence and reproduction of the "ruling class", through the existence of "divine lords" and "gods"; these were served by specialists called priests, and by other components at the service of the new system: military chiefs, civilian leaders and officials.

From this series of markers, gathered from Childe's theses, formulated in several of his works on the subject, and summarized here, we will extract the ones we consider essential:

a. The emergence of specialized technology, which significantly raised the level of development of the productive forces effective in neolithic society; this altered the existing equilibrium between material forces of production and the relations of production, which did not include the new type of human element —the specialist—, conductor of the new productive forces.

b. The consequent social division into classes, differentiated through a different participation in the productive process, with the conflictive effects upon the new social relations of production and consumption, including the necessary rise of the State.

c. The rise of the city, not only as a new space for the concentration of population, but fundamentally as a new and totally different center of production, necessary working place and residential location of the new social class, which concentrates there its working instruments (temples or workshops), and its life resources (warehouses or storage areas), all of which become "public facilities" (and are no more communal, domestic or household facilities).

We are obviously not facing common changes, as can be technological ones, by which society continuously advances, or population transformations of a quantitative nature, or much less changes in the natural environment. We are dealing with changes causally chained together, which encompass all the material forces of production: a technology associated with the social sector which is intrinsically linked to it, and which alters at the same time the natural conditions over which it acts. These changes affect the population, the instruments and the means of production, and they modify the relationship between human beings, and between them and nature.

These changes do not come by chance, neither do they originate through isolated actions (such as discoveries or inventions), nor do they result of the pressure of external factors. They are determined by the development of society, through the process of dominating the environment and improving living conditions. They are the product of the total success of "neolithic barbarism" in the management of vital resources, within the specific conditions of its historical development. They appear as the result of the complete maturity of the neolithic mode of production, whose fundamental historical task was the discovery and development of the means of food production, through the domestication of plants and animals, and through the consequent management of the soil, the water and the climate. These are the concrete material conditions which allow such a domestication to be possible and productive.

We are, thus, facing the rise of new forces of production, within the neolithic or tribal formation, which are materially superior to the ones in operation, and which appear to be at the highest point of their development. Nevertheless, in order to progress, significant changes are needed in the egalitarian relations, which are conflictive with the new conditions of production. These conditions necessarily involve, among other things, the social division into classes and, in order to resolve such a conflict, a period of social revolution had to begin. It is this stage that Childe identified as the "Urban Revolution".

In order to understand the nature of this revolutionary process, it is necessary to identify the way in which the "material transformation of the economic conditions of production" took place. This can be examined through concrete scientific procedures; it would then be possible to infer, or to record, the transformations that took place within the superstructure: the juridical, political, religious, æsthetic or philosophical and ideological forms "through which people become aware of the conflict and fight it out".

Childe, for this purpose, used the recording and the identification of the consequent phenomena of the Industrial Revolution, led by the Bourgeoisie in the Western

World. For that he used the demographic indicators and the accelerating process of technological changes. This is the reason why many mistakenly took the Urban Revolution (and the neolithic one as well) for a simple accelerating rhythm of technological development, which affected demography, and which they called "revolution". In so doing, they distorted the qualitative and global meaning of the social revolutionary process, and they replaced it with the segmentary and quantitative concept of "Technological Revolution", with which some sociologists identify the stage that followed the Bourgeois Revolution, known as "Industrial Revolution". For those who see things this way, revolutions are the accidental result of technological progress, and are unrelated to the causality chain that is generated at a social, rather than technological level. We think that, even though the demographic effects are substantial and measurable, the indicators of change are a lot more significant in other components of the productive forces (like the instruments and the means of production), and in those through which it is possible to see changes in the social relations, where the study of settlements (village–city) is obviously fundamental.

1. *"Neolithic Barbarism" and its Decomposition.*

The discovery of agriculture and animal husbandry in the central Andes, and the improvements in harnessing sea resources, have had a long and unequal development. This started right after the Pleistocene, around the seventh millennium BC, and it has had one or two thousand years of experiments in different parts of the territory. Actually, there are indications that around the fifth millennium, some plants and Andean camelids were already domesticated. This is the stage of the neolithic Revolution.

The "neolithic chain" spread very rapidly, and between the fifth and the third millennium BC the carriers of the neolithic mode of production had settled in every inhabitable part of the territory.

The Marañón Andes, on which our study is focussed, are comprised between the 6th and 12th degrees of southern latitude, approximately, i. e. between the large Sechura desert, to the north, and the arid zones that separate Lima from Ica, and the Junín plateau from the Apurímac Andes, to the south. This region is characterized by a dominantly mesothermic environment, which contrasts with the cold and arid southern lands, and the hot and humid northern ones. This is where the "*puna*" (bleak tableland) ends, that characterizes the Andean highlands to the south, and the "*páramo*" (high barren plain) dominates, which has a cold but humid kind of environment, typical of the high mountain chains or "*cordilleras*". It is also the area of the widest and strongest rivers, so much on the oriental slope which drains the large Amazon river, as on the western slope which flows into the Pacific Ocean. The Andean "*cordillera*" runs parallel to the seashore, forming a relief that practically rises from the beaches up to 4000 to 6000 meters above the sea level. The "*costa*" (seashore) is a narrow strip of sandy desert, interrupted by green triangles formed by the draining of rivers from the "*cordillera*", and which agriculture has actually transformed into artificial valleys. The ocean is rich in mammals, fish and mollusks, and this allows the capture of generous amounts of protein. The "*sierra*" (mountain chain) is formed by the "*páramos*" and "*punas*" of the highlands, as well as by the inter-Andean valleys and ravines. These altogether form landscapes of great diversity and, consequently, contain resources of various kinds. But the "*sierra*" also shows different regions from east to west: it is arid close to the coastal desert, and it becomes a forest near the Amazon jungle. This, in combination with the various altitude floors, generates a remarkable variety of landscapes and ecosystems. The eastern Andes are obviously in close relation with the Amazon region.

The development of primitive agriculture, thus, must have been confronted with the search of different technical resources, in order to subdue such diverse environments to the needs of plants and animals in domestication. In fact, the archæological evidence shows that, around 7000 BC, agriculture was in process at a local level. The cultivated plants basically depended on the natural conditions, and they were mostly legumes (*Phaseolus*), growing in valleys and ravines. Sometime later, maize and cotton were introduced (or domesticated), and the consumption of mesothermic fruits increased. Simultaneously, the cultivated areas were expanded by the use of irrigation techniques of various levels of complexity. This favored, among other things, the rapid transformation of the drainage cones into extremely fertile valleys. The combination of seashore riches and the product of increasingly successful cultivation, allowed a remarkable population growth. Consequently the number of villages increased, and there was a greater dispersion of the inhabitants of the region, who started moving toward the interior valleys, though they never abandoned their coastal settlements, which provided them with seafood. In the "*sierra*", even though the success of the neolithic revolution considerably widened the life span, the resources were not so abundant and the cultivated areas did not allow such a population increase and wealth.

Around 2 000 BC, the situation had changed in terms of the population figures, of the adaptability of plants, and of the varieties being cultivated; there had also been innovations concerning productive resources. Constructions, other than residential ones, began to appear associated with certain villages, on the coast as well as in the sierra. These were temple-like constructions, and they were obviously public in nature. On the other hand, there were great concentrations in those villages that were close to the shores, where abundant sea fauna could be gathered.

The cultivation of cotton (*Gossypium barbadense*) allowed the development of textile techniques, and the creation of valuable tools for fishing (nets and ropes), transportation (bags), and other activities.

The tendency towards more permanent settlements became general, and there was a shift in the use of construction materials, from perishable ones to stone and clay, be they subterranean or semi-subterranean precincts. But the most remarkable changes regarding settlements concerned non-residential constructions. There is a somehow complex village in a place known as Alto Salaverry, on the Trujillo coast, where simple domestic constructions, which appeared to have no planning, coexist with constructions that are apparently communal in nature. These buildings have a rectangular plan, and they are associated with platforms and burials. They are better built than the domestic units, and their arrangement suggests a certain planning. Finally, there is, in their vicinity, a circular precinct, like a sunken court, the model of which became generalized later on, and which seems to have had a practical function: it served as an instrument for measuring sunlight changes through its moving shadow, during the day and year round.

Alto Salaverry was not, by the way, the only settlement of that period to have such an "observatory". Other contemporary constructions can be found in the same northern coast, as the one in the "*Salinas*" of the Chao valley, or the one in Piedra Parada, in the Supe valley, to the south. All of them share in common their association with a village.

On the coast, at Ancash and Lima, the development of these villages with associated public buildings, was even greater. On the Casma valley, one of the most spectacular settlements of this period, known as Las Aldas, has a very large complex of non-domestic constructions of a pyramidal form. One of them, the central building, is made of seven platforms excavated from the natural hill, which suggest a great investment of labor force. There, the houses were not subterranean, but they were built upon the rock. In the proximity, similar situations can be found, where the non domestic buildings, in the form of superimposed platforms, become more and more important. Of these sites, the best known is the one in "Aspero", in the Supe valley, where a complex of public construction mounds has been found, associated with residential areas. One of these buildings, called "Huaca de los Idolos", appears as a pyramid with precincts on the top, which seem to have had a ceremonial function. The buildings are arranged in such a way as to suggest a regulating intervention, apart from the fact that their existence implies a greater amount of work than that required for the construction of a domestic unit. Others, to the south, show deposits associated with the public buildings.

The beginning of a process of great change in the current social formation was obviously taking place. Around the beginning of the second millennium, there was an intensification of interregional contact. This allowed the insertion of ceramic artifacts, from north to south, and of new cultivated plants, coming from the south. Very soon, the llama and alpaca livestock were also introduced; their domestication seems to have started in the southern regions, between Junín and Titicaca. It is necessary to analyze the nature of such changes.

In the primitive community, the productive process is carried out at a domestic level. The community as a whole participates actively, and everyone has access to the same instruments and means of production, so that the division of labor results from individual ability, or from age or sex. This allows an egalitarian access to the products generated by a socially non-differentiated work, through mechanisms of reciprocity, and through a collective notion of property. As Marx says: "Property, thus, originally means the behavior of man towards his own natural conditions of production, and towards those conditions that are his, that are tied to his own existence..., that, so to speak, only constitute an extension of his body". Inasmuch as the production and use of the instruments of labor and means of production are susceptible of being controlled by every member of the collectivity, in a generalized way, property is thus communal, and there are no different classes of persons, as a result of a different participation in the productive process. With the discovery of agriculture, the simple tools produced to carry out the work are the property of individual workers, while land (which is the principal means of production created by man through his work) is collectively owned and requires —at this stage— cooperative work.

In effect, since 5000 BC settlements occur in the vicinity of fields susceptible of being cultivated, even though on the coast, the proximity of mollusk shoals is preferred. In those places where camelid hunting is successful (in the highlands), changes take place more slowly, and they are progressively tied to pasturing in some specific southern areas. When the Neolithic starts to appear more clearly, during the third millennium, the still disperse settlements are associated with areas of natural watering, as the so-called "green disks" created by the residual waters of the phreatic mantel. It seems that the valleys were not adequate for agriculture, since they did not exist as such; they were rather water courses humidifying irregular and rocky banks. It was necessary to clean and flatten them, and to divert and channel the water. These works seem to have started during this period. Thus, cultivated fields were small and their productivity was low, and the work could be done at a domestic level, through working units constituted by a few families.

Enemies were not neighboring communities, since geographical conditions are not favorable to easy migrations (there are deserts or immense and difficult "*páramos*" or "*punas*" between valleys). The enemies to be defeated were the natural factors, so much small rodents or predatory insects, and fungi plagues or illnesses, as the even more difficult to fight: the frost that burns and destroys the crops, the droughts that prevent germination or stunt the development of plants, the floods or alluvium that destroy fields industriously prepared for cultivation, and even houses and other types of constructions, and finally the small amount of fertile land and the scarcity of hydraulic resources.

Thus, three are the essential problems the Andean peasant had to face: the climate and the seasonal changes, the fertility of the soil, and the water needed for cultivation. These problems exceeded the domestic productive capacity, and they seriously affected the life of the community, who saw its existence limited by the absence of means to overcome its hazardous dependency upon such natural factors. With respect to the relationship water-soil, it was necessary to create irrigation systems, through canalizations and flooding, which could be dealt with at a domestic level, and to develop the means of flattening and cleaning up the areas through terracing and simple fallowing work. What the empirical-spontaneous knowledge could only resolve at a very simple level, was the control of weather by means of keeping useful calendars to predict the periods of alternating climate.

The agricultural calendar is mainly solar, given the fact that it is associated to seasonal changes that occur annually according to solstices and equinoxes. These empirical observations, based upon the movement of shadows and other indicators, such as the "first rain", do not require any specialization, and they can be kept in the collective memory: they allow a rudimentary programming of agricultural activities. The presence of such circles as those at Alto Salaverry, or at the Salinas de Chao, indicates that a technically more sophisticated, and certainly more precise way of predicting seasonal changes, was taking place at a semi-specialized level. With the use of a central element —for instance, an obelisk, or a simple vertical pole— the projection of its shadows, forming rays that vary in the course of the year, would allow to set a fairly simple solar calendar, with a high degree of accuracy. This necessity is common to every peasant.

But in the Andes, the problems faced for weather prediction are greater. The foresight of frosts in the sierra must take place days in advance, in order to regulate the sowing tasks, according to the different types of cultivated plants and their growing ranges, depending on the altitudinal floors and on the hydraulic variations of each of them. Moreover, even though the seasons are measured there in terms of periods of drought and periods of rain, seasonality is in fact relative. What is more likely to occur is the alternation of major cycles, and the most typical classification of agricultural weather, is one of rainy years alternating with dry ones, the predictability of which greatly exceeds the solar cycle. The nature of the observations, the record of regularities, and the appropriated measuring tools, require a level of specialization higher than the one offered by the daily accumulation of empirical-spontaneous knowledge.

At this point of the historical process, the degree of complexity in the production determines the fact that the whole population is no more in a condition to take equally care of the productive tasks. This is due specially to the fact that the development of instruments of production and forms of labor, needs to be conducted by initiated persons, or specialists, in whose hands this part of the productive process now rests, and for which the rest of the population is neither technically nor socially fit.

The kind of specialization we are referring to, does not depend upon the capacity or the will of people, as happens with the crafts specialization (pottery, weaving, carving, metallurgy). The existence and development of these artisans depend mainly on external factors, such as individual skill (artists), the abundance of raw materials, or the possibility to rely upon a surplus economy. This is a specialization that is technologically and socially necessary; i.e. this kind of production can only be conducted by specialists that are identified as such, not so much for their skill, but for their technological knowledge, and whose tools and forms of production are not accessible to others.

This kind of specialization also has a direct incidence on the production of goods that guarantee social reproduction, in such a way that society could not progress without it. We are talking about workers, whose existence must not only guarantee the improvement of the living conditions of the non-specialized workers, but even guarantee their survival in certain specific cases. Their products are thus not required by the consumers' will; they are not "dispensable" but, on the contrary, they must be socially indispensable. The existence and reproduction of the specialized workers as a class, do not depend upon the existence or lack of surplus, the demographic pressure, or their bellicosity and will of power, but upon the nature of their social being itself. Thus, this is a class of people that is socially necessary by itself, independently from the external factors we mentioned above, which only contribute to the reinforcement, acceleration, magnification or retracement of its development, but which do not determine its existence.

The second half of the second millennium BC was a period of remarkable improvement in that direction. Little before that, between 1800 BC and 1500 BC, pottery had arrived from the northern Andes, and apparently from the neighboring Amazon region as well. It was very successful, as can be seen from the accelerated diffusion to all points of the Marañón Andes territory. Pottery confirmed the possibility of settling far from the direct sources of water supply, and it also greatly facilitated the storage, preparation and consumption of food. Almost simultaneously textile technology improved through the creation of more advanced looms. And finally, construction techniques acquired a remarkable level, through the development of complex calculations on the resistance of materials, and of geometry. This way, it became possible to design and build larger and higher architectural complexes, and to improve the articulation of accesses and services. Hydraulic technology was favored the same way, by improvements in the construction of water gathering systems that were much more sophisticated than those which could be dealt with at a domestic level, allowing the watering of totally uncultivated areas.

There is an association of stable villages with constructions of masonry and clay (*adobe*), and with public ceremonial buildings, consisting of truncated and terraced pyramids, like the ones on the coast, or with precincts which had a highly elaborated hearth (including a system of underground air supply) like the ones in Kotosh or Tantamayo in Huánuco, those of La Galgada in Pallasca, and others in Callejón de Huaylas, in Cajamarca, and (although different in form and size) in El Paraíso, Lima.

The coast played a most important role during this period, not only because of sea products, but essentially because the transformation of the drainage cones into valleys took place at this point, allowing a great economic flourishing. This success had to take the dry years into account —when there is insufficient water coming from the rivers, that at times dry out— as well as the wet years —when it even rains in the desert, and destructive alluvium comes down the river beds, invading cultivated areas, and sometimes changing directions. The inability to make predictions in that sense still brings today fatal consequences. In 1925 and in 1983, the "rainy years" provoked terrible disasters, destroying entire cities, fields, irrigation systems, roads, etc., thus forcing the populations to migrate, or to spend several years of hard reconstruction work and famine: they were declared areas of national disaster. What happens on the coast, is naturally linked to situations of rain or drought in the "*sierra*".

The humid phases in the region are called "*Niños*". It is a phenomenon that is originated in the ocean, where two submarine streams flow: a cold one, called "Humboldt", that flows from south to north, and a warm one, called "*El Niño*", with a north-south direction. Their effects on the Andean weather are far-reaching. The former affects the environment, provoking the formation of the Peruvian coast desert, and influencing it in such a way as to inhibit the rain along its western strip (including the maritime front of the "*cordillera*" that has, for this reason, an almost lunar landscape). The latter is an equatorial stream, which favors the formation of clouds and their precipitation; during the summer (around December, at Christmas time), when it flows to the south, it is said to be "the Child's advent", in the form of rain, and the river flows increase. But "*El Niño*" does not move the same way each year: there are years when "it gets lost". The problem is the difficulty to record these irregularities and, above all, to predict the real magnitude of the "rainy years".

Though this is just the beginning of a research project on this topic, during a meeting with Jorge Marcos on the coast of Ecuador, in the summer of 1985, and with the help of Presley Norton, in Salango (Manabí), there was an agreement to formulate a hypothesis Marcos had long been searching for, to explain some specific aspects of the Valdivia culture. We, on our part, wanted to explain certain elements of the Chavín culture: the traffic of *Spondylus princeps* along the Andes.

Spondylus princeps (*mullu* in quechua) is a bivalve shell, of white and intense red tones and a spiny surface; it is found in the warm waters of "*El Niño*" stream, or in the sea affected by it. It has a great thermic sensibility and it can move very rapidly, so that it can easily follow the comings and goings of the stream. It seems that when "*El Niño*" moves southward and rises to the surface, the *mullu* shell follows the stream, and when it recedes or descends, the *mullu* sea shell disappears, and it must be looked for at a depth of 20 to 40 meters, and also near Manabí to the north, where it generally can be found.

When the Spaniards reached the Andes, they found that the *mullu* was one of the most important, if not the most important, exchange item between the central and south Andean peoples and the north. There was, in the hierarchy structure of the State, a public official associated with the traffic of *mullu*, called "*Mullu-chaski-camayoq*". Furthermore, the Andean gods demanded that the tributes be paid in *mullu*, an element related to the water. The archæological data show that a huge amount of *mullu* was actually required in the central Andes in all cultures, at all times. And these sea shells were necessarily imported from the Guayas and Manabí coasts, in Ecuador, since these do not live in the cold Peruvian waters.

The first known *mullu* in the central Andes date to the second millennium BC, even previous to the emergence of pottery, and they were found in the form of necklace beads and other small manufactured items. By the end of the second millennium and the beginning of the first, their consumption increased considerably, so that there is hardly a site of this period where *mullu* shells are not found, either as manufactured items, broken or in their natural state. Their presence from then on is generalized.

The hypothesis that came out in the meeting, on February 17, 1985, in Salango, has been brilliantly exposed by Jorge Marcos:

> According to the actual evidence, we suggest that the ritual use of the *Spondylus princeps* in the Andes, was based on the fact that it served as a tool for weather prediction, used by specialists through time along the equatorial coast, in order to determine rain and drought cycles, and such important events as the "*El Niño*" phenomenon, that could so dramatically affect food production in Andean America.

It is in fact a bio-indicator of the irregular rain cycles in Guayas and in Manabí, and it became by extension a fundamental instrument of weather forecast for central Andean specialists, regarding the frequent and catastrophic alterations produced by "*El Niño*". Thus, while it served as a predictive instrument accessible to every peasant in Guayas or Manabí, it was a costly and even luxury, but fundamental item for the Andean peasants.

In order to obtain *mullu* shells, or to get in touch with those who could get it from the sea, and to make predictions (according to its abundance or absence), the central Andean populations had to cross the immense desert of Sechura and Olmos, or the hard mountains of Ayabaca. Be it by land or by sea, they necessarily had to come across a series of barriers, such as the wide forest rivers: the Chira, Tumbes, Guayas, which were filled with huge crocodiles "eating humans", whose aspect and terrible grunts are enough to frighten those who —in the south— are not used to this kind of monsters. They grunt like jaguars or lions; they have eyes like falcons, eagle or feline claws, and powerful tusks —and many of them— like the ones of the serpents, only larger. They are several meters long, and they take the color of the riverside foliage or the dark beach sand. Any person who wanted to have access to the *mullu* from the south, had to cross this barrier. This is probably why after 1000 BC, or a little later, an intermediate exchange area was set for getting *mullu*, on the Azuay and Cañar "*sierra*", in Ecuador, which could be more easily reached. To obtain *mullu* was thus a task that went beyond the limits of domestic or communal units, even though in the beginning, it might have been obtained in small quantities, in the form of manufactured goods, by sea from the north. In addition, its predictive characteristic must have become a much appreciated one.

There is in Chavín a monument dating to the last few centuries of the second millennium, or the beginning of the first millennium BC, where the whole complex of these factors and elements appears to be represented. There are two monsters, one male and one female, carved in an obelisk that is almost three meters high. The structure of their bodies is one of a crocodile or an alligator, and it is constituted by a vast amount of figures, certainly recalling the myth: falcon eyes, jaws with four tusks like those of a serpent, and feline-like claws. By their heads are some interesting naturalistic elements: a feline, a fish and a falcon (the earth, the sea, the air?), and the representation of a *mullu*. Apart from that, there are some tropical plants (like the ones in the equatorial areas) and another sea shell, the "*pututu*" (*Strombus galeatus*), which is used as a trumpet of long-range sound, and which is also found in equatorial waters. It is obvious that this was part of the Chavín religion.

But speaking of Chavín, whose edification started during the second half of the second millennium BC, it is fair recognizing that we are facing a monumental complex which is qualitatively new. It is characterized by a set of public —non domestic— buildings, whose functions are of cult and services, and which are surrounded by complementary residences. This is no more a village, like those of Alto Salaverry or Salinas, where there are a few public buildings apart from the houses; it is exactly the opposite, and this is a qualitative difference. Apart from being sumptuous and expensive, in terms of the labor power invested, the constructive spaces are made with the necessary participation of experts and architects of specialized rank. But moreover and above all, it is worth emphasizing the functional aspects of such buildings: all of them are apparently organized as observatories, workshops and warehouses. Descriptions are abundant in the literature, so we will not deal with them here, but we must nevertheless point out the generalization in the use of terraces, platforms and precincts, designated for the observation of stars. This specialized task will be related to the temples, and will last up to the seventeenth century, according to a story of the Andean mythology that was found in the Huarochirí area, in the Lima highlands, that says:

> And we know that, ever since old times, people follow the same ritual to order the *waqsa*. This is as follows: A man from the *ayllu Qaqa siqa* served as master of ceremonies during these rituals. Ever since old times, there was one or two of them, and they were called *yanka* in every community. They say that he observed the course of the sun from the wall, a very well aligned wall. When the sun reached a certain point on the wall, he would tell people "now", or if it did not reach that point, he would say "we shall go tomorrow" [translated from Quechua into Spanish by Urioste].

Naturally, by the sixteenth century other means had replaced these costly and primitive observatories, and the "*yanka*" specialists were no more at the center of the system, which was then dominated by lords of war and others associated with the palaces rather than the temples.

Chavín was not an isolated or unique ceremonial complex during this crucial period, between 1500 and 1000 BC. The Marañón Andes were covered, on the coast and in the "*sierra*", with similar complexes, some of them larger, some of them smaller. There were two or more such temples in each valley, with associated precincts, and some of them were very large. All of them had a foreseen cosmic orientation, and elements that were useful for making different types of observations. If this formal tradition started in the beginning of the first millennium, and there are large complexes close to the coastal valleys, like Las Aldas in Casma, we could say that their conductors, forecast specialists, priests, had certainly succeeded to prevail due to the success of their predictions, and to the enlargement of cultivated areas through the use of extensive irrigation systems. It is clear that they did participate in these operations, not only because the flourishing of the so-called Formative temples is parallel to an intensive program of canalizations in the valleys, and to the increase in the number of villages, which grew together with the new cultivated fields; but it is furthermore obvious that such operations had a supra-communal character, since they encompassed entire valleys, and they involved a cooperation well beyond the local production units. Besides, because of their magnitude and their extension, they required a specialized knowledge concerning water falls, the span and level of the canals, the nature of the ground, of the slopes, etc., which involved, in effect, precise calculations. In this respect, the construction of Chavín, with its subterranean canals of different types —some of which were designated to create "mysterious" noises—, are a sample of this kind of specialization. There is no doubt that the work of the temple priests considerably increased the production, in terms of quality and of the extension of the cultivated area. It also guaranteed certain security margins to face the climatic irregularities. The "oracle" priests used the *mullu* and the observatories as the concrete instruments of their productive work, and they created a new product for the agriculture tasks: highly specialized working tools, which the peasants needed without being able to generate them by their own means.

This is when society divides itself in two different classes of producers: those, the majority, who continue to take part in the productive activities of communal–domestic character (food production and crafts), and those who must abandon this kind of work, to devote themselves to a long and costly specialization (of economically unproductive time), and then to productive activities absolutely different from those of food production or domestic craftsmanship. This new class of people has different instruments of production; it requires a different productive time; its physical working space and necessary means are also different. Therefore, it is not linked to the other class of workers through common results or interests; it has, on the contrary, very different interests, its work results in very different products, and therefore it provides different life resources. For the members of one class to have access to the products generated by the members of the other class, it is necessary to establish the rules of the game that will allow it. These could not be set at the level of a "Neolithic" or egalitarian type of community.

2. *The Initial Class Society*

The relationship between one social class and the other is determined by the kind of participation each of them has within the productive process, consequently by the specific possession of and access to the means of production it needs, and finally, by the access to the final product of the working process.

Each worker is the owner of the instruments of production he uses in his work, as long as such instruments are the result of his own work. But if such instruments are originated by the work of someone else, a necessary mediation of a dependent nature is generated between this worker and the other, given that the former will not be able to do his work freely without the necessary intervention of the latter. His productive capacity will depend on the instruments available, the possession of which is not accessible to him without the participation of third parties. Thus, his work will necessarily be alien to such a participation, to the point where the absence of such a mediation constitutes a serious danger of production failure.

Specialists generated through their work highly specialized means and instruments, which not only guaranteed the success and expansion of food production (which is basic to the reproduction of the species), but which also regulated the technical working process through calendaric, hydraulic and other specialized techniques. Thus, they possessed such instruments of production; on the other hand, the actual food producers, the peasants, only possessed their labor force and simple tools, which they themselves could generate. Their productive capacity was successful at a domestic level, but it could not be so within the new conditions, when they simply become executors (instruments, in the last instance) of an activity depending on more advanced, and thereby superior, means of production.

Land, which is the main object of labor, was directly associated to labor power, in such a way that this power was the one that transformed land in a means of production, according to its needs and possibilities. And at this stage of development, land had a value only as an extension of labor power.

Consequently, if labor power (which includes workers and their simple instruments, and the land transformed into cultivated fields) was alien to specialists, who owned the essential instruments of production at this stage of development, these obviously became owners of the social totality; this was the juridical form of this type of social relations, which finds its social expression in the direct property of labor power, through the acknowledgement of the subjection of food producers to the means of production generated by priests, which were explicitly formulated through their ideological representation: gods and devils. Therefore, social relations acquire a character of exploitation relations, while specialists become owners of the work of others, of part of the social work upon which they do not directly act.

For this to happen, old social relations, that have an egalitarian nature, must be broken. The development of new productive forces, with their social conductors —the specialists who live in the temples— needs, in order to continue and to progress, a substantial transformation of the whole productive process. This means that, first, old forms of production must be substituted by new ones. Old forms have a domestic character (family or community levels), and they are based on the simple cooperation of the producers of consumption goods. New forms of production go beyond the domestic level, and they require the complex extra-familial and supra-communal cooperation; this implies inter-tribal relations, determined by the new conditions of production, such as the territory or the working resources, and independent from kinship or affinity.

The rupture of these social relations —necessary requisite for the development of the new productive forces— involved significative changes within the organization of the family, within the social organization of labor, and consequently within the whole group of institutions that regulated this form of relationship. The fact of keeping them or choosing to change, is the logical component of the critical conflicts tribal society had to face, in order to become a class society, besides having to assume a new kind of dependent relationships with the new class of specialists who lived in the temples. This demand for change, necessary and compulsory to the new class —not to the direct producers of consumption goods— represents the necessary factor for confrontation or struggle between both classes. This could have taken different forms, from inter-tribal conflicts originated by opposing and divergent interests, to the configuration of gods as demanding, powerful and terrible entities, with sanguinary servants endowed with supernatural and other powers.

Religion was, undoubtedly, an essential element of this process of class struggle, with gods and other forms that emerged, not so much from the will or imagination of priests, as from the concrete conditions of their participation in the productive process.

In order to understand this, one must take into account the fact that the priests-specialists' work instruments consisted in complex technological systems, which they are not only unable of producing individually (like the astronomic observatories, the recording of cosmic movements, or the measure units of the regularity or singularity of phenomena), but, at the level of producing results for their use, they are not even under their control, since they depend on natural forces, the origin and nature of which cannot be perceived by material recording instruments (rain, frost, drought, *niños*, etc.). From this we can say that the phenomena by themselves, and above all those which can be perceived (the stars, the water, the sea, the mountains, etc.) become the instruments of knowledge, and the objects of specialists' labor. Their dealing with such objects is of a trascendent nature for their purposes, and it is governed by non-comprehendible causes at this stage of development. Those natural forces, which are at the same time objects of labor and instruments of knowledge, acquire this specific condition through the notion of divinity (gods and devils). This is an identification of their character and form, and it resolves their representation at a social level, granting them the "supernatural" dimension that these forces obviously have at this level for explaining natural phenomena. The external aspect these forces, or supernatural beings, have will depend on traditional factors, or the form in which they are captured, on their singular effects, etc.; their description and explanation belong to the field of myth and liturgy. Thus, these divinities were not invented or foreign to the living conditions of society; these gods and devils had a real existence and a social transcendence, and they were therefore necessary and effective. Their identification or recognition, and their cult did not depend on peoples' will or empathy, but on the specific effects of their existence and powers, which priests were able to know directly, thanks to their specialization.

These conditions were very clear in Chavín and in central-Andean cultures that were contemporary. Besides the multiplication of ceremonial centers, and the clear definition of specialists with respect to land workers, specifically in terms of settlements, religion is an eloquent reflection of the conflictive character of the necessary rupture, caused by the new class. We are not yet able to describe the specific nature of these conflicts, but it is obvious that the priests of Chavín had to have recourse to cruel and highly representative divinities, using power symbols of strength and death: gigantic tusks, claws in hands and feet, hair in the form of serpents, cadaverous or ferocious heads sculpted and nailed to the temple walls, etc. The obscure labyrinths of the temple mean that there was a restrictive access to secret rituals, for only a few initiated ones, and to the sumptuous character of the temple were added mysterious noises and a paraphernalia, not easily comprehensible for the non-initiated ones. It was a religion of terror, that surely was the guarantee of a certain form of power, concentrated in the hands of the specialists.

This is the condition under which new norms of social relations emerge. These, among other things, involved the dependence of a group of people with respect to another. This dependency found a concrete expression in the conveyance of the product of the work of one group to the other, through a new economic form: tribute.

Tribute is a form of economic relation which consists in the compulsory conveyance of consumer or capital goods, as a compensation for services, supposed or effective, not measurable and hence not susceptible of a concrete equivalence of any kind. This is the way in which property relations between specialists and producers of consumer goods are dealt with. For as much as labor power is the fundamental good that generates all the material goods, tribute in the form of labor is the main form this economic relationship takes: workers owe their labor to those who provide them with the essential instruments of production. The form, content and magnitude of the access to labor power that the temple specialists have, will depend on external factors, such as the nature of labor requirements (agriculture, public constructions, services, etc.), or the relative potential of human or natural resources available. This is the reason why the formal aspect of tribute has a singular character and is not susceptible of generalizations.

For as much as tribute is the form of appropriation of labor power by specialists, in a concrete way, it represents the appropriation of the workers themselves, and/or the goods produced by them. The different forms this appropriation takes are singular as well, and they depend on variables that are equally contingent: they can vary from slavery relations or different types of servitude, to tribute in species of different kinds and character, including the organization of compulsory regular work, and redistribution or asymmetric reciprocity forms. What is neither singular nor contingent is the appropriation —in one or several modalities— of the labor of one social class by another, in the form of compulsory tribute, as we have define it here.

Whatever the etiology that shaped the way in which this problem was faced in Chavín, the temple priests received tributes, from land workers and communities distant by hundreds of kilometers, that we shall for now identify through the offerings that are found in the temple. To Chavín —which is not a region susceptible of generating surplus— came llamas from the highlands, deer hunted at heights superior to 4000 meters (the huemules or "*tarucas*"), deer from the inter-Andean gorges (the "white-tails" or "*luichos*"), and little deer from the low forests (mazamas). They received cuyes or guinea pigs (*Cavia porcellus*) that were domestically grown, vizcachas (*Lagidium* sp.), ducks and other fowl of diverse provenance; fish and mollusks of the northern and southern seas; beautiful terra cotta wares, in the style of the northern coast settlers (Cupisnique), of the Cajamarca (*sierra*), and of the Ancash and Lima coast; finely elaborated bone and shell items, as well as fine crafts made of select stones, also of diverse provenance. At the beginning of the first millennium BC, the priest-specialists of Chavín, with their ferocious gods, had access to all of these goods, which they naturally were not able to produce themselves. The only possible explanation is that this was originated through a relationship based upon the power accumulated by these lords.

Power is the possibility to have access to the resources necessary to guarantee the survival and reproduction of consumers. Hence, the relations of distribution of consumer goods are subject to the norms that regulate power. Within the egalitarian societies —where all members are able to have access to such resources, in the production of which they participated in the same way— power is dissolved within the community, taking different forms that are adequate to the necessity of facing external contradictions with other communities, or with the environment. This dissolution of power takes place according to age or sex, or through kinship and status rules. Within class societies, an entity superior to the community is needed to regulate the extra-communal class relations. This social entity is identified with the State.

The State is the form of organization of society that corresponds to the necessity of establishing and imposing the norms of relationship that must exist between the different sectors of society. Consequently, it appears and becomes necessary when society is constituted by sectors whose interests are not equal, and their relations require arbitrage. It is thus, the typical form of organization of class society, in which interests derive —in form and content— from a different participation in the productive process and, consequently, from a different access to the means of production and consumer goods.

If this form of organization responds, in effect, to the different, and normally conflictive interests of social classes, it must regulate them according to the character and social condition of each one of them. Such a form, again, does not depend on the will or the number of individuals that form a certain class, but on the specific and determinant relation of the means of production they control, according to their participation in the productive process, and to the nature of the mode of production to which they belong. This means that the form or content of the State will necessarily respond to the unequal correlation of the social classes involved, according to the interests of those which possess the means of production (that constitute the determinant factor concerning the nature of the specific mode of production), and not to the interests of either one of the actual classes. This, in turn, determines the fact that such a class necessarily constitutes itself into the dominant one, economically, socially, and politically: economically, because it possesses the means of production, which are the mobilizing axis of society; socially, because it becomes the strategically most important social element of the whole productive process; and politically, because it is able to establish the rules of relations with the other social sectors, according to its own interests and living forms, and for its own survival and social reproduction.

Therefore, in the transition from an egalitarian to a class society, the specialists' class becomes the dominating class. This is due to the fact that the new social order that is born with it, is structured in relation to a mode of production that is supported —beyond agriculture and other forms of production of a domestic nature— by the specialized production of this group of individuals. This means that agriculture and the other productive forms, become structurally dependent on the specialized forms and are alien to them. This alienating process advances to the limit where it submits the whole agricultural productive process to the development of specialized production.

It was the duty of the Urban Revolution to make this state of things possible, at a social cost that is yet unknown, but that clearly intensified violence, and created henceforth the institutionalization of warfare; at this stage, the aggressive and defensive tasks seem to have been solved at a non-specialized level. No manifest military forms existed, but there is, on the other hand a strong increase in cannibalistic practice, human sacrifice, ritual decapitation, and other means of aggression or imposition of ideological models based on violence. If data are still insufficient, everything seems nevertheless to indicate that cannibalism appeared, in the Marañón Andes, around the beginning of the second millennium BC, and that it progressively became generalized, so that in Chavín times, around 1000 BC, the practice had extended to the whole area. The evidences are fragmented human remains, partly burnt, and found in domestic refuse contexts, often with high frequencies in relation to the remains of domestic or hunted animals. We have noted that there is a coincidence between the appearance of specialists in the social scene, and the beginning of these practices, which, in other areas (like the Andean north), seem to have started much earlier. At the same time, we find ritual practices that indicate human sacrifice in Chavín, including the use of human heads in offering contexts, and their artistic representation, as well as the cremation of corpses in similar contexts. In a very well-known place called Cerro Sechín, in the Casma valley, there are scenes of great violence carved on the walls of a temple, that show pared bodies, heads, vertebra, viscera, arms and legs, as in an orgiastic butchery, with richly dressed characters acting as executioners or priests (or warriors?), in a context that seems to correspond to the second half of the second millennium BC. The period within which the Urban Revolution took place is, thus, an era of great violence, even though there are no indications that priests had to protect themselves from eventual enemies. This would indicate that violence was exerted upon peasants, who lived close or far away from the temples, in villages or hamlets that were not very different from the preceding types.

The novelty concerning settlements was of course the housing of the new class of specialists. We are facing a very well defined differentiation of the working space, and consequently of the residence. It does not mean that peasants decided at a certain point, by a gregarious instinct, or because of demographic pressure or fashion, to increase the size of their settlements, and that these later automatically became cities. Food production allowed the continuation of village life (in the size and form adequate to the type and abundance of available resources), with a settlement pattern directly associated to cultivated fields or to the extraction sources of raw material (beaches or coves, valleys or forests, etc.). The work of specialists requires a different space; the most important characteristic of this kind of space is related to the fact that sedentarism appears at the higher levels of locality —around the working place, which is stable and permanent, and whose location and characteristics are the result of decisions taken by the social user himself. His residence, thus, can be fixed the same way. In fact, while the settlement pattern of land workers depends on the natural instruments of labor —since their activities take place outside the residential space— the pattern corresponding to specialized workers does not require that they live outside the physical limits of their working place (the temple). They can actually remain there all the time, without having to move at all. This necessary condition of nucleation (between the working place and the residence) is the origin of urban centers, which appear and develop, not as centers of greater density, but as production centers, different from the rural ones, which are qualitatively and quantitatively different from those. Their ulterior growth and development depend on the success of their products, on the possibility to capture the surplus generated by their environment (which can be richer or poorer), etc. This is the organic and causal origin of the city.

Between 1500 and 500 BC, the central Andean territory completely rearranged its *modus vivendi*, and a new social order, of a classist character, took over in a generalized way. Chavín is just the most significant example of this process, but similar conditions prevailed to the north and south of the Chavín cultural area, within the geographical limits of the Marañón Andes. Later, the phenomenon extended to the southern, semi-arid area of the central Andes, up to the Titicaca highlands, with very similar characteristics to those examined here, and with particular traits responding to the analysis of derived phenomena, which we shall not deal with here.

3. *Conclusions*

The examination of the theses presented here, and their comparison with the data from the territory of the Marañón Andes, allow us to arrive to the following conclusions:

A. The decomposition of the village community, or "Neolithic Barbarism", began as a consequence of the development of advanced procedures in food production, that simultaneously and necessarily included the use of new instruments of production, the presence of specialists for the production and handling of such instruments, as well as certain modifications in the territorial sphere of their application. These are, thus, novel alterations within the complex of productive forces, which involved global changes in the corresponding labor process, and in the technical and social relations.

B. The former social relations of production were of a domestic and communal character, with labor and property forms of the same nature, all of this deriving from a similar access to the instruments and means of production. For as much as the new productive forces involved a new social division of labor, that divided workers into classes with a different access to the instruments and means of production —and consequently a different form of property over the resulting products— a transformation in the current social relations became necessary; these can no more be substituted by norms of equivalent reciprocity or egalitarian rights.

C. It was thus necessary to start a period of social revolution that resolved the conflict existing between the new social forces of production and the communities of workers, devoted to the direct production of consumer goods, which sustained them.

D. The Urban Revolution had to establish the new rules of social relations, in conditions such as to permit the survival and reproduction of the new productive forces (specialists, their instruments of labor, and the new means of production); it had to withdraw every obstacle to their development, and to fix the nature and content of property, i.e. the use and access of both classes to the production and consumption goods.

E. As a result of this process, which occurred in conditions that indicate different levels of violence, new social relations stemmed from a new form of property, of a classist character, and from a new organization of society, based on the State, which could guarantee the compliance of the new forms of coexistence.

F. The new forms of property, which are nothing more than the juridical expression of changes in the social relations of production, that impelled the revolutionary process, established that the specialists class (producers of the new means of production, and who were the main impellers of the new productive process upon which the success of the concrete action of labor power depended) became the owner of the labor power —including workers, their instruments and means of production. On the other hand the direct producers of consumer goods remained as owners of their labor power, alien to the instruments and productive means, of which specialists were the sole proprietors. Thus, in the last instance, specialists became absolute masters of the social totality, making use of labor and production goods, according to their own interests and needs.

G. State, which becomes the regulating entity of the established relations, is organized in relation to the specialists class, for as much as it must guarantee the survival and development of the new social order. Evidence indicates that it reinforced its institutional character through the use of the same instruments of labor used by specialists, i.e. the concrete natural forces, socially made explicit through ideological configurations, such as gods and their mythology. At this stage, State does not seem to be separated from Religion.

H. So much the State as the gods are due to the same causality sources that gave birth to social classes, to property and to the reorganization of family and labor, in much the same way as they determined not only the social and economic, but also the physical separation of the social groups divided into classes. This finds an expression in the emergence of the city, as a qualitatively different kind of settlement from the previously dominating village form. In effect, the origin of the city appears to be directly and necessarily associated to the development of the new productive forces, that can permanently live and work in the same physical space (for example, the temple), without having to move at all. Moreover, the city is more a center of production than a residential space, and residence is only a consequence of the necessary proximity to the center of the worker's daily activity. In the case of peasants, the village and the cultivated fields, or the sources for the extraction of goods, are two separate spaces.

Finally, we would like to point out that V. Gordon Childe's thesis concerning the necessity of a revolutionary process in the transition to the first classist formation, stated in relation to the Old World experience, finds a complete confirmation through the examination of evidences concerning the central Andean experience.

The theses sustained here, which are based in a historically necessary causal chain, differ from those which served Childe in his original statements, in various aspects; they also differ from those of whom have treated the problem in a segmentary way, from an analytical perspective that tries to explain only some parts of the process in terms of the origins of the State, of the differences in the structure of the Asiatic mode of production, or of the emergence of the cities, among others. The origin of these differences is not, of course, in the empirical sources on which we rest our comparison, but in the starting point, that is in the thesis that sustains the necessary historical condition of the Social Revolution, originated through a process which is independent from the will of persons, the participation of whom is determined by the place each of them has in the confrontation of the historical conflicts of their time.

FUNERARY RITES AT TWO HISTORICAL MOMENTS IN MESOAMERICA: MIDDLE AND LATE FORMATIVE

MARI CARMEN SERRA and YOKO SUGIURA
Universidad Nacional Autónoma de México

Tombs, mausoleums and burials have attracted the attention of last century antiquarians, precisely because of the idiosincrasy of the wide range of information they can render concerning past societies. This encouraged the search for and excavation of burials and their dead. In fact, they eloquently express not only the chronological and cultural aspects, but the social and ideological ones as well, thus becoming an outstanding archæological indicator. Thus, funerary evidence reflects the direct or indirect culmination of man's conscious behavior (O'Shea 1981).

Over the years, the approaches to research on funerary rites have changed and during the 70's, radical modifications started to take place. In fact, the appearance of the book *Approaches to the Social Dimensions of Mortuary Practices* in 1971, opened a new dimension to the analysis of burial rites and practices. In this work, Binford (1971) and Saxe (1971) emphasize, from a processual methodological point of view, the fact that mortuary evidence has a great potential to explain the social system to which the individuals have belonged during their life span. At the same time, the authors criticize the formalist approach under which burial practices were formerly analyzed. Their argument gave birth to further studies mainly concerning the processual analysis of the social aspects of the dead; the aforementioned studies emphasize the intents to correlate social configuration with the deliberate and systematic management of death within a theoretical anthropological framework. On this respect, one of the problems that bothered many archæologists, was to explain, through mortuary evidence, the social complexity levels, under the assumption that the variables found in a burial site are so differenciated that they divide the universe of mortuary practices; thus, this grouping or division represents the different social ranks. Besides, the principles that make up the various state groups or ranks are the same that conform the social relations (Saxe 1970). From this standpoint, the main thing is, in the first place, to identify and define the groups or social ranks and, later on, place them at the adequate complexity levels within an evolutionist scheme.

Brown (1981) suggests three fundamental arguments for the analysis of social ranks:

1. The first one refers to the principle of energetic expense (Binford 1971; Tainter 1973a, 1975a; O'Shea 1981). It is based on the assumption that the social rank of the buried individual is correlated with the amount or volume of energy channeled towards mortuary rites. In other words, a higher social rank implies a higher energetic expense and this, in turn, is reflected upon the degree of complexity, the treatment of the dead, the location and construction of coffins as well as the extension and duration of the funerary rites themselves. Social power is thus symbolized by the exclusive access to the specific burial locations. The importance of the spatial factor in the mortuary analysis follows from the fact that the spatial structure and arrangement of the burial site, as well as the spatial pattern occupied by the dead within that specific area help us to detect not only interpersonal but intra and inter-communitarian relationships as well; that is, the social system to which the dead had belonged when alive (Goldstein 1981). Nevertheless, the various problems inherent to the

archæological material may bring about some trouble when evaluating the energetic expense.

2. The second argument deals with the identification of tools and attributes as symbols of authority and power within a society, which Binford called socio-technic elements, such as, for example, some type of garments, manufactured weapons and other tools, especially offerings, the material of which is not ordinarily used in dwellings (Brown 1979).

It is assumed that the "social person" of the deceased manifests itself in the funerary rites by the way in which the symbols of his authority are arranged, especially when those symbols have been buried or deposited regardless of their differentiation, within a society, by age, sex and personal ability (Peebles 1974), it is thus easier to establish ranking.

In contrast with what archæological literature generally states, Brown claims that mortuary richness *per se* is not such a strong variable as that which grants the right to a certain burial position within a socially relevant space. For example, within an evolutionary perspective, the interpretation given to infant burials —especially with regard to mortuary richness— has frequently been used as an argument to indicate the presence of inherited power. Contrary to the above, Brown points out that the location analysis of the infantile burial is more useful to detect social implications and, consequently, it is not worthwhile to emphasize the material evidence *per se* associated with the dead, which rather constitute a very important element to explain the funerary habits combined with the space factor.

Brown's third argument is the demographic composition of the skeletal population, although here we also face some difficulties due to the fact that, in most cases, the sample of the population is not representative to define and differentiate social groups.

Briefly, funerary evidence has revealed its particular capacity, within archæology, to explain a historical process. It is precisely our objective to define the mortuary indicators which may help us to detect the differences between a chiefdom and one which is on the threshold of a state organization. In retrospection, a considerable number of studies carried out on this field have been especially aimed towards isolating the social ranking which will enable us to determine the presence of an important social variable within a hierarchical society.

Let us bear in mind that, to Binford (1971), the analysis of funerary attributes helps to determine the dead's social status and, consequently, to point out the type of social system to which he belonged during his life. In accordance with the above, Serra, Sugiura, and Espinosa (1976) have made an attempt to analyze the funerary practices of the Mesoamerican Formative, applying a "cluster analysis". By means of the above-mentioned statistical method, we tried to define social groups differentiated by sex and age in 213 burials excavated during the 4th season at Tlatilco (1962–1969). The Tlatilco data are of a unique character and exceptional in the Middle Formative of the Basin of Mexico, due to the high number of burials containing rich offerings found in a well-defined space.

According to the argument that the dead's social status may be identified by the "indicator-objects", as well as the amount of funerary offerings, the analysis of the Tlatilco burials constitutes an unusual case in Mesoamerica.

The information obtained has been analysed through the application of the modified program of the *Aggregative Hierarchical Clustering Program* elaborated by Donald C. Oliver, of Harvard University (Serra, Sugiura, and Espinosa 1976). During the initial stage of this analysis, the burials were divided into three groups: masculine, femenine, and infantile. The offerings were divided into: pottery and non-pottery. We were thus able to control both the scattering and the variability produced by *the presence of these factors.* Later on, during the second stage, we considered the number of presences and absences of offerings as "similarity coefficient". Finally, we assigned a given weight to the variables, based on archæological criteria; that is, according to the potentiality of the variables which indicate the "social person" of the deceased, a specific weight was given to each element, which ranged from 1 to 10 kgs. The resulting data were classified according to four methods: *single linkage, complete linkage, average linkage within the new group and average linkage between merged groups*, from which the dendrogram of the "*longest link*" or *complete linkage* has rendered the most significant results.

With regard to the "offering" analysis at a ceramic–type level, the final ordering of the dendrogram point out four clusterings in the masculine and the femenine burials and 2 in the infantile ones. It should be noted that the variables which define each group are separated by sex and age; for example, in the masculine case, the first group, which represents the higher number of burials, is clustered based on the presence of a single variable: the polished brown type, whilst the second intermediate group consisted mainly on two ceramic variables: the polished brown type and the red over brown, while the third group, with a reduced number of burials, was characterized by the highest number of variables, such as Figurine D, black polished pottery, burnished brown and polished brown, and burnished red. The polished brown variable, *per se*, works also as an indicator, since it clusters a unit which may be isolated in the femenine and infantile cases. The second femenine group is similar to that of the masculine burials. The difference between these two groups lies in the number of variables which define the third group, in view of the fact that in the femenine burials there is a higher number of elements, besides the following types: polished brown, burnished brown, polished black and Figurine D; we also found the fugitive white type and Figurine K. Besides these three groups, there are others which are isolated due to the absence of offerings, both with regard to masculine as well as femenine burials.

When observed retrospectively, we have noticed that the analysis of the pottery offerings at the ceramic-type level, even when the most significant groups had been isolated, is not accurate enough as to define "the social being" of the deceased, except where exceptional types, such as lacker and fine white clay —having a specific implication— are concerned. For the Tlatilco analysis, in our opinion, it is better to apply the same method to both —pottery and non-pottery— offerings with regard to form and function, since this variable can best symbolize the ranking differences of the deceased.

The analysis of the non-pottery variables rendered very different results. The outstanding presence of non-pottery offerings in masculine and infantile burials is significant; and even objects such as "metates" and grindstones, which could be regarded as femenine symbols, are abundant in masculine burials. On the other hand, pottery objects show an opposite trend. Within the non-pottery elements, only lithic variables show no difference between sexes. The objects which symbolize political authority, such as mirrors, jade belts, jade earrings and crystal-rock beads are distributed within the third group isolated in the pottery analysis, regardless of sex or age.

Now, the Tlatilco data we have handed in the present study pose a serious problem due to the impossibility of incorporating other variables of importance, such as skull deformation, dental mutilation, pathology, nutritional level, orientation and position of the skeleton which, together with the offerings, may be considered as potential indicators of social rank or status. Neither did we have location data for each burial though, at a glance, no specific space seems to have been used to define social status of the buried individual.

Although the analysis of the burial offerings at Tlatilco is still beings carried out, we can now mention some outstanding points:

1. Both, masculine and femenine burials have been divided into four groups, at a ceramic-type level.

2. As far as femenine burials are concerned, we found out that women occupied an important social position, similar to men's. The same phenomenon was also observed in the "El Arbolillo" burials (Vaillant 1935), wherein women were buried with a higher number of offerings. At "El Arbolillo", as distinct from Tlatilco, the males have a lesser amount of offerings than females.

3. Infant skeletal population is quite reduced; nevertheless, it has been grouped basically into two units. We should point out the existence of one specific case, burial no. 74 at Tlatilco, which presents numerous offerings, indicative of a high social ranking such as mirror, plate, seal, rattle, as well as other pottery objects. However, as we discussed earlier in this paper, it is only after incorporating other variables we are able to define whether social status is inherited. The fact that burial no. 74 is an infantile one containing all the elements that symbolize a high social position, we cannot yet determine whether the "social person" of the said individual occupied an inherited position. A considerable amount of similar offerings has been found in the infantile burials at "El Arbolillo" (Vaillant 1935).

4. Burial analysis must be multidimensional. Due to their frequency and richness, the offerings have been considered as one of the outstanding factors which are more relevant for the understanding of the social organization. However, in order for the results of the analysis to be valid, they must be supported by other variables and, therefore, we should not look for a simple correlation between the offerings and the social organization. For example, the absence of an attribute as the exclusive access to the specific location of a burial with the data from Tlatilco is significant. At Tlatilco, there is no sharp difference in the amount of energy expenditure channeled in *the burial.* All the dead have been deposited directly in the mortuary pits dug directly into the ground, without either coffins nor any other kind of construction. All have been buried without showing any variation as far as the funerary treatment is concerned.

In the "El Arbolillo" case (Vaillant 1935), a difference has indeed been detected in the way of burying the dead. Out of the 49 burials located at the site, a third part, i. e. 16 burials, are deposited in coffins made of slabs, whilst the remainder was buried in pits dug directly into the ground. Apparently, the highest number of burials with coffins appears in the Late Arbolillo I Phase, decreasing considerably later on. The "El Arbolillo" case points, then, that the ranking distinction of the buried individual is reflected not only in the amount and richness of the offerings but also on the way in which the dead are taken care of, indicating various levels in the energy output to carry out the funerary act. Thus this implies that the factors which indicate social status in mortuary practices change with regard to both time and space; therefore, it is of utmost importance, first, to isolate the variables which are inherent or adequate to each site.

As we have already seen, our objective is to detect, in mortuary practices, differences between a chiefdom and a state society. To that end, we have chosen two specific historical stages which represent both moments: the Middle Formative and the Late Formative. Tlatilco is the representative sample of the first instance, not only because of sufficient number of burials for statistical analysis, but also due to its well defined time and space. To these, we must add the presence of rich offerings and other data relative to each burial. In order to support our explanation and clear out, in a comparative way, the likenesses and differences between the burials of the Middle Formative, located at various sites in the Basin of Mexico, we have tried to incorporate the data from "El Arbolillo" (Vaillant 1935) under the same methodological procedure.

Unfortunately, the data submitted by Vaillant are quite fragmentary and, consequently, not very appropiate to carry out this type of analysis; we have, therefore, only added —as a suggestion— some highlights taken from Vaillant's study.

In order to study the burials at the beginning of state formation, we chose Ticomán (Vaillant 1931) and Tlapacoya (Barba de Piña Chan 1980). At Ticomán, a total of 61 burials have been found, eleven of which are extremely deteriorated. With regard to the remaining 50, over half of them have mortuary offerings. As far as skeletons are concerned, those of males are slightly more predominant, with a total of 21 burials, whilst female burials amount to 17. There are 9 infant burials and 13 skeletons could not be determined.

The Ticomán data present problems similar to those found at "El Arbolillo"; however, we should point out the following aspect: both quantitatively as well as qualitatively, the offerings indicative of the deceased's "social status" show a marked trend towards male predominance, while at Tlatilco it is just the opposite.

Another sample used for the analysis comes from Tlapacoya, located, like Ticomán, in the lacustrine bank of the former lake of the Valley of Mexico. To the northeast of the Tlapacoya Pyramid, 13 burials were located, corresponding to the Late Formative and coming from 4 different stratigraphic shafts. As the sample is excessively reduced, no generalizations could be drawn therefrom. Of the 13 burials, only no. 3 was a multiple one, with 3 skeletons: two males and one female, associated to a considerable amount of offerings, consisting mainly of lithic tools placed between both male skeletons. Burial no. 5 is also a masculine one, with five offerings; with the exception of these 2 cases, the remaining burials located herein contain no offerings, presenting, at most, one or two offerings (namely pots). Thus, in the Tlapacoya case, there is also a marked predominance of the male sex.

On the other hand, at the same site, but on the area of the pyramid, southwest of the already mentioned burial area, 3 tombs were excavated. These present extremely rich mortuary equipment (Barba de Piña Chan 1980: 105). According to the description of the author herself, the mortuary objects found in these tombs differ from the offerings found in the aforementioned burials. All this leads us to believe that they were manufactured specifically for that purpose, or they were obtained through exchange. Tomb no. 1 (1.20 x 0.9m.) is built with mud walls, provided with a lid of basaltic slabs. Inside the tomb 69 offerings were found, 64 of which are pottery objects of various shapes, 4 are figurines and there is also an engraved slate plate. Unfortunately, the sex of the individual buried herein could not be determined.

Tomb no. 2, located at greater depth than no. 1, is 1.42 m. long x 1.10 m. wide and 0.85 m. high. Its constructions technique differs from the first tomb, as it is made up of basaltic slab walls lying against the fill-in material of the structure itself, with a top of the same material. The skeleton found herein belongs to an adult male, placed on a bed of lacustrine plants. The tomb's bottom was covered with red paint. Inside there were 12 offerings, 7 ceramic objects, 7 obsidian knives showing no use-wear traces, 2 serpentine drills, a seashell with red paint and jade beads. The last tomb (no. 3) was found at the same depth than the first one and it shows the same construction technique. It is 1.25 m. long x 0.98 m. wide and 0.95 m. high. The skeleton was extremely deteriorated; therefore, neither sex nor age could be determined. It was covered by a great amount of offerings. There were 74 pieces of pottery, 5 obsidian knives and one slate plate. On the tomb's floor there were organic remains, probably of lacustrine origin and some red paint remains. Thus, it is quite clear that in Tlapacoya there coexisted two different burial mechanisms.

The first one was related with low social rank. The dead were deposited to the northeast of the pyramid and, as a general rule, their interment (burial) represented a minimum energy output, since they were simply placed in mortuary pits dug into the ground, and accompanied by an offering. We must point out, among the individuals buried in this area, the two adult males already mentioned who had a considerable amount of offerings. With regard to this burial, the author (Barba de Piña Chan 1980) assumes that they were professional craftsmen.

The second burial mechanism is represented by three tombs, probably belonging to the priestly class or, perhaps, to the ruling class. The idiosincracy of this social rank can be judged by the rich offerings, by the construction of the tombs themselves, the way in which the bodies were deposited and by the exclusive location of the burial inside the pyramid. In spite of the fact that the sex of only one individual could be defined out of the three tombs, we may suggest that towards the Late Formative males assumed and consolidated their higher social position *vis à vis* women, which did not formerly occur. On the other side, in a ruling center such as Tlapacoya, there are strong differences as far as treatment of the dead is concerned, always in accordance with their social status. The above-mentioned factors tell us that a more complex social structure had —by then— taken place, with marked differences between the ruling group and those who were ruled. The latter group was subdivided into professional craftsmen and the remaining individuals. On the other hand, although Ticomán is placed in a historical time similar at Tlapacoya, it presents differences. Due to its somewhat rural character, funerary rites at Ticomán do not reflect the existence of extremely definite social groups as in Tlapacoya; it rather seems that the Ticomán people followed the patterns which were predominant at that time. The difference may only be observed in the consolidation of the male's position within society. All this suggest that, in order to explain a definite historical stage, it is necessary to clearly bear in mind the synchronic variation of contemporary communities. Only when we are able to specify the likenesses and differences between the various contemporary groups, will we be able to understand the different social systems which work at a given moment.

It is also advisable, as we have already mentioned, that the analysis of funerary rites envisages a multidimensional approach and that the results be supported by the elements from other archæological dimensions, such as the spatial pattern of intrasite structures, as well as spatial distribution of sites at a given region. Thus, mortuary evidence is linked to various levels of archæological studies.

Although social divisions are not always clearly and directly reflected in funerary rites, the archæologist must evaluate the capacity of mortuary data to reflect a social structure, always with the assistance of other complementary data.

REFERENCES

BARBA DE PIÑA CHAN, B.

1980. *Tlapacoya*, Biblioteca Encilopédica del Estado de México, México.

BINFORD, L.

1971. "Mortuary Practices: Their Study and Potential", in: *Approaches to the Social Dimensions of Mortuary Practices*, J. A. Brown (ed), Memoirs of the Society for American Archæology no. 25.

BROWN, J.A.

1981. "The Search for Rank in Prehistoric Burials", in: *The Archæology of Death*, R. Chapman, I. Kinnes, and R. Randsborg (eds), Cambridge University Press.

CHAPMAN, R. and K. RANDSBORG

1981. "Approaches to the Archæology of Death", in *Ibid.*

GOLDSTEIN, L.

1981. "One-Dimensional Archæology and Multi-dimensional People: Spatial Organization and Mortuary Analysis", in *Ibid.*

O'SHEA, J.

1981. "Social Configurations and the Archæological Study of Mortuary Practices: a Case Study", in *Ibid.*

PEEBLES, C.S.

1974. Moundville: the Organisation of a Prehistoric Community and Culture, Ph. D. Dissertation, University of California, Santa Barbara.

SAXE, A.A.

1970. Social Dimensions of Mortuary Practices, Ph. D. Dissertation, University of Michigan.

1971. "Social Dimensions of Mortuary Practices in a Mesolithic Population from Wadi Halfa, Sudan", in: *Approaches to the Social Dimensions of Mortuary Practices*, J.A. Brown *et al.* (eds), Memoirs of the Society for American Archæology no. 25.

SERRA, M.C., Y. SUGIURA, and G. ESPINOSA

1976. Clasificación numérica de las ofrendas en entierros de Tlatilco: algunas consideraciones sobre la estructura social del Formativo Temprano, (paper presented at the XLII International Congress of Americanists, Paris).

TAINTER, J.A.

1975. The Archæological Study of Social Change: Woodland Systems in West-Central Illinois, Ph. D. Dissertation, Northwestern University.

VAILLANT, G.

1931. *Excavations at Ticoman*, American Museum of Natural History, Anthropological Papers no. 32 (2), New York.

1935. *Excavations at El Arbolillo*, *Ibid.*

AGRICULTURE AND THE FORMATION OF THE TEOTIHUACAN STATE

EMILY McCLUNG DE TAPIA
Universidad Nacional Autónoma de México

Introduction

Although the work of V. Gordon Childe considers aspects of sociocultural and technological development in the Old World, especially in Mesopotamia and Egypt, the relevance of his theoretical position in relation to other areas where the cultivation and eventual domestication of plants (and, in some cases, animals) and the development of complex urban societies took place is apparent.

In particular, Childe's position concerning the production of food surpluses and their importance for development is explicit in his written work, both in academic as well as in popular books. We refer here to his book *Man Makes Himself* published first in 1936 and re-edited with some modifications in 1951 in which we can appreciate frequent references to the concept of surplus and its implications for the development of certain social institutions throughout the text. At present, there is an apparent tendency among archæologists to consider surplus production, which is viewed as characteristic of and a pre-requisite for urbanism, as a phenomenon which occurs after the establishment of agricultural societies. Childe, however, establishes a fundamental position with respect to this as seen by the following observation:

> Food production, even in its simplest form, provides an opportunity and a motive for the accumulation of a surplus. A crop must not be consumed as soon as it is reaped. The grains must be conserved and eked out so as to last till the next harvest, for a whole year. And a proportion of every crop must be set aside for seed (Childe 1951: 71).

Furthermore, the concept of surplus is not only associated with agricultural production, but with animal raising as well:

> The storage of grain, the conservation of live-meat "on the hoof" is much simpler, especially in a warm climate, than the preservation of stocks of slaughtered game. The surplus thus gathered will help to tide the community over bad seasons: it will form a reserve against droughts and crop failures. It will serve to support a growing population. Ultimately it may constitute a basis for rudimentary trade, and so pave the way to a second revolution (Childe 1951: 71).

On the other hand, it should be recognized that surplus production is most relevant insofar as it permits certain sectors of a [growing] population to dedicate at least part of their time to non-agricultural activities. Childe apparently saw the roots of this process of differentiation of activities in a level of social organization corresponding to what is currently referred to by some authors as the tribe, in which subsistence activities are generally thought to be essentially cooperative:

> The Neolithic economy as a whole cannot exist without co-operative effort. The heavier labor of clearing patches in a forest or draining a marsh must be a collective undertaking. The digging of drains, the defense of the settlement against wild beasts or floods, must again be communal responsibilities...implies some social organization to co-ordinate and control the communal activities (1951: 81).

Finally, Childe points out that the production of a food surplus is basic to any endeavor which an agricultural society might undertake, including the intensification of its productive system:

> The works in question were collective undertakings, they benefitted the community as a whole, and were beyond the power of any individual. And generally their execution required capital in the form of a stock of surplus foodstuffs, accumulated by and at the disposal of the community. The workers engaged in draining and embanking must be fed; but while so employed they were not direclty producing the food they consumed. As the reproductive works of a community become more ambitious, so the need for an accumulated stock of surplus foodstuffs would increase. Such an accumulation was a pre-condition of the growth of the village into a city, by conquering ever more of the territory surrounding it from marsh and desert. Conditions of life in a river valley or other oasis place in the hands of society an exceptional power for coercing its members: the community can refuse a recalcitrant access to water and can close the channels that irrigate the fields... The social solidarity needed by irrigators can thus be imposed owing to the very circumstances that demand it... So when the social will comes to be expressed through a chief or a king, he is invested not merely with moral authority, but with a coercive force too; he can apply sanctions against the disobedient (1951: 90).

The importance of these passages cited from Childe rests in the references to the role of surplus foodstuffs as basic to sociocultural and technological development. We can see Childe's apparent position with respect to the relationship among the origin of surplus production (a logical consequence of the domestication of plants and animals), its later role in sustaining non-economic development, and the development of intensive agriculture (particularly irrigation) as the means by which surplus production could provide subsistence for a large sector of the society in question.

Agriculture at Teotihuacan

The significance of the preceding comments is clearly demonstrated by studies of the development of urbanism in Mesoamerica and in particular, in the case of the first prehistoric Mesoamerican urban center, Teotihuacan. The role of agriculture in the development of the Teotihuacan state, especially in its most intensive manifestations, has provided material for discussion during more than 20 years. We will leave this discussion aside for the moment, since there exists a substantial amount of literature dealing with the relationship between "civilization" and "state" and cities (urban centers) as manifestations of civilizations or growing states. At this time, we propose to turn attention to some of the more elemental aspects of the discussion: the role of agriculture in the formation of the urban center of Teotihuacan, its role in the articulation of the Teotihuacan state, and the relationship of these to economic symbiosis in central Mesoamerica during the Classic period (150–750 AD: Middle Horizon).

The Teotihuacan Valley: Settlement Patterns and Agricultural Activities

The Teotihuacan Valley is located approximately 40 km northeast of modern Mexico City. The most notable archæological site in the region consists of the urban center of Teotihuacan, to the northeast of the town of San Juan Teotihuacan. The prehispanic city covered an area of approximately 20 km^2 (Millon 1970: 1077), and its maximum population is estimated to have been between 75,000 and 200,000 inhabitants, with an average falling around 125,000 (Millon 1970: 1080). The urban area extended over a total area of 26–28 km^2 (Millon 1964: 344) during the period between 150 and 750 AD. However, not all of this area was occupied at any one time, but rather there are indications for settlement concentrations during some phases of its development. The surrounding valley covers an area of approximately 505 km^2 bordered on the south and southeast by the Sierra Patlachique; on the north by Cerro Gordo, Cerro Chiconautla and Cerro Malinalco; and opening onto the Tepeapulco–Apan plain in the extreme northeast and the Texcoco plain in the southwest (fig. 1). Elevation of the valley floor varies between 2240 and 2300 m above sea level.

The Teotihuacan region was first settled during the Middle Formative period (Altica phase, *c.* 900–500 BC: First Intermediate I–II). Settlement patterns characteristic of the early phases of occupation in the region have been determined and described by Sanders (Sanders *et al.* 1975: 294–317), who distinguished a number of settlement types: hamlets, villages, hilltop centers and exploitation sites. The distribution of sites during Formative period cultural phases are indicated on a series of maps (Sanders *et al.* 1975, II). Sites tend to be concentrated in the Sierra Patlachique from the time of initial settlement of the region until the beginning of the Patlachique phase (*c.* 150 BC–AD 1), although a gradual increase takes place in the settlement of the lower piedmont, a pattern which is manifest by the Tzacualli phase (AD 1–150).

In general terms, settlement during the Middle and Late Formative periods (First Intermediate I–II) in the valley is characterized by its piedmont orientation. At the end of this period, however, the alluvial plain and the Lower Valley (especially the village of Cuanalan) become important. By the end of the Terminal Formative (300 BC–AD 150: First Intermediate III–IV), during the Tzacualli phase, site orientation had changed markedly and focused on the lower slopes and the valley floor where a large settlement founded during the previous Patlachique phase was rapidly expanding. Also during the Tzacualli phase there is an increase in the number of sites on the north slope of Cerro Gordo and in the area designated by Sanders as the North Tributary Valleys.

Based on the distribution of sites in the piedmont with indications of Formative period occupations (phases Altica, Chiconautla, Cuanalan/Tezoyuca, *c.* 900–150 BC: First Intermediate I–III), and the availability of certain natural resources, we propose a subsistence pattern based on the exploitation of wild plants (cacti, edible "weeds" and fruits), forest resources (especially, wild game such as deer and rabbits), and agricultural products (maize, beans, and cultivable edible "weeds") obtained by the extensive temporal cultivation of suitable areas.

At the same time, it seems reasonable to suppose that the construction of agricultural terraces and some form of control of the intense seasonal flow of nearby barrancas could have begun some time during the Terminal Formative. Since few sites corresponding to this period have been excavated and there are few data with respect to the subsistence resources utilized by their inhabitants (with the exception of some faunal remains from Tlachinolpan, corresponding to the Late and Terminal Formative periods), it is impossible to be more specific at this time.

The occupation of the valley floor and, in particular, the alluvial Lower Valley, suggests the growing importance of agriculture and, probably, the turning of attention to the greater Basin of Mexico, both in political as well as in economic terms. During the Late Formative (First Intermediate II), the only settlement of the valley floor occurs during the Cuanalan phase (*c.* 500–150 BC) at the site of Cuanalan in the Lower Valley and three additional sites (2 hamlets and a village which appears in the area of the future urban center).

During the Patlachique phase, the settlement of the valley floor expands rapidly, and by the Tzacualli phase (Terminal Formative, First Intermediate IV) it reaches urban dimensions in terms of population size and density. As the principal site in the valley expands, outlying sites are greatly reduced in size and become marginal in function. The relative absence of villages on the valley floor or in the lower slopes of the surrounding piedmont suggests two possibilities: 1) that agricultural lands in the alluvial plain were cultivated by city residents and the inhabitants of piedmont settlements engaged in local production, complemented by other subsistence activities such as the exploitation of forest resources and the channeling of these to the city; and 2) that the inhabitants of the lower piedmont also cultivated lands on the valley floor. By this phase, Cuanalan had been abandoned and other contemporary sites were situated in the lower piedmont instead of the valley floor. This is the moment in time when population growth in the Teotihuacan Valley becomes important and the "agricultural" carrying capacity of the region is relevant.

Carrying Capacity and Agriculture in the Teotihuacan Valley

There is an incorrect tendency among archæologists to base calculations of human carrying capacity on the potential agricultural productivity of a particular region. However, agricultural potential refers to the maximum productivity of all crops in an area, and when this concept is used, calculations are generally based on the potential productivity of one (or, at most two) plants. It would be more satisfactory to consider agricultural potential as one of a number of components which together make up the total ecological potential of a region. Ecological potential includes the potential productivity of wild floral and faunal resources as well. As such it is potential productivity of all of these subsistence resources which establishes the human carrying capacity of a region.

Another problem with the concept of carrying capacity most frequently used in relation to prehispanic Mesoamerican populations is the tendency to base calculations of agricultural potential on actual productivity under modern conditions. While it is obvious that modern environmental conditions can provide some useful information for the reconstruction of the past, there is evidence which suggests that conditions in the Teotihuacan Valley were somewhat different during the period of prehistoric occupation (especially insofar as the availability of water resources and the degree of erosion are concerned). These differences must be considered in relation to agricultural potential and carrying capacity.

The most recent efforts to determine the number of persons which could be sustained by maize agriculture (and amaranth to some extent; Lorenzo 1968: 66–69; Charlton 1970: 331–334; Sanders, Parsons and Logan 1976) establish that under essentially modern but technologically simple conditions of maize cultivation in the Teotihuacan Valley, approximately 40,000–50,000 persons could be maintained. This includes the use of rainfall cultivation (*temporal*), floodwater irrigation and permanent irrigation in all suitable areas of the valley. After many years of debate over the subject of irrigation in relation to Teotihuacan agriculture, there are now numerous data, both indirect and direct, which support its use in the valley. It has been mentioned elsewhere (McClung de Tapia 1979a, 1979b) that there is little doubt of the importance of irrigation as an intensive technique in the Teotihuacan Valley during the Classic period in order to increase agricultural productivity to support the urban population. What we have questioned and continue to question is the role of this and other intensive techniques in relation to the development of the state.

As we mentioned previously, population growth during the Tzacualli phase was significant, and the settlement attained urban proportions. Based on the statistical analysis of ceramic concentrations from surface collections, Cowgill (1974) suggests an average Tzacualli phase population of 50,000–60,000 inhabitants, a figure substantially higher than that originally proposed by Millon (1973: 521) of 25,000–30,000. This implies that the urban population (not to mention the rest of the valley's inhabitants) would have had to obtain a proportion of their foodstuffs from sources outside of the valley, if we accept 40,000–50,000 as an approximation of the carrying capacity of the region. A number of considerations are important here and time does not permit a detailed discussion of the arguments surrounding the use of the concept of carrying capacity, the "real" carrying capacity of the region, the size and density of population during different occupation phases and the necessities of that population. The major point for the moment is that the city of Teotihuacan undoubtedly had to resolve problems related to the obtention and distribution of subsistence products very early on in the history of its development. Part of the resolution of this problem probably took the form of more intensive agricultural production within the valley itself. However, at the same time, the obtention of foodstuffs from other areas implies aspects of social, economic and political organization, some of which we try to suggest here.

Regional Symbiosis and the Teotihuacan State

Another important concept which has come out of the study of Mesoamerican culture is that of Regional Symbiosis, mentioned by Palerm and Wolf (1972: 194) and developed by Sanders (1956) and Sanders and Price (1968). One of thc outstanding characteristics of the Mesoamerican territory was the marked diversity in biophysical environments, and, consequently, the corresponding diversity and localized nature of certain resources. Based on the diversity of human necessities and the development of relations of mutual dependence among populations as a response to satisfying these needs, Mesoamerica was conceived of as divided into a series of "Symbiotic Regions", zones of contrasting environmental character, generally composed of a highland and a lowland component. In the case of prehispanic Mesoamerica, symbiosis has been defined as an important ecological process in the sociocultural evolution of the region, directly related to the efficient obtention of raw materials and technology, including subsistence products. The symbiotic region which has received the most attention because of its role as a key area for prehispanic sociocultural development is the so-called "Central Mexican Symbiotic Region" (fig. 2). Although the geographic definition of the region as well as the specific products and resources which circulated within it are based on data from the 16th century, immediately before the Conquest, there are indications for the existence of such a network during the Classic period (AD 150–750; First Intermediate V and Middle Horizon). Although this area has never been the object of a specific research project designed to determine the validity of Sanders' hypothesis, the concept of symbiotic relations among populations in Central Mexico, focused especially towards the Central Highlands and more specifically towards the Basin of Mexico, provides a context within which we are able to postulate the economic development of an urban center of Teotihuacan's magnitude.

The evidence for certain types of exchange at Teotihuacan is supported not only by traditional artifacts such as foreign ceramics, obsidian of known provenience, etc. but also by archæobotanical remains which indicate the presence of plants not indigenous to the Teotihuacan Valley or the Basin of Mexico in general. In view of the difficulty in proposing the importance of an exchange network in the formation of the city of Teotihuacan as well as the state when it has not even been proven that such a network existed, we find ourselves in a position similar to that of those archæologists who insisted on the importance of irrigation for the formation of the Teotihuacan state while being unable to demonstrate that the teotihuacanos effectively utilized irrigation systems. Over the years, data have accumulated indicating the use of irrigation early on in the development of other regions (Flannery *et al.* 1967; García Cook 1981, etc.), as well as in the Basin of Mexico (Sanders and Santley 1977; Nichols 1980), in the Teotihuacan Valley during the Postclassic (Millon 1954, 1957; Charlton 1980), and, finally, in Teotihuacan during the Classic period (Nichols 1981). Similarly, it will be necesary to accumulate a large quantity of circumstantial evidence which tends to support our hypothesis. Fortunately, recent research has begun to provide new evidence which tends to support both the concept and the reality of a Central Mexican Symbiotic Region during the period of Teotihuacan development and expansion.

Based on the study of settlement patterns and potential land use in the Cuautitlan Region (fig. 3), defined as part of the Basin of Mexico Settlement Survey directed in large part by Sanders (1976; Sanders, Parsons and Santley 1979), a relationship is postulated between the exploitation of agricultural lands and the location of settlements during the period of Teotihuacan domination in the Basin of Mexico. Nichols (1980) observes that the region is practically abandoned during the Tzacualli phase, as occurs elsewhere in the Basin at about the same time (Blanton 1972; Parsons 1968, etc.). During the Classic period, the area is re-populated and a pattern develops in which approximately 70% of the population is concentrated in nucleated villages or regional centers of Teotihuacan type (Sanders 1976: 15, in Nichols 1980: 234). In addition, in the same region, there is evidence for the development of irrigation as a measure against the high risk of rainfall (*temporal*) cultivation in the area. The political demise of Teotihuacan is paralleled by the reduction of population centers in the Cuautitlan region. Without going into further detail here, Nichols (1980) presents a convincing argument for the possible role of this region in the production of agricultural surplus for the urban population at Teotihuacan.

Another source of recent evidence in support of the concept of a symbiotic region is the re-analysis of carbonized maize specimens, most of which were recovered from the fill of the Pyramid of the Sun from a Late Tzacualli context [excavation carried out as part of the "Teotihuacan Mapping Project", Rene Millon, director (University of Rochester).] Benz used Multiple Discriminant Analysis (Benz, McClung de Tapia, and Iltis, n. d.) to compare populations of modern races of maize with the archæological specimens in order to determine the afiliation of the Teotihuacan material. The results indicate that the archæobotanical remains are divided into three racial groups, all so-called Mexican Pyramidal types: *Palomero toluqueño, Cónico*, and *Arrocillo*. The variability demonstrated by these races therefore was in existence in Central Mexico at least as early as AD 100. *Palomero toluqueño* is adapted to altitudes between 2200 and 2800 m, and although its present distribution is limited to the Valley of Toluca (it is reported to be almost extinct), presumably it extended across a broader area in the past (Wellhausen *et al.* 1952: 51). *Cónico* is also adapted to the same elevation interval, and represents the principal commercial maize currently cultivated in the Central Highlands. It occurs in the states of Mexico, Tlaxcala, Puebla, part of Michoacan up into Jalisco, and Hidalgo up to Queretaro (Wellhausen *et al.* 1952: 84). Finally, *Arrocillo* is of the most interest here because of its particular distribution limited to the northern Sierra de Puebla, and because of its adaptation to lower elevations than those characteristic of the Basin of Mexico in general, between 1600 and 2000 m (Wellhausen *et al.* 1952: 52). Although elevation in these instances does not represent an unflexible limiting factor, it at least suggests that this race might not do as well in the Basin of Mexico or the Valley of Teotihuacan in particular. For the moment, the occurrence of Arrocillo among the archæobotanical material suggests the possibility of a broader relationship with the Puebla–Tlaxcala region, which may have involved exchange or other means of obtaining maize to supply the city of Teotihuacan. Other hypotheses concerning the relationship between Teotihuacan and the Puebla–Tlaxcala region during the Late Formative period will not be discussed here; however, at least the presence in the city of a type of maize which could have been imported from another region and whose ecological requirements would be difficult to maintain in Teotihuacan (at least without the use of irrigation) suggests an interesting hypothesis worth investigating in more depth.

Conclusion

We have touched a number of points up until now and it is possible that some of the threads have become tangled. We have tried to point out the importance of food surpluses which were so significant for Childe, and to suggest their relevance for urban development at Teotihuacan as well as the Teotihuacan state. The types of surplus should be distinguished: agricultural surplus to be stored by the population and utilized over the course of the year, including seed for the following season; surplus which is destined to support non-agriculturalists in the society; and surplus which is exacted by the state in the form of tribute or taxes to be utilized according to the criteria of a centralized authority.

In the case of Teotihuacan there are indications that it was necessary early on to obtain food surpluses from outside of the immediate region, in order to provide for the basic needs of the urban population, and it is probable that this situation somehow influenced the growth of the state. The need to articulate already established economic relationships in a local symbiotic network and broaden them in order to supply a monstrous urban center would foster the development of an efficient economic organization, supported and, ultimately, controlled by the political system.

Our position rests on the assumption that Teotihuacan was founded where it is for religious–ritual reasons, possibly perceived in relation to certain economic advantages. The availability of specific resources such as obsidian, agricultural lands (although not risk-free), basalt, and clay deposits, together with an access to communication routes to such important areas as Puebla–Tlaxcala and the Gulf Coast, would be advantageous insofar as the region's future potential was concerned. Recognition of the possibility of intensifying agricultural productivity through the use of irrigation would make the region even more favorable. The substantial population increase in the city during the Tzacualli phase, owing mainly to immigration from other parts of the Basin of Mexico, would increase the need for surplus foodstuffs, especially given the apparent inability of the region to produce a sufficiency. Since the shortage of subsistence resources had to be resolved during the initial phase of Teotihuacan's urban growth, it propitiated the development of mechanisms for the acquisition of food from outside of the valley, which in turn probably stimulated the centralization of control over agricultural production elsewhere in the Basin of Mexico.

This position differs somewhat from others in that it suggests that the need to organize the acquisition of subsistence products from outside of the Teotihuacan Valley stimulated and/or reinforced a developing centralized authority structure rather than arguing that the necessity of providing for a growing population stimulated the development of intensive agricultural systems, including irrigation, which in turn stimulated the centralization of authority as a sideline to the managerial requirements of irrigation.

At the moment, research which is being carried under the direction of Linda Manzanilla at Teotihuacan is partly focused on the recovery of storage facilities in association with a temple complex dated to the early period of Teotihuacan urbanization, in order to provide data relevant to the integration of subsistence production and distribution in its economic and political contexts. In addition, there is a preliminary suggestion of possible storage structures (based on the high proportion of maize pollen in samples from circular structures excavated in the Merchant's Barrio by Evelyn Rattray during the Classic period. Nevertheless, there is no other evidence for the acquisition and distribution of subsistence products in the city, during the period of transition to urbanism or during its height.

FIGURE 1. THE TEOTIHUACAN VALLEY ECOLOGICAL DIVISIONS (From Sanders 1965)

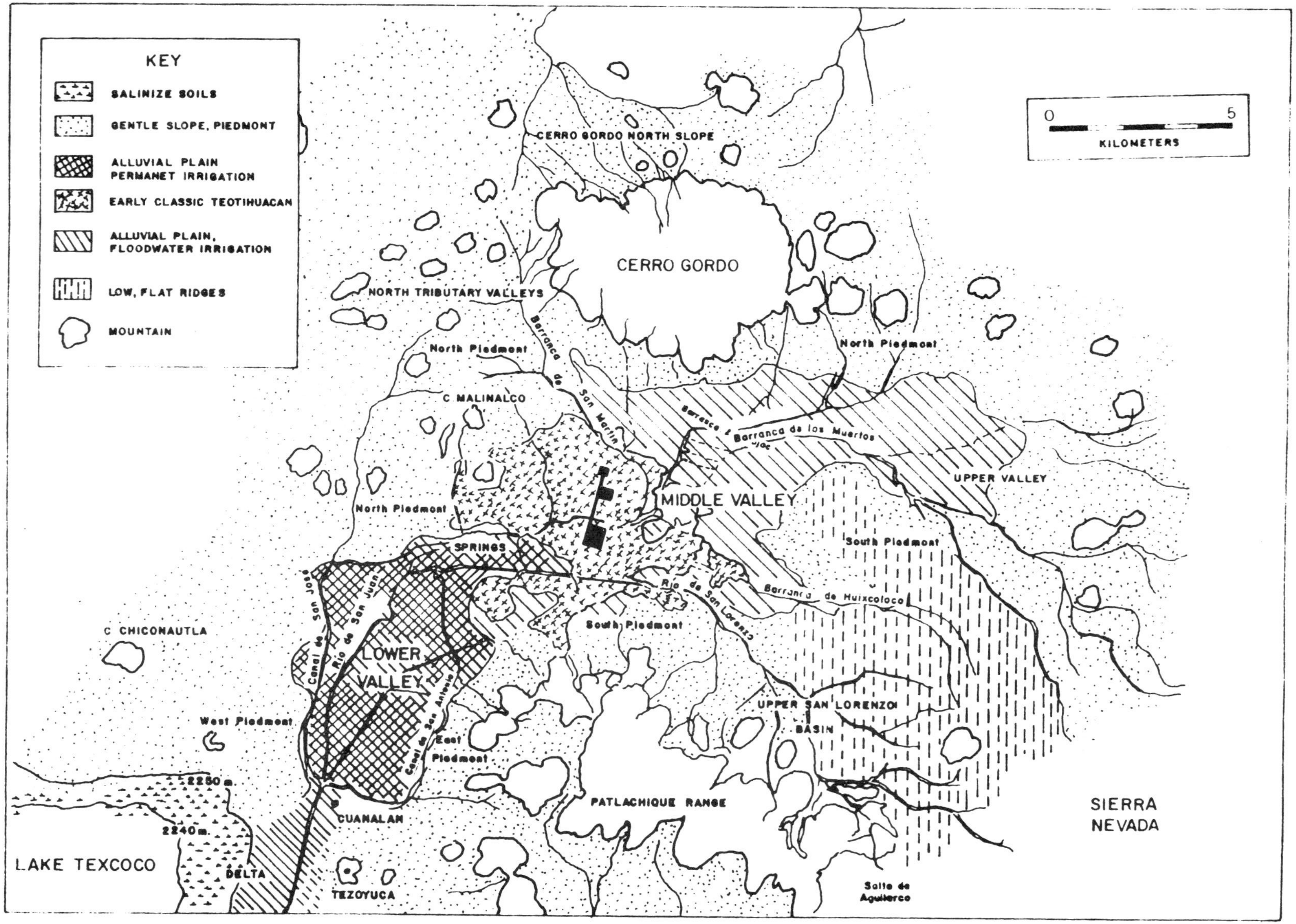

FIGURE 2. The Central Mexican Symbiotic Region during the late horizon Period. (from Sanders and Price 1968: 192).

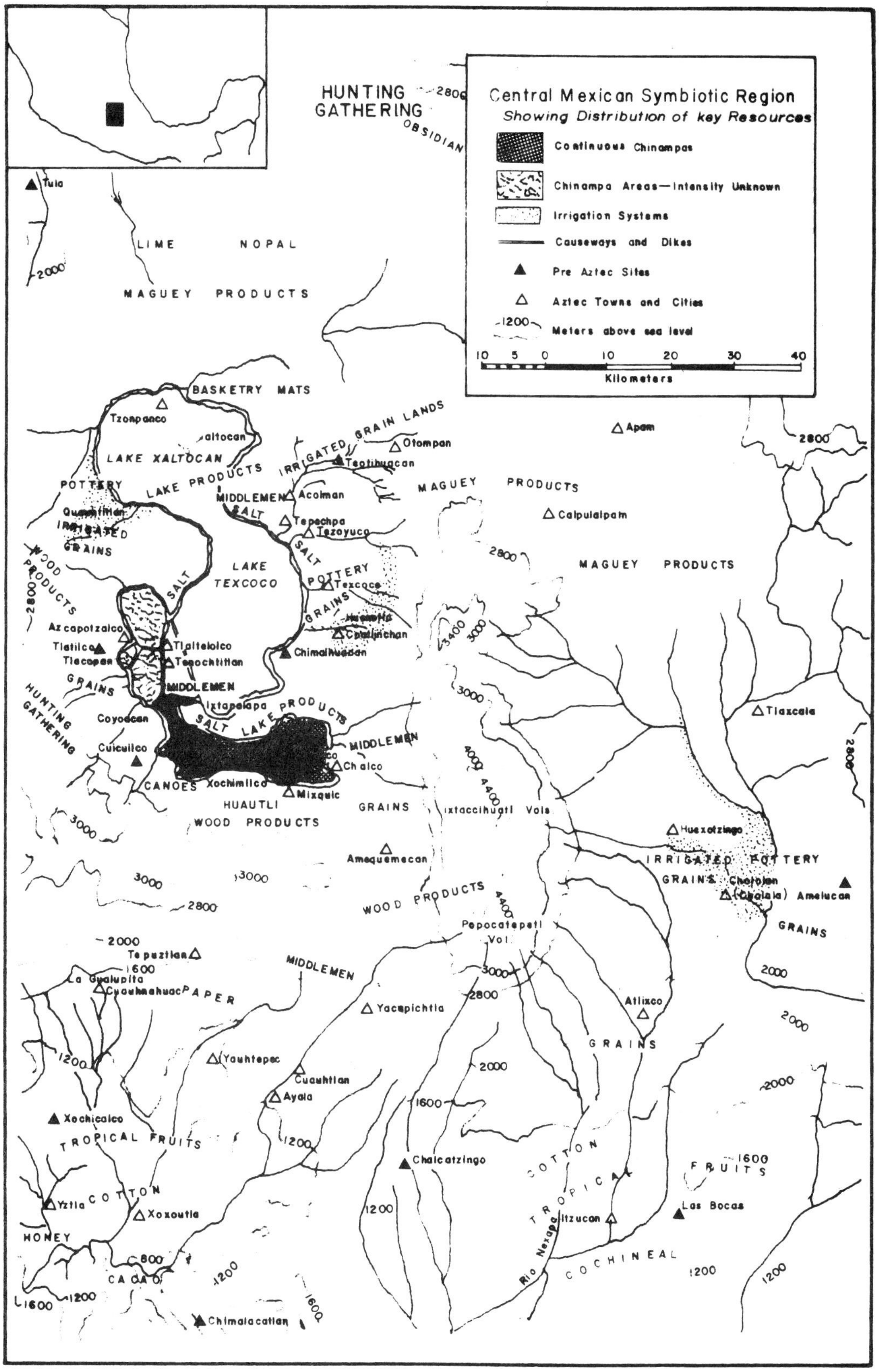

FIGURE 3 . THE BASIN OF MEXICO , SHOWING SURVEY AREAS
(from Logan and Sanders 1976 : 57)

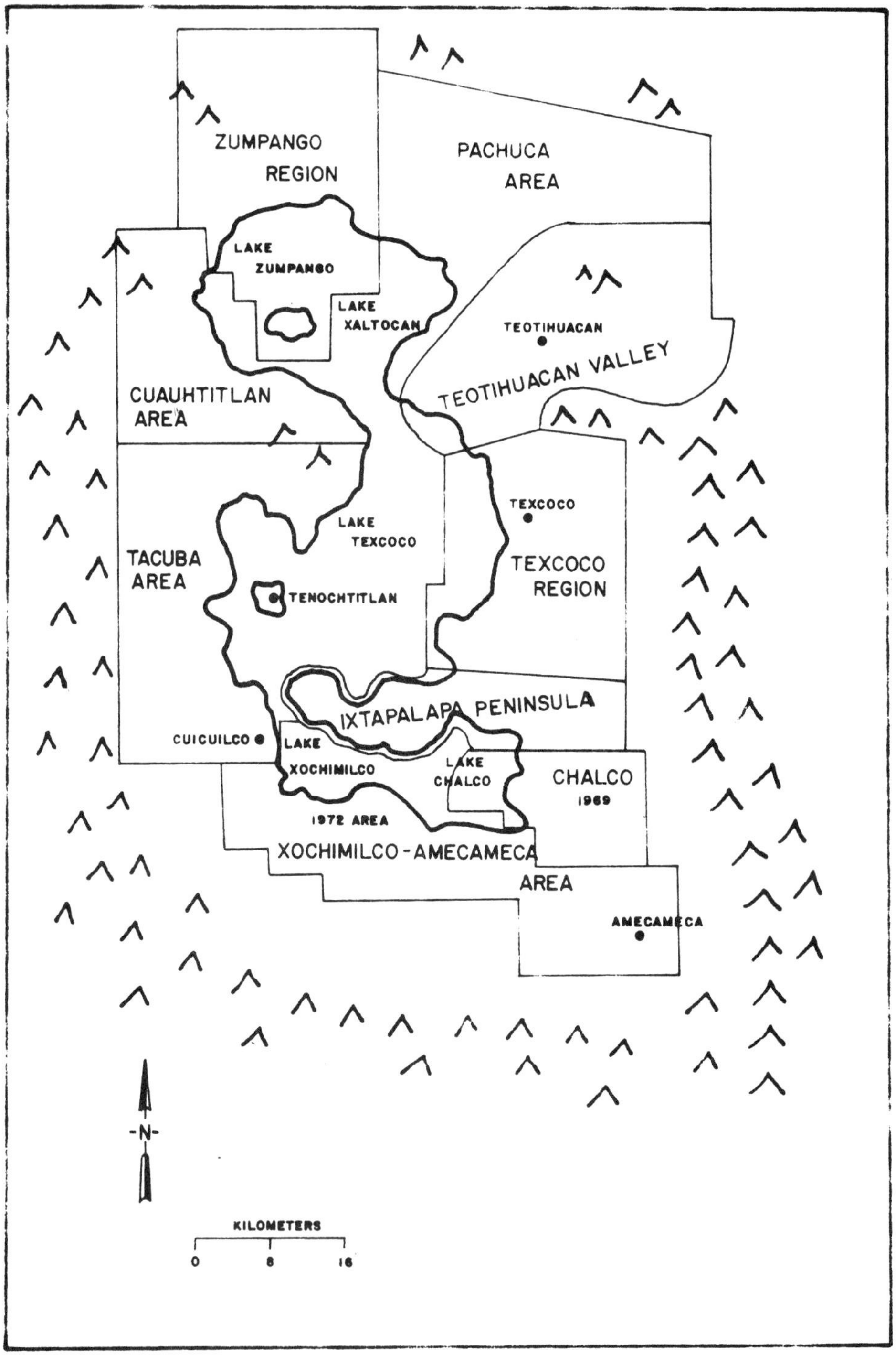

REFERENCES

BENZ, B.F., E. McCLUNG DE TAPIA, and H.H. ILTIS

n.d. "Studies in Archæological Maize", in: On the Origin of the Mexican Pyramidal Complex, Manuscript.

BLANTON, R.E.

1972. "Prehistoric Adaptation in the Ixtapalapa Region", *Science* vol. 175 (4028): 1317–1326.

CHARLTON, T.H.

1970. "Contemporary Agriculture in the Teotihuacan Valley", in Sanders *et al.: The Teotihuacan Valley Project: Final Report*, vol. I. Department of Anthropology, Pennsylvania State University: 253–383.

CHILDE, V.G.

1951. *Man Makes Himself*, Mentor Books.

COWGILL, G.L.

1974. "Quantitative Studies of Urbanization at Teotihuacan", in Hammond, N. (ed): *Mesoamerican Archæology: New Approaches*, Austin, University of Texas Press.

FLANNERY, K.V., A.V.T. KIRKBY, M.J. KIRKBY, and A.W. WILLIAMS

1967. "Farming Growth and Political Systems in Ancient Oaxaca", *Science* 158: 445–454.

GARCIA COOK, A.

1981. "The Historical Importance of Tlaxcala in the Cultural Development of the Central Highlands", in Sabloff, J. (ed): *Supplement to the Handbook of Middle American Indians* vol. I, Austin, University of Texas Press: 244–276.

LORENZO, J.L.

1968. "Clima y agricultura en Teotihuacan", in Lorenzo, J.L. (ed): *Materiales para la arqueología de Teotihuacan*, Instituto Nacional de Antropología e Historia, Serie Investigaciones no. 17, México: 53–72.

McCLUNG DE TAPIA, E.

1979a. *Plants and Subsistence in the Teotihuacan Valley, AD 100–750*, Ann Arbor, Xerox University Microfilms.

1979b. *Ecología y cultura en Mesoamérica*, Dirección General de Publicaciones, Universidad Nacional Autónoma de México, México.

MILLON, R.

1954. "Irrigation at Teotihuacan", *American Antiquity* 20 (2): 177–180.

1957. "Irrigation Systems in the Valley of Teotihuacan", *American Antiquity* 23 (2): 160–166.

1970. "Teotihuacan: Completion of Map of Giant Ancient City in the Valley of Mexico", *Science* 190 (3962): 1077–1082.

1973. *Urbanization at Teotihuacan, vol. I. The Teotihuacan Map*, Part 1, Austin, University of Texas Press.

NICHOLS, D.L.

1980. Prehistoric Settlement and Land Use in the Northwestern Basin of Mexico, the Cuautitlan Region, PhD Dissertation, Department of Anthropology, Pennsylvania State University.

1981. "Prehispanic Irrigation at Teotihuacan, New Evidence: the Tlajinga Canals", Paper presented at the Symposium *Teotihuacan nuevos datos, nuevas síntesis, nuevos problemas*, October 1981, Universidad Nacional Autónoma de México, México.

PALERM, A., and E. WOLF

1972. "Potencial ecológico y desarrollo cultural en Mesoamérica", in Palerm and Wolf (eds): *Agricultura y civilización en Mesoamérica*, México, Sepsetentas no. 32: 149–205.

PARSONS, J.R.

1968. "Teotihuacan, Mexico and its Impact on Regional Demography", *Science* 162 (3856): 872–877.

SANDERS, W.T.

1956. "The Central Mexican Symbiotic Region: a Study in Prehistoric Settlement Patterns", *Viking Fund Publications* no. 23: 115–128.

1976. "The Agricultural History of the Basin of Mexico", in Wolf, E. (ed): *The Valley of Mexico: Studies in Pre-Hispanic Ecology and Society*, Albuquerque, University of New Mexico Press: 101–159.

SANDERS, W.T., and B.J. PRICE

1968. *Mesoamerica: The Evolution of a Civilization*, New York, Random House.

SANDERS, W.T., M. WEST, C. FLETCHER, and J. MARINO

1975. *The Teotihuacan Valley Project: Final Report* vol. II, The Formative Period Occupation of the Valley, Parts I, II, Occasional Papers in Anthropology no. 10. Department of Anthropology, Pennsylvania State University.

SANDERS, W.T., J.R. PARSONS, and M.H. LOGAN

1976. "Summary and Conclusions", in Wolf (ed): *The Valley of Mexico*: 161–178.

SANDERS, W.T., and R.S. SANTLEY

1977. "A Prehistoric Irrigation System near Santa Clara Xalostoc in the Basin of Mexico", *American Antiquity* 42 (4): 582–588.

WELLHAUSEN, E.J., L.M. ROBERTS, E. HERNANDEZ X., and P.C. MANGELSDORF

1952. *Races of Maize in Mexico*, Bussey Institution, Harvard University, Cambridge.

CONQUEST, TRIBUTE AND THE RISE OF THE STATE

TERRY STOCKER
University of West Florida

Flannery (1972) reviewed the theories proposed for the evolution of civilizations. He discusses and dismisses the "prime mover" theories. He concludes that a system analysis might well be the establishment of a series of rules by which the origins of some complex system, such as civilization, could be simulated.

In this paper warfare as a prime mover will be reviewed with new data. Carneiro (1970: 734) states, "warfare is surely a prime mover in the origin of the state", though "it cannot be the only factor. After all, wars have been fought in many parts of the world where the state never emerged". Another major variable is population circumscription. The state by definition requires a high population level and if that level is reached in a geographic area where there can be no release through massive out migration there will be war for limited subsistence resources. The war will either result in a leveling off of the population or subjugation of one group by another. This in turn leads to an intensification of social stratification.

Flannery (1972: 405) says: "Carneiro is probably right in assuming that no early state was without war". However, he questions whether warfare results in or results from the state. He argues that most evidence cited by Carneiro is from periods after which the state was formed, and the *evidence is still ambiguous* for the Formative periods.

Background

There are three major problems concerned with the analysis of warfare as a variable in the rise of the state. The first is ethnographic analogy. I assume Flannery wrote in haste when he questioned whether warfare results from the state. The ethnographic record leaves no doubt that warfare does not require a state level of social organization (Carneiro n.d.).

The second problem is overcoming previous labels of major sites of the Formative period. Flannery correctly notes that most evidence cited by Carneiro dates to periods long after state formation, and he also correctly notes that the evidence for warfare during the Formative periods is ambiguous. To resolve some of the ambiguity, I choose as a major example the site of El Paraíso, Peru. Some individuals use this major Preceramic center as a focal point of debate regarding theories on the rise of the state. They have incorrectly characterized El Paraíso as having one building which they interpret as a temple (see Quilter and Stocker 1983).

El Paraíso has betwen seven and nine stone structures comprising about 100,000 m^3 of mass and covers 58 ha (Moseley 1975). Jackson and Stocker (1982) proposed that the one "unit" referred to as a temple was a granery. I have included a photograph of this building for the reader's evaluation (fig. 1). Engel excavated this building because it was the highest in the site and was less covered with overburden. It is one of the smallest building on the site. Fifteen thousand tons of rock and fill were used in Engel's (1966) reconstruction. I implore the reader to realize they are looking at a *Preceramic* building with 37 rooms, the walls of which are a meter in width, and reached heights of 10 meters.

When I saw this building in 1979, I was astounded. I have then visited archæological sites in 30 countries. To realize that it was built before the advent of ceramics, in the New World, is without a doubt one of the most remarkable feat's in human existence. If in fact it was a temple, we definitely should look to religion as a major variable in the rise of the state (cf. Wheatley 1971).

One necessary variable in the rise of the state was food storage which could feed the populace during events of shortage. If the famine was too severe the stored food would be used to support only the ruling class and presumably warriors. The three year famine of the Aztecs is a paramount documented example. That is, in the final stages of the famine much of the populace in the Aztec capital sold themselves into slavery to people of the Gulf Coast, but the ruling elite and warriors remained in the city.

Stored foods require protection, thus there is an interlink between the military and stock piled foods. The link between stored foods and the military is most manifest at boundary fortifications where the military might come under seige and would be required to withdraw into a fortification and maintain themselves. Stocker and Kylar (1984) documented large granaries at the Aztec fortification of Oztoman, Guerrero which maintained a crucial position at the southeast corner of the Tarascan empire.

Not all storage sites necessarily served the military. Morris' work at Huánuco Pampa, Peru revealed large storage facilities, but no military structures. The interpretation is that the stored supplies were turned into *chicha* and sub-imperial leaders presented "their followers" that *chicha* in return for required labor. However, the crucial aspect of understanding this site and the "absence" of a military presence is that Huánuco Pampa was well within the boundaries of the Inca empire. Furthermore, the ruling elite surely recognized that should an army be sent near Huánuco Pampa they would be privileged to the stored supplies.

Had the site of El Paraíso been correctly presented (not necessarily interpreted) Carneiro might have used it as a primary example of population circumscription. Populations of coastal Peru must inhabit the 50 some river valleys which run from the highlands to the coast. The terrain in between is among the driest in the world, and even in modern times is not inhabited. Should one of these river valleys suffer from the effects of the El Niño, (destruction of sea resources due to red tide and land resources due to massive rainfall), the inhabitants would have only one option: raid the closest river valley with a food source.

Recent documentation by Feldman (1983) demonstrates that the effects of El Niño can be localized to certain valleys. "The isolation of the north and the highlands is also affecting Lima. Produce, which travel by truck, cannot make it to the markets of Lima, which are beginning to experience spot shortages of fresh fruits and vegetables. People are beginning to complain. It remains to be seen if Peru's government...can survive El Niño" (Feldman 1983: 18).

This is an obvious post-industrial example. Imagine what it would have been like without modern transportation. In Preceramic times the population of El Paraíso built a tremendous site. They built buildings of such immense dimensions that a functional interpretation is required, and religion is far from suggested, and further from demonstrated. I suggest, the Preceramic Period inhabitants of El Paraíso were well aware of periodic catastrophic effects of El Niño, and they prepared themselves for an eventual encounter of El Niño itself and/or an encounter with other groups which might have suffered from it. The population of El Paraíso prepared themselves with one of the hallmarks of the state: monumental architecture. Let us not forget this hallmark of the state came before the invention of ceramics, in this case.

This brings us to the third problem in determining the role of conquest and tribute in the "Formative" state. What would be the evidence of conquest in the archæological record? There are many places in the world with obvious fortifications where archæologists cannot demonstrate battles or conquest or tribute. The Hopewell region of Ohio is a primary example (Stocker n. d. a).

I have devoted considerable energy in attempting to evaluate this problem with regard to Aztec empire building. Were it not for the written record, conquest as ***the*** major variable in the expansion of the Aztec state would never have been known. Aztec history spanned some 200 years and they conquered 250 major centers. These centers had their own tributaries; therefore, they in essence conquered approximately 1000 to 2500 centers.

The Aztecs placed governors and some of their own population at only eight of these conquered centers. I have conducted research at two and there is definite archæological proof of the Aztec presence (Stocker 1981; Stocker and Cobean 1984; Stocker and Kylar 1984). There is no evidence of an Aztec conquest at centers without governors; nor is there any evidence of Aztec presence at what I consider to be tributaries of the site at which governors were placed. ***Therefore, without the written record, how could we demonstrate conquest?*** We could not. This likewise means that conquests for the earliest states cannot be documented in the archæological record.

I hypothesized that if archæologists were to go to the crucial centers of the Aztecs that they would also find evidence of the earlier conquest by "foreign" cultures. I (Stocker and Lamb 1985; Stocker and James n. d.) did in fact discover Olmec "presence" in two of the areas where the Aztecs had governors. The criticism I have met is that I cannot demonstrate that these are Formative Period conquests. Again, the haunting problem of "ambiguous evidence"!

My reply? Without the written record we could not determine that the Aztecs had conquered so many centers. Without written records we would have interpretations of the Aztec state similar to what we have for the Toltecs and Teotihuacan. What does this say about those interpretations of the Toltecs and Teotihuacan? I must state the obvious. Warfare as the major variable has not entered into an explanation of the development of these states, because it cannot be seen in the archæological record.

The Hypothesis

To date, Mesoamerican archæology is concerned with what can be seen in the archæological record. Therefore, we borrowed geographer's models to explain the relationship of sites. We borrow models from ecologists to explain how "states arise in socioenvironmental contexts where effective management is either necessary or especially beneficial" (Brumfiel 1983: 262). Both are predicated on being able to see some tangible thing in the archæological record. A site or a certain type of site in a certain ecological niche.

In Mexico and the United States there are plaques or monuments to significant battles from the historic period, battles which were crucial in the ultimate formation of both states. To satisfy the members of this conference, I will point only to the Alamo. I could provide hundreds of other examples and several more are listed below, but the point is that this is often impossible for prehistory, but with certain critical data it can be done in specific cases (Stocker 1983).

According to Brumfiel (1983: 278):

> Perhaps the conclusion to be drawn from this discussion is not that anthropologists need less ecology in their efforts to understand state formation, but that they need to focus more sharply on the interplay of ecological variables and political dynamics. An-

thropologists need to understand the goals of political actors as generated by different types of prestate political systems, and they need to be able to define how ecological variables present obstacles and opportunities to political actors pursuring their goals. From such studies, it should eventually be possible to derive a general theory of state formation, one that specifies the necessary ecological and political conditions under which state formation occurs.

In the remainder of this paper, I want to focus on the political actors. I further focus on what cannot be seen in the archæological records as it related to political actors. I believe that political scientists may have a better understanding of the essence of civilization than do anthropologists, ecologists or geographers. Of course, anthropology in its initial years set itself apart from other social sciences by defining itself as the "science" of the small "traditional" (primitive) societies. This ghost still haunts us. To attempt to arrive at the creation of the state from this epistemological background has been difficult. Political science has something to offer because it has always been concerned with the state.

What we call progress is a mysterious marriage of creativity and plunder. Civilization has flowered when human beings have devised ingenious new ways to organize production and social life, but this has been done usually with stolen goods. The Greek city-states, the Roman Empire, and Elizabethan England come to mind. Undergirding their extraordinary achievements in art, philosophy, literature, and statecraft was military power and conquest. Rome lived off grain exacted as tribute from Sicily, Spain, North Africa, and Gual. The civilization of modern Europe was built on shiploads of gold and bullion extracted from mines of the New World, the slaves, diamonds, copper, and vegetable oils of Africa, and the cotton, rubber, and spice of India and Malaysia —all taken at gunpoint. (Barnet 1980: 113).

Political science and historical literature abound with such quotes (Schwab 1984; Chambers 1979). I leave the reader with one more example which encapsulates U. S. history.

The American nation, born in one war (1775–83), defended in a second (1812–15), expanded in a third (1846–48), held together in a fourth (1861–65), given global power by a fifth (1898–1901), sixth (1917–18), and seventh (1941–45), and taught the limits of that power in an eighth (1950–53) and ninth (1965–73), has been affected deeply throughout its history by military affairs. Despite our distrust of standing armies, the United States can be said to draw its very existence from the endurance of Washington's army. (McElheny 1986: 35).

In the next section, examples will be provided which demonstrate the same thing for the Aztecs. How do we then make the leap into Prehistory? How do we make the leap to the Formative Period? I suggest that we simply take all of history as a model. I refuse to believe that there was ever a civilization or state that was not founded on conquest and tribute. To refuse such an idea because archæologists cannot provide the battles or the battle casualties and the resulting conquest and tribute is limiting our interpretations of prehistory. How often do tribute items become "translated" into trade items in the archæological record?

To date, there seems to be one type of model to explain the processual development of historic groups and another for prehistoric groups. History is my model of prehistory.

My hypothesis would be stated something like this. If the state is characterized by social stratification and that stratification intensifies and expands through time, then there must be an increase in the productivity of food and status supplies to satisfy that growth. If there is no demonstrable qualitative (or quantitative) change in technology to supply the demand of food and status items, then it must be obtained by force elsewhere (outside).

The Aztec Case

Brumfiel (1983) provides a thorough review of the theoretical positions advanced to explain the growth of the Aztec state. She summarizes Aztec state formation with four discrete logical steps (Brumfiel 1983: 276).

1. Intensification of competition within and between petty kingdoms which broke the power of local rulers and led to Triple Alliance hegemony with the Valley of Mexico.
2. Centralization of power through organizational reforms that reduced the economic and political power of subordinate rulers and Tenochtitlan's nobility.
3. Consoloditation of power through the initiation of public works with the valley and expansionary conquests beyond it.
4. Development of bureaucratic complexity characterized by the specialization of administrative personnel by task and by the presence of three or more levels of decision makers in many of the administrative branches.

The Population/Stratification Factor of Conquest and the State

I will expand on points 1 and 3. It must be remembered and cannot be overstated, that the Aztecs in the beginning of their history were among the tributaries of the center of Atzcapotzalco. They successfully rebelled and overthrew Atzcapotzalco, and then went on to rule a very large part of Mesoamerica.

> Thus were assassinated the highest in office in Azcapotzalco, and from that time on the city lost all its importance. It had once been the greatest and most populous in all the land, royalty had resided in it, and it had conquered thirty tributary states long before the Aztecs had even dreamed of coming to this land (Durán 1964: 71).

Rebellion was essential in the Aztec state organization and I assume many other Mesoamerican states. I (Stocker 1982; Stocker and James 1986) argue that rebellion was essential in order to insure a supply of sacrificial victims. Were it not for the fact that they needed sacrificial victims they could have put governors throughout their empire, much like the Inca. They did in fact put governors at crucial centers where they could not afford rebellions and a possible loss of territory (Stocker and Kylar 1983). This is the basic difference between the military organization of the Inca and the Aztec, although some of the differences may relate to a degree of political centralization (cf. Gorenstein 1966).

Aztec history spanned some 200 years and approximately 250 major centers were conquered. The tribute from these centers certainly was the major means of supporting a major attribute of complex society: class stratification. Again I ask the question, how often are tribute items translated into trade items, especially in the Formative Period archæological record?

Collier (1982: 17) extends Carrasco's (1971) theory that Aztec ascendance involved no fundamental changes in the principles of central-Mexican stratification. But it was a fundamental change for the Aztecs themselves because they went from a chiefdom "type" society to a state. ***This is documented with written records, and it is documented that it was done by means of conquest and tribute.*** However, the Aztec empire did not present something socially different from that which had gone before in the Postclassic Period, but it was substantially larger. It was obviously larger because it was much more sucessful in military endeavors, and we must assume population growth.

I assume the Aztecs were militarily successful because they had adept military leaders (as well as warriors), and analogies can be drawn to Alexander the Great, D. David Eisenhower and the present day Israeli state. I am still in the process of examining the Aztec selection criterion for centers to conquer, and this will take a considerable research endeavor because almost nothing is written on the conquered centers and analysis requires indepth field work. But this information and analysis is not required to understand that the Aztecs were militarily adept. The first Aztec king, Acamapichtli, had four conquests, and to paraphrase a quotation above, maybe the Aztec state drew its very existence from the endurance of Acamapichtli's army (also see Kurtz 1978).

Until a more definite analysis of selection criterion for conquest can be presented, there are two points to be considered. First, the Aztecs were not always successful, and quite often the losses were considerable. For example, the loss to the Tarascans during the reign of the 7th king, Axayácatl, was detrimental to the Aztec empire and one which would direct their military endeavors for the remainder of their history (Stocker n. d. b). Thus there were other "complex societies" which confronted them, and were militarily superior. The significance of these defeats is not immediately manifest, it may be that groups like the Tarascans were more adept at defense than offense. Second, the military not only supported social stratification by supplying necessary elite regalia, but it likewise amplified it.

We need now to examine in detail how warfare made the state possible. As with less organized societies it provided a means of reducing the home population first through deaths and second by movement into defeated areas. This could occur on a diffuse level such as when the Spanish migrated to Mexico after the defeat of the Aztecs. Or it could be a focused reduction and movement of an entire population. A primary example is the Aztec annihilation of the centers of Oztoman and Alahuitzlan, and subsequent repopulation of each of these by 2000 people from Central Mexico (Stocker and Kylar 1984).

More importantly these types of repopulation migrations serve to reduce the expanding elite which multiplied in the main center. This is another concept which I believe should be borrowed from political science (Mills 1956). As the elite expand societies become much more stratified. The Aztecs are a primary recorded example. Each succeeding king was progressively separated from the public masses, and in fact from other elites. By the time of Moctezuma II, no one was permitted to watch the king eat.

The growth of the elite meant that there had to be a simultaneous increase in production of status goods. This of course gave rise to increasing military campaigns which is clearly demonstrated in the *Codex Mendoza* (see Stocker and Kylar 1984). But no matter how great the increase in status items, there would be an increase in competition for a limited number of elite offices. Thus repopulation of a site (annihilated in warfare) could serve to syphon off the expanding elite in the main center. That is, the brother of a king could now be governor. In fact, by the time of the last Aztec king, the nobility had grown to the point that Moctezuma II, had to restructure the bureaucracy and access to state office. Nobles of questionable birth were relieved of their office and "many" were executed (Carrasco 1971).

Thus, the military supported social stratification by supplying necessary elite regalia. It likewise amplified social stratification with chieftains and governors and other ruling officials for captured sites.

The Material Factor of Conquest and the State

A sound interpretation of the individual motivation for going to war was to gain status in Aztec society. However, there was also the motivation of booty. Durán (1964) is full of incidences of looting of defeated villages. "By the time the army returned to Mexico with their splendid loot, men were so eager to go to war that one seldom saw a soldier in the city. All wished to go to war since they fared so well" (Durán 1964: 118).

Archæologists should keep such passages in mind when they excavate "exotic-foreign" ceramic wares at any given site. Quite often these are referred to as "trade wares", but the proof of trade is no more substantiated than booty. This also results in trade being offered as a primary variable in the rise of the state (see Stocker 1983).

The Aztec military was almost self-supporting. A detailed analysis of the tribute indicates that tribute from conquered towns was a major means (if not the major means) of supporting the Aztec military. Shields are the most depicted item in the *Codex Mendoza.*

> Another tribute consisted of shields made of fire-hardened rods so strong and heavy that a sword could not dent them... Big thick bows were also paid as tribute, together with arrows of different forms and of many types. Finely worked stones in the form of balls for the slings were also sent, and so were numerous slings of maguey fiber... Other contributions were white and black blades for the swords, flints for the arrows and the darts. Let one try to imagine all the products of this land and he will know what was sent as tribute to Mexico! (Durán 1964: 130).

The Energy Factor of Conquest and the State

War provided free slaves and slaves provided "free" energy. Slavery was common in Aztec society and throughout Mesoamerica and Central America (Sherman 1979), but there is little definite information with regard to their functions. In part, this may be due to the fact that many of the chronicles were composed so long after the conquest. However, there is enough fragmentary information to assume that slaves were a major work force of the Aztec state.

Some slaves were purchased in the markets. An amazing thing about this practice is that there is no mention (or at least I have not located one) as to how slave traders came into possession of slaves. Many slaves were captured in war, and I assume that the individual making a capture which was not to be sacrified could sell the capture, but again there is no mention of such transactions in the Aztec chronicles.

A great number of slaves came from annihilated villages. Only the men and women were annihilated by the Aztecs. The children were taken as slaves.

What did the slaves do in Aztec society? There are few passages with reference to this. However, a very lucid insight is provided in Sahagún's text on the Soothsayers which details the date of the days of birth. The day of Tezcatlipoca was the day of the slaves, "Those who existed by the hones for sharpening and grinding stones" (Sahagún 1979: 34).

Our modern world marvels at the great stone statues "made without steel tools" by the Precolumbian populations. The stone walls of the Inca which were so well fitted that a knife blade cannot fit between them. "How did they do it?" My answer is slave power. I also propose that the ***initial work*** on the great Olmec heads was done by slaves under the watchful eye of craftsmen.

Slaves have a very subtle and quick way of getting lost in history. Of course, what would be the lot of a slave? Not to be remembered in history, that much is certain. The Aztecs referred to slaves as the divine dead. What is more, slaves cannot be seen in the archæological record.

Parenthetically, I have given a tremendous amount of thought to the matter of what life must have been like in a tribute paying village. It must have been quite burdensome, I would think. However they were free. Free to rebel.

Summary

I believe that military conquest and subsequent tribute was the major means of achieving a state level of social organization among what have been competing chiefdoms. It would have provided material tribute for status items; as well as armament. It also provided food sources for the elite (Zorita 1963) and a stockpile in granaries for times of shortage. It provided an ecological balance by the reduction of lives and a means of syphoning off some of the population through migration into defeated areas. This worked especially well for removing an expanding elite which required so much of the tribute.

I believe that it is the expanding elite competing for a limited number of offices which caused the decline of many state organizations. The Inca might be a primary example. They may have been the one Native American social organization which could have defeated the Spanish. However, upon the arrival of Pizarro, there were two rival sons that were heir to the Inca Empire and they split it in military battle. But this is just the tip of the proverbial iceberg. Each son was supported by a factionated elite which numbered into the thousands.

The populace can support only a given number of elites, and then the system begins to "fall under its own weight". At the time of Spanish arrival the Aztecs did not have this problem. They had suffered two recent major defeats, and many of the high class warriors (and therefore nobles) were killed.

The major contention of this paper is that the "facts" presented for the Aztec state are historic, and that such historic facts are not preserved in the archæological record. The battle of Atlixco and the loss of Aztec lives has never been demonstrated in the archæological record, and never will. If we cannot do it with the written record, what can we say about the causal factors responsible for the rise of the Toltecs, Teotihuacan and the Olmecs?

We must either make a leap in reasoning about conquest, tribute and the rise of the state since we cannot recover demonstrable evidence or we will be limited in an appreciation of war as a cultural process. Until now war is summarized as a limited cultural activity. We seem to have dichotomized human existence into a historical epoch where warfare is an obvious cultural process and prehistory where warfare is a limited cultural activity.

> One of the principle activites of both Olmec and Classic Maya lords seems to have been warfare and the humbling of captives as shown on Olmec "altars"...and numerous Maya stelae (Coe 1977: 186).

Plate 1. The excavated building at the Preceramic site of El Paraíso, Peru.

REFERENCES

BARNET, E.

1980. *The Lean Years*, New York, Simon and Schuster.

BRUMFIEL, E.

1983. "Aztec State Making: Ecology, Structure, and the Origin of the State", *American Anthropologist* 85: 261–284.

CARNEIRO, R.

1970. "A Theory of the Origin of the State", *Science* 169: 733–38,

n. d. Chiefdom-Level Warfare as Exemplified by the Chiefdoms of Fiji and the Cauca Valley. ms.

CARRASCO, P.

1971. "Social Organization of Ancient Mexico", *Handbook of Middle American Indians* vol. 10.

CHAMBER, J.

1979. *The Devil's Horsemen*, Atheneum, New York.

COE, M.

1977. "Olmec and Maya: A Study in Relationships", in: *The Origins of Maya Civilization*, R. Adams (ed), Albuquerque, University of New Mexico Press: 183–195.

COLLIER, G.

1982. "In the Shadow of Empire. New Directions in Mesoamerican and Andean Ethnohistory", in: *The Inca and Aztec States 1400–1800*, G. Collier, R. Rosaldo and J. Wirth (eds), Academic Press, New York.

DURAN, D.

1964. *The Aztecs*, Orion, New York.

ENGEL, F.

1966. "Le Complexe Préceramique d'El Paraiso (Pérou)", *Journal de la Société des Américanistes* 55: 43–96.

FELDMAN, R.

1983. "El Niño: Recent Effects in Peru", *Field Museum of Natural History Bulletin* 54: 16–18.

FLANNERY, K.

1972. "The Cultural Evolution of Civilization", *Annual Review of Ecology and Systematics* 3: 399–426.

GORENSTEIN, S.

1966. "The Differential Development of New World Empires", *Revista Mexicana de Estudios Antropológicos* 20: 41–67.

JACKSON, B. and T. STOCKER

1982. "Peru's Preceramic Menu", *Field Museum of Natural History Bulletin* 53 (7): 12–23.

KURTZ, D.

1978. "The Legitimation of the Aztec State", in: *The Early State*, H. Claessen and P. Skalnik (eds), The Hague, Mouton: 169–189.

McELHENY, V.

1986. "Swords into Plowshares. An Interview with Merritt Roe Smith", *Invention and Technology* 1 (3): 34–43.

MILLS, C.

1956. *Power Elite*, England, Oxford University Press.

MOSELEY, M.

1975. *The Maritime Foundations of Andean Civilization*, Menlo Park, Cummings Press.

QUILTER, J. and T. STOCKER

1983. "Subsistence Economies and the Origins of Andean Complex Societies", *American Anthropologist* 85: 545–562.

SCHWAB, R.

1984. *The Oriental Renaissance*, Columbia Univeristy Press, New York,

SAHAGUN, B.

1979. *The Soothsayers. Florentine Codex*, Book 4, Santa Fe, School of American Research.

SHERMAN, W.

1979. *Forced Native Labor in Sixteenth-Century Central America*, Lincoln University Nebraska Press.

STOCKER, T. and R. COBEAN

1984. "Preliminary Report on the Obsidian Mines at Pico de Orizaba, Veracruz", in: *Prehistoric Quarries and Lithic Production*, J. Ericson and B. Purdy (eds), Cambridge University Press: 83–96.

STOCKER, T. and D. JAMES

n.d. A Structural-Semantic Analysis of Monument II, Chalcatzingo, Morelos, ms.

STOCKER, T. and E. KYLAR

1984. "Aztec Warfare, Sacrifice and Cannibalism", *Explorers Journal* 62: 126–133.

STOCKER, T. and S. LAMB

1985. "Mysteries of the Mexican Codices", *Explorers Journal* 63: 2–7.

STOCKER, T.

1981. "Obsidian Technology in Mexico", *Explorers Journal* 59: 176–181.

1982. "Chichimec Military Potential: The Teotihuacan Case", International Congress of Americanists, Manchester, England.

1983. Figurines from Tula, Hidalgo, Mexico, Ph.D. Dissertation on file at the University of Illinois.

n.d.a. The Hopewell Culture in New Light. ms.

n.d.b. Aztec Military History. ms.

WHEATLEY, P.

1971. *The Pivot of the Four Quarters*, Chicago, Aldine Company.

ZORITA, A.

1963. *Life and Labor in Ancient Mexico*, New Brunswick, Rutgers Univeristy Press.

Acknowledgements

I profited from comments by Robert Carneiro, Jeff Quilter and Susette Doyon–Bernard.

EPILOGUE: LIFE ENDS BETTER WHEN YOU ARE LIVELY AND STRONG

JOSÉ PÉREZ
Escuela Nacional de Antropología e Historia

> The sun is the same in the relative way, but you're older
> Shorter of breath and one day close to death.
> Every year is getting shorter, never seem to find the time
> Plans that either come to naught or half a page of
> scribbled lines
> Hanging on in quiet desperation is the English way.
> (*Pink Floyd* 1973).

I. Vere Gordon Childe died on October 19, 1957 near Katoomba in the Blue Mountains (New South Wales, Australia) at the age of 65. His jacket, eyeglasses and compass were found lying at the edge of the cliff by Mr. Newstead, driver of the rented car. Fearing a fatal accident he notified the local police.

The police report stated that "...the loss of his glasses was the reason why Professor V. G. Childe plunged to his death from 900 feet at Govett's Leap". The Sydney Daily Telegraph and the Sydney Morning Herald accepted and echoed the official version.

With this apparently accidental death, history closed controversially on probably the most important archeologist of the twentieth century. Theories have been advanced from the most widely differing personal and political standpoints with each person collecting fragments of evidence and fitting them together like a jigsaw puzzle in a variety of ways. But, the crux of the matter lies in the question of whether his death was or was not the result of a voluntary act.

Trigger says that:

> ...various explanations have been offered of the circumstances of Childe's death. He was afraid of illness and loss of independence in old age, but his half-sister Ethel said that he was hoping to return to England, even though for many of his colleagues he had apparently packed up all his belongings. Some people have tried to link his death with disillusionment with the Soviet Union or with Marxism which, it is claimed, he had finally seen to be a bankrupt philosophy. The evidence is overwhelmingly against this.

Based on the testimony given by Peter Gathercole, we said before that the death of V. G. Childe occurred accidentally during a geological field trip. Today written evidence tells us something different.

II. In March 1980, Glyn Daniel published two crucially important documents in connection with this matter. The first is a letter from Childe to W. F. Grimes, who succeeded him as director of the Institute of Archeology of London University. It runs as follows:

The Carrington,
Katoomba
Blue Mountains, N.S.W.
20.10.57

Dear Grimes:

The enclosed packet contains papers that in time might be of historical interest for the Institute. But today they could cause distress and even lead to a libel case. After ten years they will be less irritating. I earnestly beg you that they be deposited in the archives and not opened until January 1968, supposing this year arrives.

Yours sincerely,

V. Gordon Childe.

The second document, which we may regard as his real will, was most probably written the same day, although it is important to note that the date of the letter to W. F. Grimes is wrong since it is the day after Childe's death. For reasons of space the complete text has not been produced.

III. Because of the mysterious circumstances surrounding Childe's death, it opened up a field to argument and conjectures; this is precisely why, having his will now available, we must analyze the text. It opens with what we can call a general statement of the problem of old age, whose main argument lies in the survival of the fittest and consequently, in the negative role of old people in biological and social terms. "The progress of medical science has burdened society with a horde of parasites: people of independent means, pensioners and other retired people that society has to maintain and even look after". He attacks the supremacy of the old, the gerontocracy as "the worst possible form of leadership", in his own words; but he goes even further, saying of the old "they exploit youth, which they expect to produce for them and take care of them". These are lines steeped in frank naturalist and social Darwinism that make absolutely no allowances.

We know that Childe has to retire from his position at London University, having reached the age of 65; it might be supposed that his disillusionment stemmed partly from his administrative rule. But, apparently this is not so. He dismisses university professors over 65 —though admitting there are exceptions— with the words: "I doubt they can ever produce new ideas. Obligatory retirement from academic, judicial and administrative posts has of course contributed in some measure to opening the doors to younger men, and has saved students and subordinates from inefficient teachers and incompetent administrative bosses. The old system in British universities has provided me with warning examples of distinguished professors who mumble through classes ten years out of date and who waste departmental funds on obsolete equipment". As an example of what he says he mentions Arthur Evans.

The solution he proposes is drastic. "I have always thought that a healthy society should rid itself of such parasites by offering euthanasia as the highest honour or even by imposing it in extreme cases, but certainly not condemning them to misery and hunger by inflation".

To become more personal, I feel that he has nothing to contribute to the study of prehistory since his memory is now beginning to fail him. He cannot see a solution to the question of "the cradle of the Aryans" which for him was still of crucial importance. He rejects the possibility of becoming a counselor of some cultural or university society, at the same time admitting that he lacks "the will to face the discomfort and anxiety of traveling in the USRR or China".

He is certainly worried and distressed by the economic situation that England is going through at this moment, confessing that "I have become too dependent on a lot of material —even luxuries— to do a job I am trained for... I have never saved money, and if I had, inflation would have eaten it up".

"Of course, with my pension I would not be able to keep up my standard of living, without which life would be intolerable and it is necessary to prevent myself from becoming like an invalid, a burden to society". It is precisely at this point that he shows his decision to take his own life. "I have always thought of ending my life before this happens".

He approaches the subject of suicide straightforwardly, praising it as the human quality *par excellence* and the supreme and most dramatic expression of freedom of choice. Let Childe speak for himself: "The British prejudice against suicide is completely irrational. To end one's life deliberately is something that distinguished *homo sapiens* from other animals, better than the ceremonial burial of the dead. I am not trying to hurt my friends by laughing at this prejudice. An accident can happen very easily and naturally on a mountain precipice". It is obvious from the last paragraph that he had already chosen the manner and place of his death; however, doubt will hover around for many years to come.

He briefly surveys childhood memories, but without showing any affection for his homeland. "I have visited the country of my birth and I have discovered that I like Australian society much less than European, with no hopes of being able to do anything to improve it, as I have lost faith in my old ideals... There is nothing left I wish to do here; nothing I feel I must or can do". He ends his letter —his farewell and his will— with a sentence full of dramatic tenderness: "Life ends better when you're lively and strong".

Childe's personal position sheds light on his suicide. He was a man alone but not lonely who had not cultivated, as far as we know, family ties. He was the only child of his father's second marriage and, it would seem, never had a close relationship with his half-siblings. On the other hand, those who were close to him on the eve of his suicide agree that there was nothing to show that he was depressed; consequently, it seems true that he was "lively and strong".

IV. In the last months of his life, living in Australia, V. Gordon Childe produced *The Prehistory of European Society, The Bronze Age, Farewell* and *Retrospection.* These works are useful in examining some aspects of Childe's work. His phrase "as I have lost faith in my old ideals" will fan the flames of controversy.

The first question he deals with in *Farewell* is that of absolute chronology. Until this time Childe had been guided in this field by typology or crossed dates. He saw the possibilities of radiocarbon clearly, at that time in its first stages of application. For Childe, absolute chronology was an imperative need, since his diffusionist patterns to explain European society historically had to rest necessarily on dates that could be trusted. He probably had some doubts as to his historical reconstructions, at the same time being aware of the adjustments that radiocarbon dating would bring.

In the same article he says that "Of course the universal laws of social development are not so many and are less trustworthy than the Marrists believed before 1950". A few pages further on he returns to the subject and says "The Marrists have resort to 'the uniformities of social evolution' which if indeed seem to make intelligible the development of each individual culture to which this concept was applied, on the other hand fail completely when they try to explain the differences between one culture and another and, by the way, erase or dismiss the differences observed as irrelevant. In this way prehistory becomes a-historical". It must be emphasized that Childe was particularly interested in underlining and explaining cultural differences. In *Retrospection* he tells us that having evaluated Marx more thoroughly, "Now at last I have cleared my mind of trascendental laws that determine history and of mechanical causes, whether economic or environmental, that automatically model its course". Clearly, for him Marxism in one thing and Marrism another; in this respect he does not fall into error and characterizes and defines the latter with absolute clarity. There is nothing to suggest that at the end of his life he rejects Marxism.

V. We have to change our outlook as we move from Childe the archeologist to Childe the humanist. In *The Bronze Age* he left us what we might call his *utopia.* He entered archeology from philology, in search of the Indo-Europeans, and ended his academic life with *The Prehistory of European Society*, a historical explanation of why Europe is like it is and not otherwise; an explanation of how this divergence from the East and America might have happened —let us remember his interest in explaining cultural differences.

The Bronze Age in Egypt and Mesopotamia coincides with the beginning of the Urban Revolution, that is, with the establishment of totalitarian regimes that systematically stripped the peasants of surplus produce, society became divided into opposing classes: on one hand the king and the nobility, and on the other the peasants and specialized artisans. The accumulation of surplus made possible the rise of the metal industry; as it could not have been accumulated in any other way, the "necessary logic" of Eastern totalitarianism is spoken of. However, the production relations that facilitated the establishment of a metal industry checked the progressive development of the same. It generated a group of experts in applied science, but relegated them to the lowest, illiterate and exploited classes.

The Aegean bronze industry developed after the eastern ones and was more progressive because as it arose within a social and economic structure that was not class-based, the craftsmen in metal were not relegated to the lowest class and, in addition, were free to move from one town to another and to choose their clients. "Consequently the new population of craftsmen and specialists had, from the start, a social hierarchy and a tradition of freedom that actually survived even when the division of classes reduced the peasantry to a low, exploited class. Precisely because they could prosper from the surplus accumulated in the East and benefit from the metal industry established in relation to it without having themselves to accumulate the enormous surplus needed to start up such an industry the societies of the Aegean were able to enter the Bronze Age without undergoing a division of classes. The new generation of full-time specialists required to maintain a Bronze Age could grow in the Aegean as in fact happened, from the peasantry in the heart of a barbaric tribal society. The artisans came to occupy an important place in the societies, and even won an important reputation.

Meanwhile, north of the Alps the Bronze Age began a thousand years later, and among poor independent barbarians; this is why they appropriated for themselves the surplus accumulated by the class-based societies of Minoan Crete and Mycenæan Greece. European villages were served by itinerant merchant–craftsmen, members of an economic association who pooled their experience and forged an international tradition. These anonymous metal–workers can be considered the ancestors of the scientist who shared their findings in international society, from Galileo, Newton and Pascal until 1945.

According to Childe, the history of Europe differs from that of the New World "because of the proximity of Egypt and Mesopotamia, the only places where social and economic conditions for the initial establishment of a metal industry pre-existed. The priority of the East in this aspect, however, offers an explanation, that is equally historical, for the divergences of European history from that of the East; it exempted the Europeans from paying the heavy price of starting this industry from nothing, at least until a tradition particular to Europe in applied sciences was established among societies that remained barbarian, but with enough stability to survive the breakdown of tribal organization". He admits that his historical interpretation may be mistaken, but does not repent having published the result, which is "The Prehistory of European Society". Furthermore he stresses precise chronologies (remember his insistence on the possible results of radiocarbon dating) since much of his argument depends on the age of Unetice culture.

What we have called the utopia of Childe the humanist, is a society that gave scientists a privileged place, and in which knowledge was exchanged and spread with complete freedom beyond national frontiers (internationalism). It is an open and pluralist Europe where the scientist, or intellectual, is given full recognition since he fulfills a useful and progressive role. If we think back on his will, it would seem that British society never gave Childe his recognition. The year 1945 is mentioned as the beginning of the "cold war"; a historical moment that did not afford the very least possibility of such an utopian political concept as that advanced by Childe. He never confused Marxism with Marrism, nor did he ever let himself be deceived by what we now call "socialist realism". Suicide seems rather to be a protest against the European political scene of the time; Childe was not disappointed with Marxism but with European society.

The statement "I have lost my old ideals" refers to the impossibility of seeing his socio-political dream come true, that is the culmination of European historical development. This utopia stretches from *How Labour Governs* to *Prehistory of European Society*, with a certain reminiscence of Plato's *Republic*. Childe identifies with and reflects himself in his utopia, presenting himself as the archetype: he knew several European languages, had traveled in many countries, read obscure archeological texts from a variety of sources and in remote languages; he pooled the information and spread it internationally. He was the 20th century personification of the description that he himself gave of the merchant craftsmen of Europe in the Bronze Age.

A final comment, aside but not unconnected: while we analyze from America the life and work of Vere Gordon Childe we cannot forget another lucid, utopian and heartbroken suicide: the Andean, José María Arguedas.

UNIV DUNELM

UNIVERSITY LIBRARY
DURHAM

DURHAM UNIVERSITY LIBRARY
3 0104 00462411 4

ACCESSION No. AP1 125/49

This book should be returned on or before the last date shown below, unless it is previously recalled by the Library when it must be returned immediately.

Fines are incurred on overdue books (see Library Regulations).